Dynamics of Multibody Systems

International Union of Theoretical
and Applied Mechanics

Dynamics of Multibody Systems

IUTAM / IFToMM Symposium, Udine, Italy,
September 16–20, 1985

Editors
G. Bianchi and **W. Schiehlen**

Springer-Verlag
Berlin Heidelberg New York Tokyo

Prof. Giovanni Bianchi
International Centre
for Mechanical Sciences
Palazzo del Torso
Piazza Garibaldi, 18
I-33100 Udine, Italy

Prof. Werner Schiehlen
Universität Stuttgart
Institut B für Mechanik
Pfaffenwaldring 9
D-7000 Stuttgart 80, FRG

ISBN-13:978-3-642-82757-0 e-ISBN-13:978-3-642-82755-6
DOI: 10.1007/978-3-642-82755-6

Library of Congress Cataloging in Publication Data.
IUTAM/IFToMM Symposium (1985 : Udine, Italy)
Dynamics of multibody systems.
At head of title: International Union of Theoretical and Applied Mechanics.
1. Dynamics – Congresses. I. Bianchi, G. (Giovanni), 1924.
II. Schiehlen, W. O. (Werner O.) III. International Union of Theoretical and Applied Mechanics.
IV. International Federation for the Theory of Machines and Mechanisms.
V. Title.
QA845.I98 1986 531'.11 86-3770
ISBN-13:978-3-642-82757-0 (U.S.)

2161/3020-5 4 3 2 1 0

Preface

A first Symposium on Dynamics of Multibody Systems was held August 29 - September 3, 1977, under the chairmanship of Prof. Dr. K. Magnus in Munich, FRG. Since that time considerable progress has been made in the dynamics of multibody systems, a discipline rendering essential services to the fields of robotics, biomechanics, spacecraft control, road and rail vehicle design, and dynamics of machinery. Therefore, the International Union of Theoretical and Applied Mechanics (IUTAM) has initiated and sponsored, in cooperation with the International Federation for Theory of Machines and Mechanisms (IFToMM), a Symposium on Dynamics of Multibody Systems, held at the International Centre of Mechanical Sciences (CISM) in Udine, Italy, September 16-20, 1985.

The aims of the symposium were to generate knowledge, to stimulate research, to disseminate new ideas, and to acquaint the scientific community in general with the work currently in progress in the area of multibody dynamics.

A Scientific Committee has been appointed consisting of

G. Bianchi (Co-Chairman), Italy; T.R. Kane, USA;
R. Kawai, Japan; D.M. Klimov, USSR; K. Magnus, FRG;
F. Niordson, Denmark; A.D. de Pater, The Netherlands;
B. Roth, USA; W. Schiehlen (Co-Chairman), FRG;
J. Wittenburg, FRG.

The Committee selected the participants to be invited and the papers to be presented at the Symposium. As a result of this process, 47 active scientific participants from 20 countries followed the invitation and 26 papers were presented. The lectures were devoted to the following main topics:

Computerized Formalisms, Modelling Techniques, Solution Techniques, Flexible Systems, Robotics, Gyrodynamics, Dynamics of Machines.

The lectures were given by leading experts in the field of multibody dynamics and represented in an impressive manner the wide scope of modern dynamics, reaching from theoretical problems to engineering applications. Because of the extensive discussions during the Symposium it is expected that a stimulating effect will lead to further research in this important field of mechanics. The publication of all papers presented may promote this development. Since some of the lectures were related to more than one of the main topics, the papers are arranged in alphabetical order in this volume.

The material contained herein reflects reproduction from original manuscripts submitted by the authors. Therefore, varying quality of the papers can be observed in this book. Only limited editing has been done as to preserve the individual flavor of each contribution.

The Symposium was financially supported by

IUTAM - International Union for Theoretical and Applied Mechanics,

IFToMM - International Federation for Theory of Machines and Mechanisms.

This help is gratefully acknowledged.

The success of the Symposium would not have been possible without the help and the excellent work of the staff of the International Centre of Mechanical Sciences (CISM). The Local Organizing Committee under its Chairman (G. Bianchi) arranged the social programme and organized the five days of the Symposium with great skill. We thank especially the Symposium Secretary (Ms. E. Venir).

The editorial work of the Proceedings was supported by the Institute B of Mechanics, University of Stuttgart. In particular, we would like to express our sincere thanks to Dr. E. Kreuzer for his very valuable contributions. Furthermore, thanks are due to the Springer-Verlag for an agreeable and efficient cooperation.

Udine and Stuttgart, November 1985

Giovanni Bianchi Werner Schiehlen

Participants

(Authors are identified by an asteriks,
Chairmen are identified by a circle)

AMBROSIO, J.A., CEMUL, Instituto Superior Tecnico, Av. Rovisco Pais, 1096 Lisboa, Portugal

* BARRACO, A., Département de Structures, E.N.S.A.M., 151 Bd. de l'Hôpital, F-75640 Paris Cedex 13, France

o BIANCHI, G., Dipartimento di Meccanica, Politecnico di Milano, Piazza L. da Vinci 32, I-20133 Milano, Italy

* BOGACZ, R., Institute of Fundamental Technological Research, Polish Academy of Sciences, ul. Swietokrzyska 21, 00-049 Warsaw, Poland

* BRAUCHLI, H., Institut für Mechanik, ETH-Zentrum, CH-8092 Zürich, Switzerland

* DESOYER, K., Institut für Mechanik, Technische Universität Wien, Karlsplatz 13, A-1040 Wien 4, Austria

* GARCIA DE JALON, J., Escuela Superior de Ingenieros Industriales, Universidad de Navarra, Urdaneta 7, E-20006 San Sebastian, Spain

* HAUG, E., Center for Computer Aided Design, College of Engineering, The University of Iowa, Iowa City, IA 52242, USA

HILLER, M., Institut A für Mechanik, Universität Stuttgart, D-7000 Stuttgart 80, FRG

* HUSTON, R.L., Dept. of Mechanical and Industrial Engineering, University of Cincinnati, OH 45221-0072, USA

* IWATSUBO, T., Faculty of Engineering, Kobe University, 657 Kobe Nadaku Rokkodai 1, Japan

* JONKER, J.B., Lab. of Engineering Mechanics, Delft University of Technology, Mekelweg 2, NL-2628 CD Delft, The Netherlands

o* KLIMOV, D.M., Institute for Problems of Mechanics, The USSR Academy of Sciences, Pr. Vernadskogo 101, Moscow 117 526, USSR

KOPPENS, W., Department of Mechanical Engineering, Eindhoven University of Technology, Postbus 513, NL-5600 MB Eindhoven, The Netherlands

KORTÜM, W.; Institut für Dynamik der Flugsysteme, DFVLR, D-8031 Oberpfaffenhofen, Post Wessling, FRG

* KREUZER, E.J. , Institut B für Mechanik, Universität Stuttgart, Pfaffenwaldring 9, D-7000 Stuttgart 80, FRG

o* LIKINS, P., Office of the President, Alumni Building 27, Lehigh University, Bethlehem, PA 18015, USA

* LILOV, L., Institute of Mechanics and Biomechanics, ul. "Acad. G. Bonchev", bl. 8, 1113 Sofia, Bulgaria

* LUGNER, P., Institut für Mechanik, Technische Universität Wien, Karlsplatz 13, A-1040 Wien 4, Austria

MÄGI, M., IKP - Linköping Institute of Technology, S-58183 Linköping, Sweden

* MÜLLER, P.C., Sicherheitstechnische Regelungs- und Meßtechnik, Bergische Universität, Gauss-Strasse 20, D-5600 Wuppertal 1, FRG

* NIELAN, P., Division of Applied Mechanics, Durand Bldg., Stanford University, Stanford, CA 94305, USA

OLEDZKI, A., Polytechnic of Warsaw, ITLIMS, Al. Niepodleglosci 222, Warsaw, Poland

PARENTI, C.V., Instituto di Meccanica Applicata alle Macchine, Facoltà di Ingegneria, Università di Bologna, Viale Risorgimento 2, I-40136 Bologna

o DE PATER, A.D., Krammer 4, NL-2641 TZ Pijnacker, The Netherlands

* PEDERSEN, P., Department of Solid Mechanics, Bldg. 404, Technical University of Denmark, DK-2800 Lyngby, Denmark

* PFEIFFER, F., Lehrstuhl B für Mechanik, Technische Universität München, Postfach 20 24 20, D-8000 München 2, FRG

* POPP, K., Institut für Mechanik, Universität Hannover, Appelstraße 11, D-3000 Hannover 1, FRG

o PRENTIS, J.M., Queen's College, University of Cambridge, Cambridge CB3 9ET, UK

* RENAUD, M., LAAS-CNRS, 7, rue du Colonel Roche, F-31077 Toulouse Cedex, France

* RIMROTT, F.P.J., Department of Mechanical Engineering, University of Toronto, Toronto, Ontario M5S 1A4, Canada

* ROBERSON, R., Dept. of Applied Mechanics and Engineering Sciences, Mail Code B-010, University of California, San Diego, La Jolla, CA 92039, USA

* SAMIN, J.C., Lab. d'Automatique et d'Analyse des Systèmes, Université Catholique de Louvain, 3, place du Levant, B-1348 Louvain la Neuve, Belgium

* SARYCHEV, V., Institute of Applied Mathematics, The USSR Academy of Sciences, Miusskaja Square 4, Moscow 125047, USSR

°* SCHIEHLEN, W.O., Institut B für Mechanik, Universität Stuttgart, Pfaffenwaldring 9, D-7000 Stuttgart 80, FRG

SCHWAB, A., Lab. for Engineering Mechanics, Delft University of Technology, Mekelweg 2, NL-2628 CD Delft, The Netherlands

* SCHWEITZER, G., Institut für Mechanik, ETH-Zentrum, CH-8092 Zürich, Switzerland

* SCHWERTASSEK, R., Institut für Dynamik der Flugsysteme, DFVLR, D-8031 Oberpfaffenhofen, Post Wessling, FRG

SHARP, R.S., Department of Mechanical Engineering, University of Leeds, Leeds, LS2 9JT, UK

* UNDA, J., Escuela Superior de Ingenieros Industriales, Universidad de Navarra, Urdaneta 7, E-20006 San Sebastian, Spain

* WEBER, R., Theoretical and Applied Mechanics Dept., Hilti AG, FL-9494 Schaan, Liechtenstein

WIELENGA, Th., Mechanical Dynamics Inc., 3055 Plymouth Road, Ann Arbor, MI 48105, USA

* WILLEMS, P., Lab. d'Automtique et d'Analyse des Systèmes, Université Catholique de Louvain, 3, place du Levant, B-1348 Louvain la Neuve, Belgium

°* WITTENBURG, J., Institut für Mechanik, Universität Karlsruhe, D-7500 Karlsruhe, FRG

WOHLHART, K., Institut für Mechanik, Technische Universität Graz, Kopernikusgasse 24, A-8010 Graz, Austria

ZHU, M.M., Department of Basic Sciences, Nanjing Aeronautical Institute, Nanjing, China

°* ZIEGLER, F., Institut für Allgemeine Mechanik, Technische Universität Wien, Karlsplatz 13, A-1040 Wien 4, Austria

Contents

Dynamics for Flexible Systems

A. BARRACO, B. CUNY, G. ISHIOMIN

Ecole Nationale Supérieure d'Arts et Métiers
Département Structures, Equipe Robotique
151, boulevard de l'Hôpital - 75640 PARIS CEDEX 13 (FRANCE)

Abstract

Two formulations of dynamics equations accounting for deformation effects are presented. Both show benefits and disadvantages that are discussed. An example demonstrates the method's capabilities.

1 - Introduction

In recent times, there was no appropriate calculation means available to engineers in order to predict and describe the dynamics behavior of any complex mechanism that they conceived. The most stringent hypotheses were therefore used such as: reducing an n-degrees of freedom mechanism to n separate one-degree of freedom mechanisms; ideal and non dissipative member connections; perfectly rigid bodies.

The first two hypotheses were recently discarded thanks to the development of multi-body systems analysis and to the production of automatic dynamics behavior simulation codes. However, such general computer codes do not account for all possible situations, and two major problems still remain: first generate dynamics equations for complex spatial kinematic loops, and secondly compute internal efforts at member connections for any mechanism. Since most real cases require the solution for these problems, the above calculation codes only apply to limited situations.

The third hypothesis, dealing with member rigidity, is usually accepted since disregarding it would induce a significant increase of difficulty to establish the phenomenon's equations. However it is sometimes necessary to take into account structural deformations to more closely describe the

Dynamics of Multibody Systems
IUTAM/IFToMM Symposium Udine 1985
Editors: G. Bianchi and W. Schiehlen

actual mechanism's behavior. Spatial manipulators usually show small velocities but they carry such heavy loads (e.g. satellites) that inertial effects induce substantial end displacements due to member flexural deformations. Industrial manipulators operate at large velocities while, at the same time, they are required to provide a very good repeatability: in those cases, both deformations in members and at member connections must be accounted for. Let us note that when using deformable members, all internal efforts at member connections can be calculated, however complex the mechanism may be.

For these reasons, we investigated the dynamics modelling of mechanisms and robots composed of both rigid and deformable portions. In the first section, the various available approaches are presented, for which advantages and disadvantages are pointed out. In a second section, an example illustrates our theories. A complex spatial mechanism is analyzed to compare outcoming results when either rigid or deformable behavior is assumed.

2 - Bibliography :

Previous works can be classified according to two criteria: first the methods used to introduce deformation, secondly the investigated mechanism. Works were started in the early 70's which dealt with mostly plane specific cases such as the slider-crank mechanism in which the lever was assumed flexible, or the well-known four-bar mechanism. Authors wrote down specific equations for those structures, considering that member masses were concentrated at given points.
Later on, the Finite Element Method provided a general frame of research and developments, the authors either developing their own code based on the F.E.M.'s theory or using an existing general code to introduce a new formulation. Little results are available for spatial mechanisms, and no comparison between results of different origins is drawn. Turcic

and Midka [1] provide a recent and exhaustive bibliography of American contributions. In Europe, as far as we know, some research was led at the University of Louvain by Samin [2], at the University of Karlsruhe by Wittenburg [3]. In France, theoretical developments are under way at the Bertin Company [4] and at the CEA [5]. More interest was put into the experimental aspect of flexible systems command by the CERT in Toulouse and MATRA-ESPACE Company.

3 - Dynamical model of a mechanical system

The dynamical model of a mechanical system incorporates both the definition of a scheme, simplified model of the real complex structure, and the elaboration of the differential equations, the solution of which provides the dynamics behavior of the model, and thereon of the real system. Let us examine hereafter these two points.

3-1 Schematic model of a real mechanical system

In order to study a given complex mechanical system, different hypotheses are set for its different composing parts: some may be rigid enough to be assumed solid while others on the contrary will show significant deformation. The structural model under study will therefore mix rigid components, called joints, and deformable ones, called members, connected by means of non dissipative hinges, possibly showing some flexibility.

Each rigid joint is characterized by the position of its center of mass, the value of its mass and the components of its inertia tensor. On a given joint, an arbitrary number of flexible members can be attached through three different types of connections: rigid connection, sliding hinge or pin joint.

As far as three dimensional deformable bodies are concerned, their elastic behavior is described using classical continuum mechanics formulations, such as beam theory, plate or shell theory, or full three dimensional continuum theory. Because of their simple geometrical shapes, the mechanical structures we analyze only require beam theory, which we indeed used.

3-2 Elaboration of the differential equations governing the movement

Deriving these differential equations has been long studied and will not be discussed herein. Different methods will be used for rigid and for flexible members, the complete structure being modelled by mixing all equations. Let us examine these three points.

Rigid member case: let us isolate a rigid part, or joint fig. 1; it is subjected to two types of external actions ; some are known, e.g. imposed forces or torques, while others are unknown and come from the interaction of contiguous members connected at the joint.

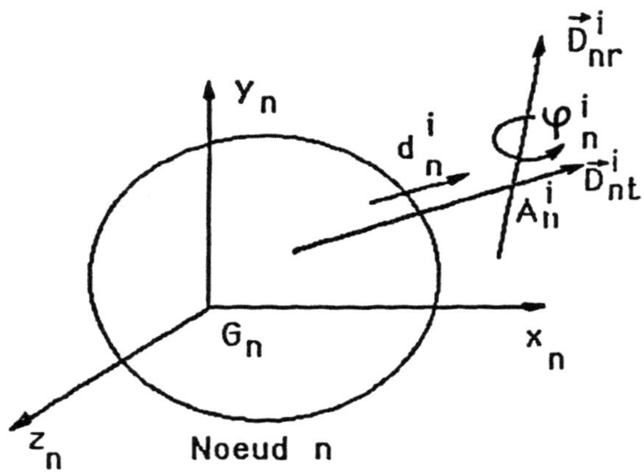

Fig. 1.: Rigid node n, connection i

Let us attach to each joint a referential frame located at point G_n, the joint's center of mass. The ith connection to a contiguous element is defined in point A^i_n as follows. Starting from point B^i_n along a given direction vector $\vec{D}^i_{nt}$ fixed in the joint's frame, we locate joint A^i_n at a distance d^i_n which can be variable and which represents a translation. In point A^i_n a second direction vector $\vec{D}^i_{nt}$ represents the axis along which the contiguous element can be subjected to a rotation of amplitude ϕ^i_n. Whether parameters d^i_n and ϕ^i_n are variable or not, we deduce all possible types of member connections

(built-in, rotation, translation, combined).
Euler's equations turn out to be in the joint's frame:

$$\underline{M}_n \cdot \Gamma_n = \underline{H}_n \cdot X_n \tag{1}$$

where matrix $\underline{M}_n$ contains mass and inertia tensor components, Γ_n includes linear and angular accelerations and X_n stands for both known and unknown external efforts. All those quantities being expressed at time t.

Equation (1) is the total form of Euler's equations. When two successive time steps t-dt and t are considered, we deduce the incremental form by linearization and substraction:

$$\underline{M}_n \cdot \bar{\gamma}_n = \underline{H}_n \cdot \bar{X}_n \tag{2}$$

where $\underline{M}_n$ is the same matrix as above computed at time t-dt, while $\bar{\gamma}_n$ and $\bar{X}_n$ represent increments of accelerations and external efforts between time t-dt and t.

Deformable member case: the dynamics behavior model is more difficult to set up in this case. The first problem arises in describing the movements of the different particles of each member: how can one define a moving reference frame when both distances and angles are continuously changing due to member deformations?

Different moving frame definitions exist which give more or less satisfaction. The best one, in our mind, would be to use as moving frame associated to the moving deformable body the frame attached to the same body considered as rigid and submitted to the same loading process.

Let us consider, figure 2, a deformable beam carrying a concentrated mass at one end and fixed by a pin joint at the other end. Let us call $(\vec{e}_s)$ the moving frame attached to the beam when assumed as rigid. Figure 2 shows two possible definitions of moving frames associated to the real deformable beam at a given time t, say $(\vec{e}_d)$ and $(\vec{e}_d')$. If the beam deformation vanishes, these two frames coincide and reduce to the rigid beam frame $(\vec{e}_s)$; whereas for significant deformations it turns out to be impossible to specify, a priori, the relative position of these three moving frames which depends on the mass distribution, the time history of the movement. In the simple case under consideration, if the concentrated mass over distributed mass ratio is large, it appears that the $(\vec{e}_d)$ frame follows move "closely" the movement of the rigid body frame $(\vec{e}_s)$.

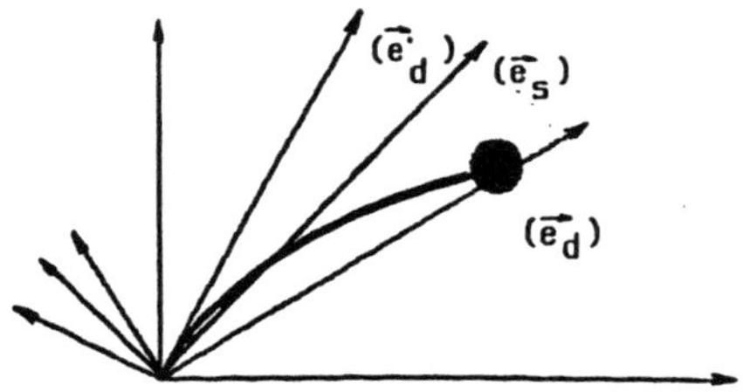

Fig. 2. Frame associated with a deformable element

Definition of the position of an arbitrary point

Describing the position of a particle in a Galilean frame at a given time t requires to know the entire member's deformation since it contributes to it. Let us consider two member's positions of interest: the present one at time t, and the initial one at time $t_o = 0$. Let us also introduce a third, intermediate, configuration that we call "reference position": it can be either a real position, previously reached by the deformable member at time t', $(t_o \leq t' \leq t)$, or a virtual position reached from the original configuration through a rigid-body movement.

Let P and P_r be the same member point in the actual and reference positions respectively, at time t. The following vectorial equation can be written for the point's position:

$$\vec{OP} = \vec{OP}_r + \vec{P_rP} \tag{3}$$

Point P_r can be chosen in many different ways as mentioned above, but two choices appear of practical interest since they lead to suh a sufficiently small $\vec{P_rP}$ vector that the geometrical transformation between the reference and actual positions can be linearized.

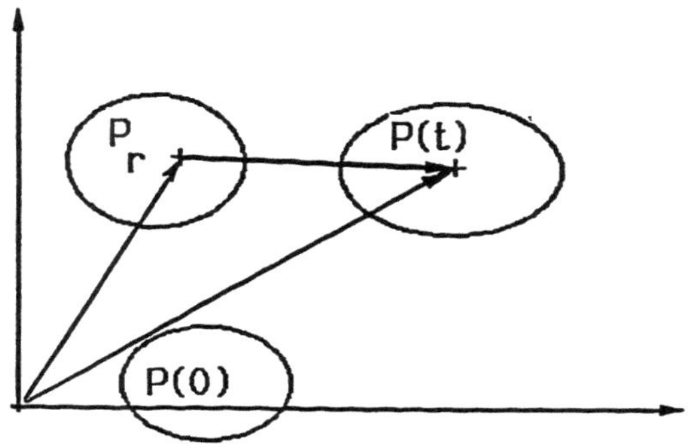

Fig. 3: Introduction of a reference frame

The first choice for the reference position, figure 4, is based on a rigid body transformation of the initial configuration in such a way that its associated frame is moved to come in the vicinity of the actual frame at time t. Under such hypotheses, the point's position parameters can be separated into two groups:

- on the one hand, six finite amplitude parameters q_s, in the most general case, defining the position of the point attached to the reference configuration.
- on the other hand, a certain number of small amplitude parameters q_d, defining member's deformations.

Let us collect both types of parameters q_s and q_d in a unique unknown vector q. The virtual work principle, when applied to the mechanical system's movement with small deformations, leads to the following equation:

$$\underline{A}_b \cdot \ddot{q} + B_b + \underline{K}_b \cdot q_d = \underline{H}_b \cdot X_b \qquad (4)$$

The first two expressions of this summation come from the virtual work of acceleration quantities: they include the well-known terms of a rigid body dynamics equation; terms involving first and second time derivatives of deformation parameters, and coupling terms of the two parameter types. The $\underline{K}_b \cdot q_d$ product, of the rigidity matrix $\underline{K}_b$ and deformation parameters vector q_d, accounts for the system's deform-

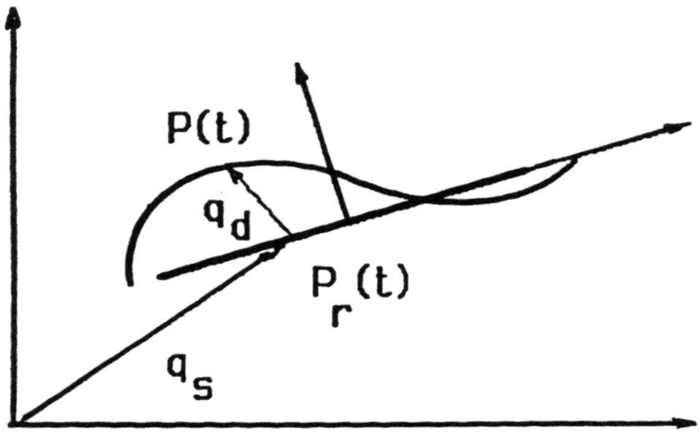

Fig. 4: First choice of the reference configuration

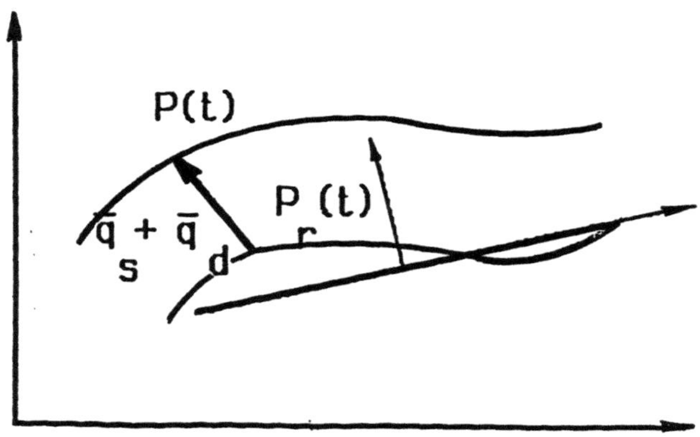

Fig. 5: Second choice of the reference configuration

ation. This method does not account for material and geometric non linearities. Finally, the right-hand side product accounts for the forces and torques vector X_b applied to individual members, some of them being given while others resulting from interactions between contiguous members are unknown.

When the q_d parameters are set to zero, there remains the well known rigid body systems dynamics equation.

The second choice, figure 5, takes as reference position the configuration actually reached at time t-dt. Therefore, the geometrical transformation from time t-dt to time t can be considered small enough, and it is decomposed into a rigid body part characterized by parameters $\bar{q}_s$ and a deformation-involving part characterized by parameters $\bar{q}_d$. Since the displacement from t-dt to t is small, the $\vec{P_rP}$ vector linearly depends on the $\bar{q}$ vector that incorporates both $\bar{q}_s$ and $\bar{q}_d$.

When the virtual work principle is written at successive times t-dt and t while considering the same reference position, the difference between these two equations suppresses that part of the movement from time t_o to time t-dt, and there remains an incremental form of the principle that reads:

$$\underline{M}_b.\ddot{\bar{q}} + \underline{G}_b.\dot{\bar{q}} + \underline{K}_b.\bar{q} = \underline{H}_b.\bar{X}_b \qquad (5)$$

In this equation, the $\bar{X}_b$ vector accounts for the increments of both known and unknown applied efforts; $\ddot{\bar{q}}$ and $\dot{\bar{q}}$ are the second and first time derivatives of the position parameter increments $\bar{q}$; matrices $\underline{M}_b$, $\underline{G}_b$ and $\underline{K}_b$ are computed for parameter values at time t-dt and represent the standard mass matrix, the gyroscopical effects matrix and the stiffness matrix. The $\underline{K}_b$ matrix contains usual first and second order terms, but moreover includes dynamical rigidity terms whose effect is to increase or diminish the standard rigidity matrix terms according to the value of the acceleration quantities to which the element is subjected.

<u>Obtaining the complete model</u>: the complex mechanism was split into separate simple parts for which we wrote down

dynamics equations corresponding to given hypotheses. To solve the entire real problem, we must complete the previous set of equations with two types of relations: first, some expressing the interaction principle, and then others specifying kinematical conditions, all to be satisfied at contiguous members connections.

All those equations together - dynamics and connections - constitute the entire model in which the unknowns are first the kinematical variables and also the connection efforts. Let us note that a complete solution, in terms of displacements and efforts, can be found only if all equations are independent, in which case the above equation system is maximum in size (largest number of equations and unknowns). If connection efforts are eliminated and kinematical relations are used, the entire system can be reduced to a minimum size, with equations involving only independent kinematical variables and known external efforts. In between these two extrema, the number of equations to be solved varies according to the chosen solution method which can be closer to either the maximum or the minimum size.

We chose to use the incremental form and write down the mininum sized system of equations. Besides, we introduce the implicit β-Newmark integration scheme as soon as possible in the calculations, i.e. just after individual element equations are written. This leads to a pseudo-static form for the equations:

- equation (2) becomes $\underline{K}_n.\bar{u}_n = \underline{H}_n.\bar{X}_n + f_{dn}$ (6)
- equation (5) becomes $\underline{K}_b.\bar{q}_b = \underline{H}_b.\bar{X}_b + f_{db}$ (7)

in which expressions $\underline{K}_n$ and $\underline{K}_b$ are called pseudo-stiffness matrices, f_{dn} and f_{db} are dynamics efforts vectors, while $\bar{X}_n$ and $\bar{X}_b$ are efforts increment vectors for joints and members. After simple geometrical considerations are used to relate the $\bar{q}_b$ variables to the nodal displacements $\bar{u}_n$, we are led to solve a matrix assembly problem such as in static analysis. We obtain:

$$\underline{K}.\bar{u} = \underline{H}.\bar{X} + f_d \qquad (8)$$

The only unknowns $\bar{u}$ are solved for, from which all parameters are computed, the new geometry is calculated and new initial conditions for the following step are established.
The program that we created manages those calculations in three steps. A first conversational part enables the user to define, modify and save his mechanism. A second part deals with the time integration of the movement in an iterative process. When numerical results diverge, any previously saved step can be used for a restart. A third part, also conversational, is available to scan through the numerous results.

4 - Example:

On figure 6 the mechanism under investigation is presented: it is composed of a planar four-bar parallelogram rotating around a fixed axis contained in that plane. It can be realized that this structure is a simplified description of an entire class of mechanical robots showing planar deformable parallelogram structures.

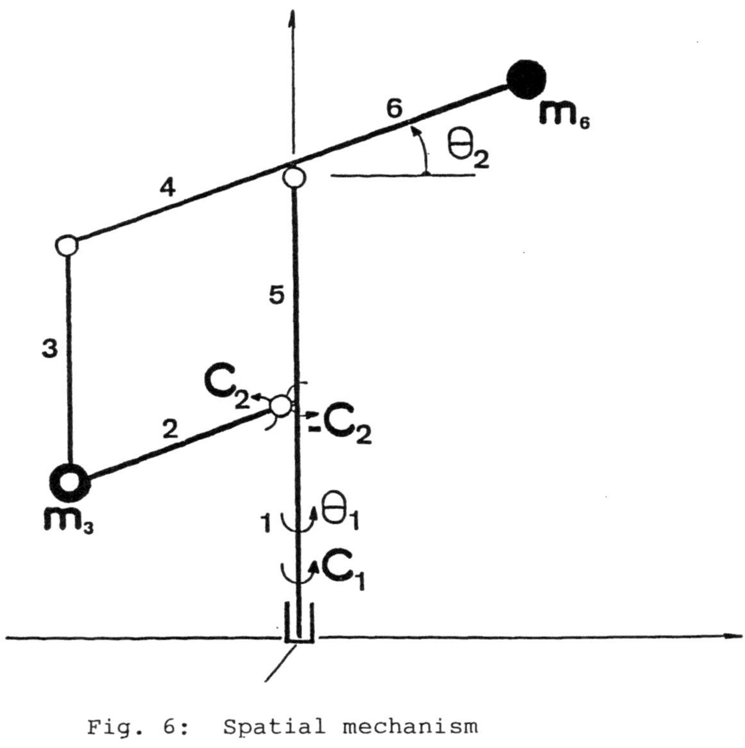

Fig. 6: Spatial mechanism

We suppose that all structural components are deformable and that concentraded masses account for the actuators and displaced end mass.
When all structural members are assumed rigid, it remains a two degrees of freedom mechanism for which θ_1 indicates the structure's plane position while θ_2 shows the parallelogram's position in that plane. When all structural members are assumed flexible and are analyzed as one single element each, thirty five degrees of freedom are necessary to represent the spatial dynamical behavior of the mechanism.
The applied efforts are two concentrated torques $C_1(t)$ and $C_2(t)$ which start at distinct times, see figure 7 . From time $t_o = 0$ to $t_1 = 1.$ second, C_1 only is acting on the whole structure which therefore starts rotating around the fixed axis. Thereon, from $t_1 = 1$ to $t_2 = 2.$ seconds, C_1 remains constant while torque C_2 starts acting so that the concentrated end mass moves up. Finally up to 3. seconds, C_1 diminishes to 0 while C_2 remains constant.
It should be noted, at this point, that each structural member could be subdivided into several elements, if variable inertia values should be taken into account for instance or if a better precision was looked for.
Two goals are aimed at, in this example: first show that it is feasible to carry out the dynamics analysis of a spatial mechanism built of flexible members; secondly compare the outcoming results to those of a classical rigid body-based analysis of the same mechanism. The numerical effects of the integration algorithm is suppressed by utilising the same algorithm and the same time step, namely $2.5\ 10^{-3}$ second, in the two analyses. This is not sufficient however to insure that no numerical perturbation will occur since the algebraic structure of the governing equations itself plays its part.
The results that we compare are the values of the θ_1 and θ_2 parameters, previously defined for the rigid body mechanism, and their counterparts for the flexible member mechanism. Figure 8 shows how different the solutions are: the angles in the deformable structure case are of smaller amplitudes, from which fact we can deduce that some portion of the applied efforts goes into deforming the structural members.

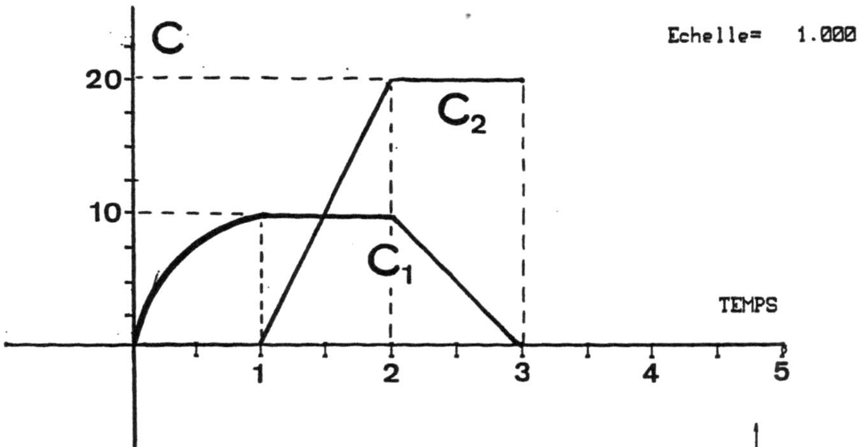

Fig. 7: External applied torques

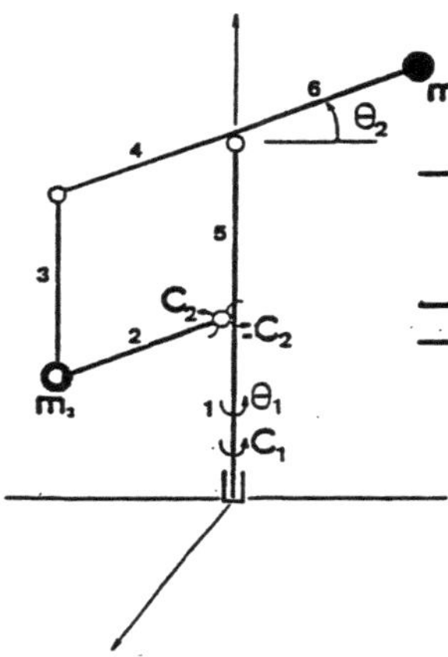

ENSAM PARIS ROBOTIQUE

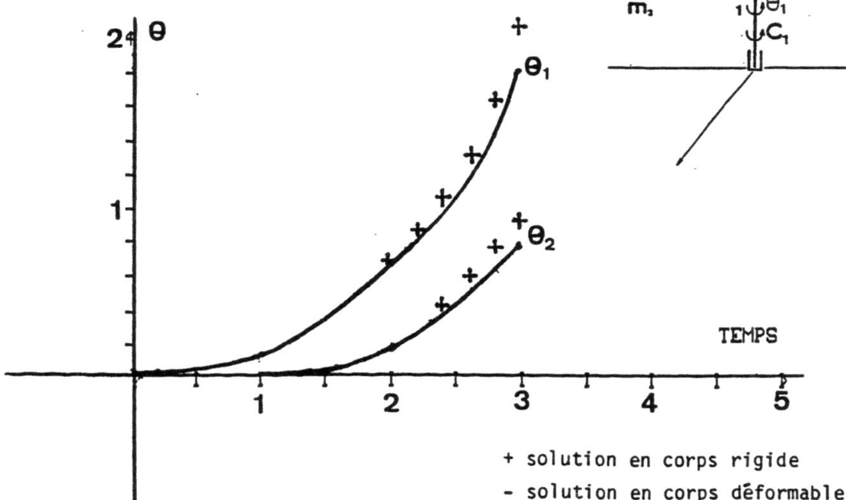

Fig. 8: Comparison between rigid or deformable solution.

ENSAM PARIS ROBOTIQUE

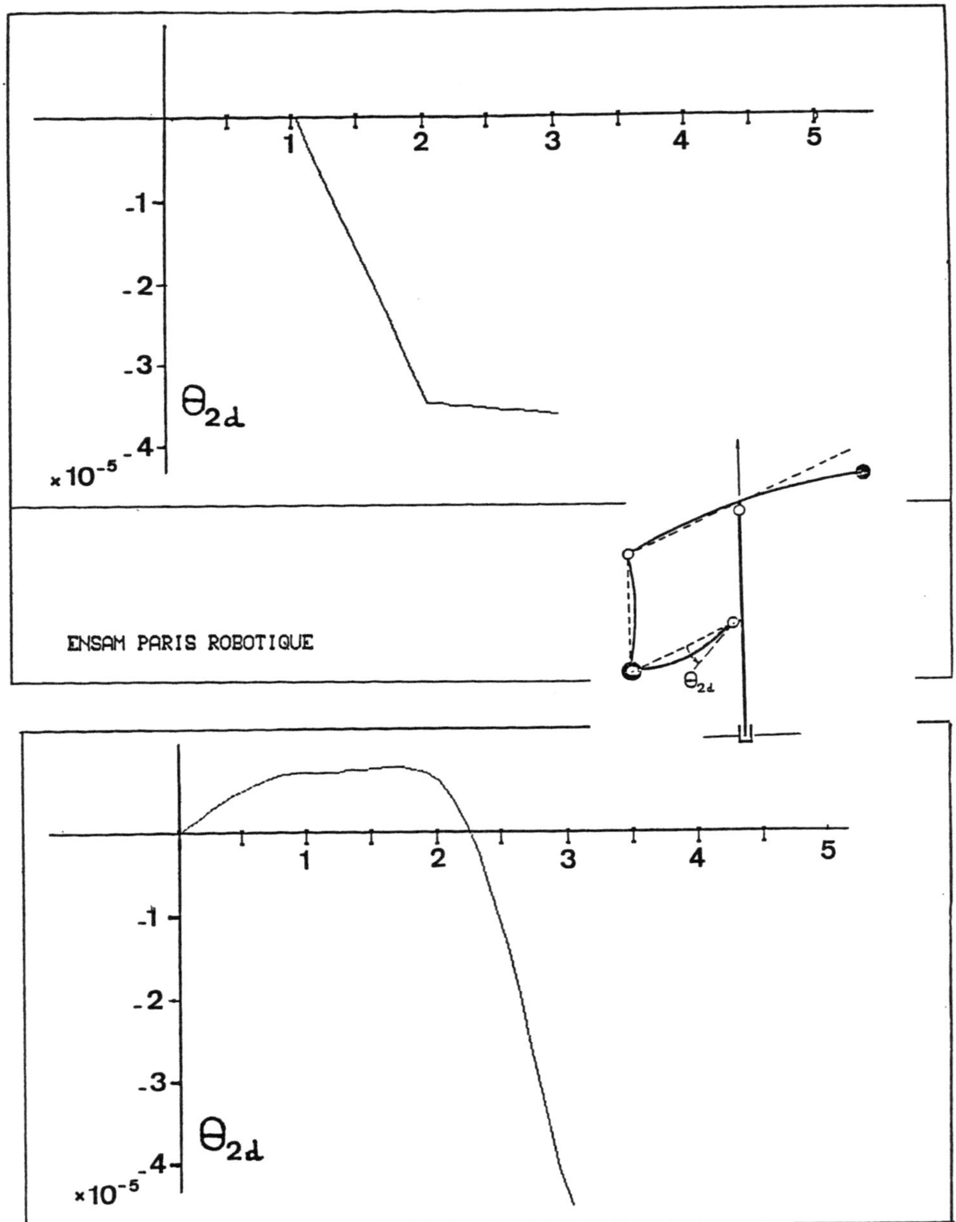

Fig. 9: Deformations in the two principal planes of flexion.

ENSAM PARIS ROBOTIQUE

Finally, figure 9 shows the end deformation variations with time for that end of member 2 where torque C_2 is applied. We have drawn deformation curves both in the flexural plane due to torque C_2 and the plane perpendicular to it. The first curve can be shown to approximately follow the shape of torque C_2 itself, with zero deformation from $t_o = 0.$ to $t_1 = 1.$ second when the rotating parallelogram does not deform. Member deformation in the perpendicular plane is due to a flexion out of the initial parallelogram's plane coming from inertial effects of rotating masses. These two types of deformations are of the same order of magnitude.

5 - Conclusion:

In a first part, we described the two approaches that we used to account for member deformations. Depending on how the reference configuration is chosen, two formulations are elaborated. In the first one, part of the unknowns represent the finite rigid body movement while the remaining part deals with the infinitely small deformation of members. In the second formulation, an incremental form of the virtual work principle is derived by subtracting virtual work equations written at two successive time steps, and the resulting unknowns are displacement increments of both solid body and deformational nature, from which small enough increments a linearized form is derived.

In a second part an example is analysed for application of the previous theory. The example deals with a relatively complex spatial structure that was mostly used to show the potential of the proposed method.

Finally, let us emphasize the complexity inherent to these methods: large number of terms to be computed, numerous integration time steps at which all parameters must be actualized. This complexity also appears in the calculation cordes developped from such methods: long calculation times on computers, large amount of data to be saved appear as the two major characteristics of any real world applications.

As a conclusion, we think that these methods become necessary when structural elements are rather deformable and when very precise results are required, or when efforts developed at member connections are looked for, whatever the type of structure.

Acknowledgement: this work was achieved as part of the national research program "Automatisme et Robotique Avancée".

6 - References

1. Turcic, Midka "Generalized equations of motion for the dynamic analysis of elastic mechanism systems". ASME - Journal of Dynamic Systems, Measurement, and Control, Vol. 106, Decembre 1984, pp. 243-260.

2. Samin, Willems, Boland "Stability analysis of interconnected deformable bodies in a topological tree". AIAA Journal, Vol 12, no 8, août 1974.

3. Wittenburg "Dynamics of systems of rigid bodies". Ed. Teubner, Stuttgart 1977.

4. Bertin CONTRAT MRT - Analyse dynamique d'assemblage mécanique.

5. Code PLEXUS - CEA - Manuel d'utilisation.

Stability of Several Degrees of Freedom Moving Subsystem Interacting with Waves of Continuous Media

Roman BOGACZ

Polish Academy of Sciences
Institute of Fundamental Technological Research

Summary

The stability problem presented in this paper is an supplement to the investigation concerning behaviour of the continuous systems with the moving load, in particular with the inertial type of concentrated and densely distributed load. The considerations deal with several degrees of freedom system which is interacting either in a few points or continuously with the travelling wavês of the continuous media. The discrete subsystem moves along the continuous system with the constant velocity. It is shown, that in the case of two interaction points between the both subsystems there exists a countable set of the bounded instability regions where solutions have an oscillating nature.

Introduction

The stability of relative motion of same continuous and mass-spring systems constitutes an important engineering problem and has extensively been investigated. For such systems some results were obtained for interaction of a continuous system with concentrated masses or densely distributed oscillators [1,2] as well as for stability of interactive multibody systems [3]. In this paper an extension of the mentioned results is performed and presented. We formulate the fundamental equations as well as the boundary conditions of the considered problem. Than we describe the method of solution. As ilustrative examples we determine the instability regions for two cases of interaction of oscillators with a travelling wave of continuous subsystem.

Dynamics of Multibody Systems
IUTAM/IFToMM Symposium Udine 1985
Editors: G. Bianchi and W. Schiehlen

The first case deals with a infinite set of densely distributed oscillators (each of them is with a few degrees of freedom) while the second is devoted to the single oscillator interacting in two points with the continuous subsystems

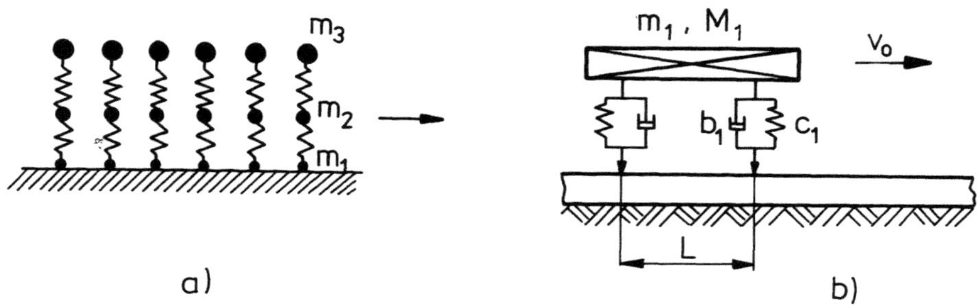

Fig. 1

Our considerations will be performed, under following assumptions:

a) relative velocities of subsystems are constant,

b) contact surfaces of continuous subsystems during undisturbed motion are planes and contact lines are straight,

c) friction on all contact surfaces is negligible,

d) viscoelastic elements which connect particular concentrated masses and lumps are massless and

e) gravitational interactions are disregarded.

Let us connect to each i-th subsystem an inertial coordinate frame $\xi_j^{(i)}$ (j=1,2,3) in the way defined by the following relations

$$\xi_1^{(1)} - \xi_1^{(2)} - v_{21}t = 0,\ \xi_1^{(2)} - \xi_1^{(3)} - v_{32}\,t = 0, \tag{1}$$

$$\xi_2^{(1)} = \xi_2^{(2)} = \xi_2^{(3)}, \quad \xi_3^{(1)} = \xi_3^{(2)} - c_2 = \xi_3^{(3)} - c_3 \tag{2}$$

where

v_{ij} - relative velocity of i-th with regard to j-th subsystem,

t - time and

c_i - some constants.

Equations of motion and boundary conditions of i-th subsystem may be symbolically written in the following way

$$\sum_{n,j,k,l,m=0}^{p} a_{njklm}^{(i)} \frac{\partial^{j+k+l+m}}{\partial \xi_1^{(i)j} \partial \xi_2^{(i)k} \partial \xi_3^{(i)l} \partial t} w_n^{(i)} = 0 \tag{3}$$

denoting its displacement by means of $w_n^{(i)}$. These equations may be also written with regard to only one coordinate frame by means of (1) and (2). Boundary conditions express; (i) a compatibility between displacements and forces (or normal stresses) of mutually interacting subsystems, (ii) the absence of shear stresses on contact surfaces and (iii) radiation conditions. The solutions of equations of motion should describe vibrations in the case of discrete subsystems and travelling waves in the case of continuous subsystems. Suppose that they have the following form:

$$w_n^{(i)} = \sum_{x=0}^{e} \Psi_{nk}^{(i)} (\xi_2^{(i)}, \xi_3^{(i)}) \exp(r_{nk}^{(i)} \xi + \lambda_k^{(i)} t). \tag{4}$$

Substituting (4) into equations of motion and boundary conditions and using (1) and (2) we obtain the relation between eigenvalues λ_k and motion parameters v_{ij}. This relation (in the case of fixed other system parameters) will be called the characteristic equation and written in the following way

$$\Phi(\lambda, s) = 0, \tag{5}$$

where

$s = (v_{21}, v_{32}),\ s \in S = V_{21} x V_{32} (v_{21} \in V_{21},\ v_{32} \in V_{32})$.

This equation is the basis of discussion concerning stability of the considered system.

Explicit determination of instability set S_I on the basis of (5) is not always possible. For instance, when the characteristic equation is either a polynomial of degree higher that fourth or a transcedental equation this determination is possible only in some special cases. In such a case the use of some auxilliary criteria makes the stability analysis possible. Each of them, for fixed motion parameters s_0 and system parameters, is equivalent to the condition

$$(\forall \lambda\ C) \Phi(\lambda, s_0) = 0 \ . \ Re(\lambda) < 0 \qquad (6)$$

where C is the set of all complex numbers the complex space. Provided that $\Phi(\lambda,s)$ is a polynomial in λ the following auxiliary criteria are generally used:

1. Routh-Hurwitz stability criterion (cf. [4], [5]) whose modified version takes into account also the more general case when coefficients of the polynomial are complex,
2. Mikhaylov stability criterion, which is based on an examination of the curvest configuration

$$\Phi_R = f(\Phi_I) \qquad (7)$$

given in the complex plane where this configuration is defined parametrically

$$\Phi_R = Re(\Phi(\lambda,s))\Big|_{\lambda = i\omega,\ s=s_0},\ \Phi_I = Im[\Phi(\lambda,s)]\Big|_{\lambda = i\omega, s=s_0}, \qquad (8)$$

where

$$\omega = Im(\lambda) \in (-\infty, +\infty) .$$

3. Generalized Mikhaylov stability criterion [5] on the basis of shapes of curves $Re[\Phi(i\omega,s)]=0$ and $Im[\Phi(i\omega,s)]=0$ allows an illustrative separation of motion parameters characterizing stability and instability domains in the space Q where

$$Q = I \times V_{21} \times V_{32} \tag{9}$$

The above mentioned procedure will be illustrated by means of the following two particular cases.

Stability of Interaction of a set of Densely Distributed Oscillators with Travelling Wave of Euler Beam under an Elastic Foundation

Similar as in [2], [6-7] for the case of oscillators possessing one degree of freedom we consider now a case of the moving subsystem of oscillators with a few degrees of freedom. In the general case we assume additional that the single oscillator has a mass m_1 on the contact with the continuous subsystem (y_0) (Fig. 2).

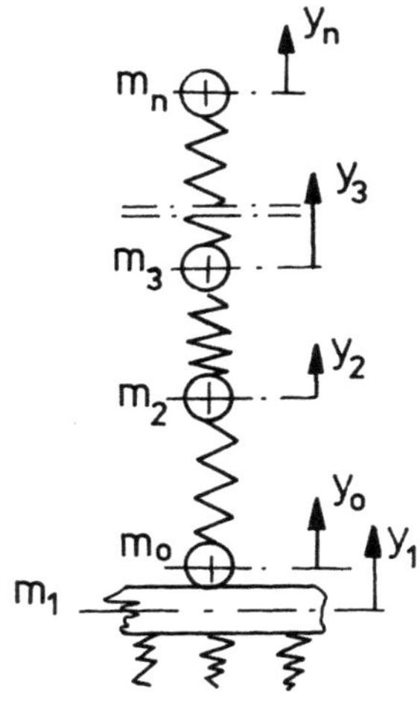

Fig. 2

Equations of motion for such an oscillator are as follows:

$$
\begin{aligned}
&m_2\ddot{y}_2 + c_1(y_2 - y_0) - c_2(y_3 - y_2) = 0,\\
&m_3\ddot{y}_3 + c_2(y_3 - y_2\ ) - c_3(y_4 - y_3) = 0,\\
&\cdots\cdots\cdots\cdots\cdots\cdots\cdots\cdots\cdots\cdots\\
&\cdots\cdots\cdots\cdots\cdots\cdots\cdots\cdots\cdots\cdots\\
&m_n\ddot{y}_n + c_n(y_{n+1} - y_n) - c_n(y_{n+1} - y_n) = 0,
\end{aligned}
\tag{10}
$$

where the contact force is given by the formula:

$$
\begin{aligned}
&m_1\ddot{y}_0 + c_1(y_2 - y_0) = p_0(x_1^*, t)\\
&y_i = y_i(x_1^*, t)
\end{aligned}
\tag{11}
$$

According to (1), (2) we describe the motion of continuous system in (x_j, t) - coordinate system conected with (x_j^*, t) by the following relations:

$$x_1 - x_1^* - vt = 0, \ x_2 - x_2^* = 0. \tag{12}$$

In the case of Bernoulli-Euler beam the equation takes the form:

$$EI\frac{\partial^4 y_1}{\partial x_1^4} + T\frac{\partial^2 y_1}{\partial x_1^2} + m_0\frac{\partial^2 y_1}{\partial t^2} + cy_1 = p_1(x_1, t) \tag{13}$$

where; EI denote flexural rigidity, T longitudinal force, m_0 mass density, and c linear elastic foundations constant.

Making use of solution in the following form of travelling wave:

$$
\begin{aligned}
&y_j = A_j e^{ik_j(x_1 - v_1 t)}; \ j = 0,1,2,\ldots,\\
&p_j = (-1)^j \ P\ e^{ik_j(x_1 - v_1 t)} = (-1)^j P\ e^{ik_j(x_1^* - v_2 t)}, j = 0,1,
\end{aligned}
\tag{14}
$$

and the relations (12) we find

$$A_1 = A_0, \ k_j = k, \ \mathrm{Im}(k) = 0 \text{ and } v_1 - v_2 - v = 0 \tag{15}$$

The condition of an unique solution (14) with respect to A_j yields a relation similar to [6]:

$$F_1(v_1^2) - F_2(v_2^2) = 0 \tag{16}$$

where for the case $m_2 \neq 0$, $m_3 \neq 0$, $m_j = 0, j = 0,4,5,..,$ we have:

$$F_1(v_1^2) = \frac{1}{\eta(R_1^2 - v_1^2)}, \quad R_1^2 = \frac{1}{m_0}(EIk^2 - T + \frac{c}{k^2}), \qquad \eta = \frac{m_2}{m_3}$$

$$F_2(v_2^2) = \frac{[\alpha_2^2 + \alpha_3^2\eta - v_2^2](\alpha_3^2 - v_2^2) - \alpha_3^4\eta}{\alpha_3^4\eta - (\alpha_3^2 - v_2^2)(\alpha_3\eta - v_2^2)}, \quad \alpha^2_j = \frac{c_j}{m_j k^2}$$

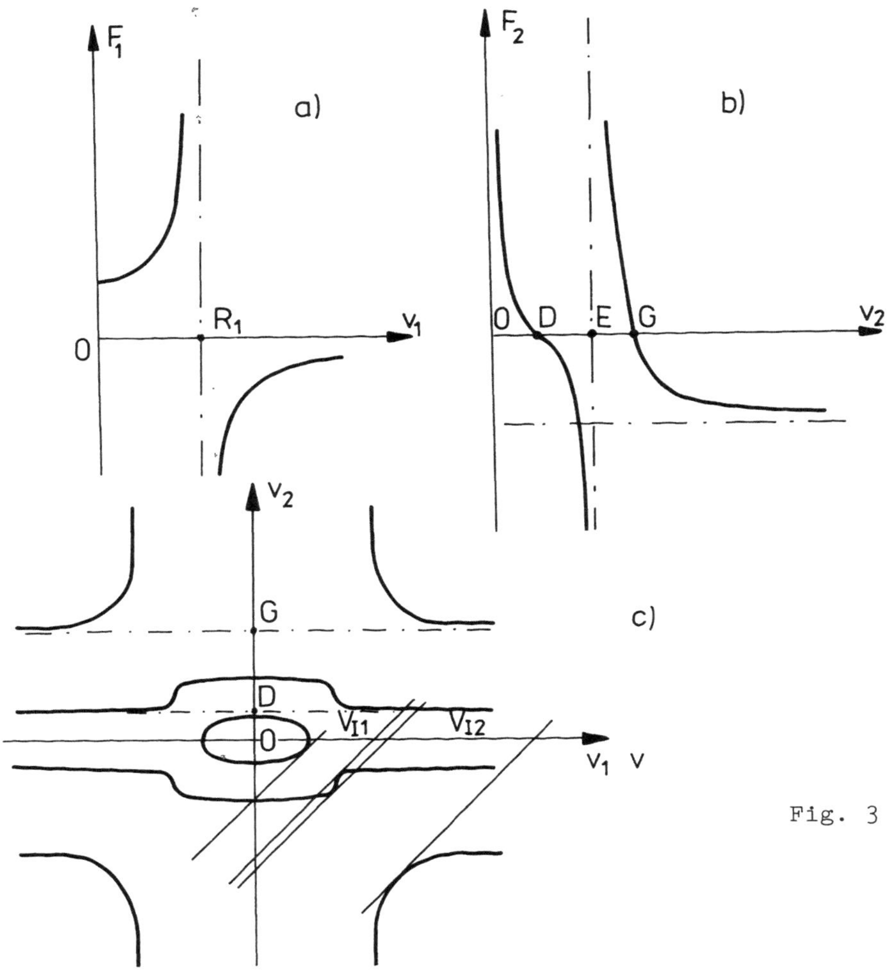

Fig. 3

Fig. 3 shows qualitatively a plot of functions $F_1(v_1^2)$, $F_2(v_2^2)$ and $F_1(v_1^2) = F_2(v_2^2)$ in v_1, v_2 plane.
Using the instability criterion (6) similar as in [2], [5,6] we determine a critical velocities v_{cr}. After comparison of results of [6] with that shoen in Fig. 3c we can conclude that each additional degree of freedom of oscillators can (not obviously) introduce additional region of instability. This depends on the value of $\eta_i = m_i m_{i+1}^{-1}$, and α_i.

<u>Interaction of Moving Lumped Subsystem with Travelling Waves of a Continuous System.</u>

We consider now similarly as in [6], [9] a subsystem possessing of at least two degrees-of-freedom Fig. 1b with mass m_i and moment of inertia M_i for i-th lump, c_{0ij} - is the elasticity of the ij-th spring. The elastic continuous subsystem is characterized by velocity of elastic travelling wave (velocity of surface wave in a general case of three dimensional system). The motion of the lump subsystem will be described in a moving coordinate system (ξ,ζ) which is related to the system of coordinates conected to the continuous system (x_1, x_2) as follows:

$$\xi - x_1 - v_0 t = 0, \qquad \zeta - x_2 = 0. \tag{17}$$

The set of questions of motion of the mass-spring system can be written in the form:

$$m_j \frac{d^2 y_j}{dt^2} + \sum_{s=1}^{n} c_{js}(y_j - y_s) + \sum_{s=1}^{n} b_{js}\left(\frac{dy_j}{dt} - \frac{dy_s}{dt}\right) = P_j(t), \quad j=1,2,\ldots,n \tag{18}$$

$$P_j(t) = c_{ojo}(y_{jo} - y_j) + b_{ojo}\left(\frac{dy_{jo}}{dt} - \frac{dy_j}{dt}\right) \tag{19}$$

where y_j is a displacement of j-th coordinate, y_{jo} denotes the corresponding prescribed displacement which is a function of cooedinates of the interaction points between the mass-spring subsystem and the continuous subsystem z_{js}, L_{js}, $j = 1,2,\dots,n$; $s = 1,2,\dots,m$, where s denotes the number of point of interaction of j-th lump. The relations characterized a system which can be written as follows:

$$y_{jo}(t) = G_j(z_{11},\dots, z_{1m},\dots, z_{n1},\dots,z_{nm}), \tag{20}$$

$$P_j(t) = F_j(P_1,P_2,\dots,P_n,L_{11},\dots,L_{1m},\dots,L_{n1},\dots,L_{nm}).$$

From the equations of motion and boundary conditions of continuous system for the oscillating moving force $P(t) = Ae^{\lambda t}$ ($\lambda=\varepsilon+i\omega$) similar as in [5] the following relation can be obtain

$$R(\lambda,L) = \frac{z(\lambda,L,t)}{P(t)}, \qquad z(\lambda,x,t) = Be^{ik(x-v^*t)}. \tag{21}$$

Substituting the solution in the form of travelling wave we obtain the charateristic equations, which will be the basis for the discussion of instability problem.

In the case of symmetric system with two degrees of freedom, without damping, with mass m, moment of inertia M interacting with continuous system at two points at distance L and for the speed greater than the velocity of elastic waves the characteristic equation takes the form

$$Q_1(\lambda) - Q_2(\lambda)R(\lambda,L) + 2Q_1(\lambda)\,Q_2(\lambda) = 0,$$

$$Q_1 = \frac{2\alpha_1 + \lambda^2}{m\alpha_1\lambda^2}, \qquad Q_2 = \frac{\alpha_2 + \lambda^2}{\alpha_1\lambda^2} \tag{22}$$

or for M = 0 after separation of real and imaginary part as follows:

$$\mathrm{Re}(\phi) = \varepsilon^2 - F_c(\varepsilon,\theta)\varepsilon + \kappa_1 - \theta^2 - F_s(\varepsilon,\theta)\theta = 0,$$
$$\mathrm{Im}(\phi) = 2\varepsilon\theta + F_c(\varepsilon,\theta)\theta + F_s(\varepsilon,\theta)\varepsilon = 0, \tag{23}$$

where

$$F_c(\varepsilon,\theta) = \frac{v^2-1}{\gamma_0} e^{-\frac{\varepsilon}{v+1}} \left(\cos\frac{\theta}{v+1} - e^{-\frac{2\varepsilon}{v^2-1}} \cos\frac{\theta}{v-1}\right),$$
$$F_s(\varepsilon,\theta) = \frac{v^2-1}{\gamma_0} e^{-\frac{\varepsilon}{v+1}} \left(\sin\frac{\theta}{v+1} - e^{-\frac{2\varepsilon}{v^2-1}} \sin\frac{\theta}{v-1}\right), \tag{24}$$

and

$$\kappa_1 = 4\alpha_1 L^2, \quad v = v_o a^{-1}, \gamma_2 = \frac{1}{2}\gamma_0 = \frac{4\rho l}{m}, \quad \alpha_1 = \frac{c}{m}, \quad \alpha_2 = \frac{cL^2}{2M}.$$

The curves representing $\mathrm{Im}\Phi = 0$ and possible instability regions according to the condition

$$\varepsilon \geq 0 \Rightarrow \frac{v^2-1}{2\gamma_0} e^{-\frac{\varepsilon}{v+1}} \left(\cos\frac{\theta}{v+1} - e^{-\frac{2\varepsilon}{v^2-1}} \cos\frac{\varepsilon}{v-1}\right) \leq 0 \tag{25}$$

can be ploted on (v,θ) plane as the functions $\theta = n\Pi(v^2-1)$ and $\theta = \frac{n\Pi}{v}(v^2-1)$ shown in all graphs at Fig. 4 for the case of string.

The influence of the values of parameters m and M on the shape of the instability regions is very important. Estimation on the basis of equation $\mathrm{Re}\Phi = 0$ is shown in Fig. 4a-4c.

The method of determination of the velocity of travelling wave as well as the relation (21) is similar as in [9].

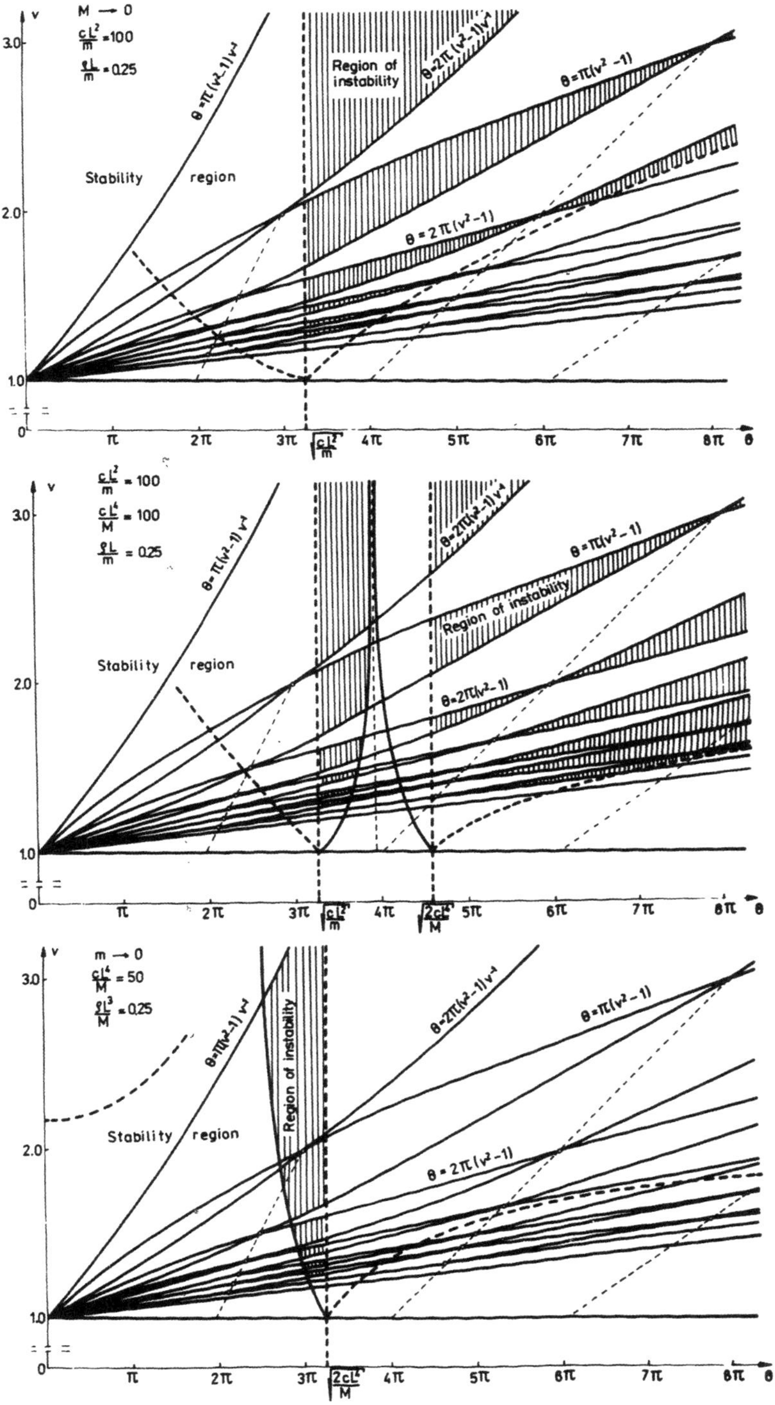
M → 0
cL²/m = 100
ϱL/m = 0.25
θ = π(v²−1)v⁴
Region of instability
θ = 2π(v²−1)v⁴
θ = π(v² −1)
Stability region
θ = 2π(v²−1)
cL²/m = 100
cL⁴/M = 100
ϱL/m = 0.25
Region of instability
m → 0
cL⁴/M = 50
ϱL³/M = 0.25
v
θ
1.0
2.0
3.0
0
π
2π
3π
4π
5π
6π
7π
8π
√(cL²/m)
√(2cL⁴/M)

a)

b)

c)

Fig. 4

The influence of damping on the critical velocities is very essential for an application but because of the limited space this problem will be discussed in separate paper.

Concluding Remarks

It follows from our consideration that in the case of interacting of the mass-spring system with the travelling wave the obtained layout of instability regions is qualitatively new. As follows from Fig. 4 for two at least points of interaction in general case a set of instability regions is always infinite but countable. The range of importance of the obtained effects for mutual interaction of the hybride systems seems to be greater than for application to mechanics.

References

1. Nelson, H.D.; Conver, R.A.: Dynamic Instability of Beam Carrying Moving Masses. Journal of Applied Mechanics, 38, 4,(1971) 1003-1006.

2. Bogacz, R.: Interaction Between a Moving Set of Nonlinear Oscillators and a Travelling Wave. Proceedings of Vibration Problems t. 9, 1, (1968), 55-77.

3. Popp, K.; Müller, P.C.: On Stability of Interactive Multibody System with Application to MAGLEV-Vehicle-Guideway Control Systems. in: Dynamics of Multibody Systems, Proc. IUTAMSymposium, Munich, 1977; Ed. Magnus, Springer, (1978), 290-305.

4. Hurwitz, A.: On the Condition under which an Equation has only Roots with Negative Real Part. Dover Publ. New York, (1964).

5. Bogacz, R.: On stability of reciprocal interaction of a continuous and discrete systems in relative motion, IFTR Reports 61, Warsaw 1977.(in polish).

6. Bogacz, R.; Popp, K.: Dynamics and stability of train-track-systems., in: Proc. 2nd Intern. Conf. on Recent Advances in Structural Dynamics, Southampton (1984) 711-725.

7. Frąckiewicz,H.; Dynamic of Concentrated masses moving along a beam resting on an elastic foundation (in polish), Rozprawy Inżynierskie, 2, 13,(1965) 397-419.

8. Mathews, P.M.: Vibrations of a beam on elastic foundation. Z. Angew. Math. u.Mech. 39, (1959), 13-19.

9. Bogacz,R.: On self-excitation of moving oscillator interacting at two points with a continuous system. Non. Vibr. Probl. 19, (1979), 240-250.

Canonical Approach to Multibody Systems Using Redundant Coordinates

H. BRAUCHLI and R. W. WEBER

Institute of Mechanics
Swiss Federal Institute of Technology, CH-8092 Zurich

and

Hilti AG
Forschungsabteilung, FL-9494 Schaan

Summary

The utility of the methods of classical dynamics as tools for formulating equations of motion is investigated. A new projection method, suitable for a general class of multibody systems, is also presented. In applying this alternative method, every body of the system is first described by its own arbitrarily chosen position and velocity variables, after which the constraints governing the system are introduced. These constraint relations are linear functions of the velocity variables. They are used to construct projection operators describing the relationship between the unconstrained and constrained system. The method allows the equations of motion to be split into two sets: one describing the motion of the constrained system, the other yielding the reactive forces. The inverse mass matrix of the constrained system is expressed in terms of an inverse matrix of lower order, and in some systems a set of explicit differential equations can be obtained. An important feature of this method is that geometric singularities are easily avoided.

1. Introduction

In current texts on dynamics of multibody systems [1,2], the methods of classical dynamics do not appear. It is the aim of this work to investigate the utility of these methods as tools for formulating equations of motion. The relationship between the equations of Lagrange and Hamilton and their generalizations, that is, the equations of Euler-Lagrange and Euler-Hamilton as stated by Hamel and others [3-7], is studied. For systems with a kinetic energy which is a quadratic function of the velocities, the equations of Lagrange and Hamilton are equivalent and simple to apply. Difficulties only arise when one wants to invert the mass matrix in order to arrive at explicit differential equations. The situation for the equations of Euler-Lagrange and Euler-Hamilton is similar, although the computation of the coefficient matrices C^h_{ki} requires additional labor.

Dynamics of Multibody Systems
IUTAM/IFToMM Symposium Udine 1985
Editors: G. Bianchi and W. Schiehlen

An advantage of these equations, however, is that they enable the velocity variables to be chosen arbitrarily, a capability shared with the method of Kane [2].

While for a single rigid body the computation of the kinetic energy and the mass matrix is simple, in multibody systems it may be tedious due to the kinematic relations to be considered. We therefore view a multibody system as consisting of free bodies. For this system the mass matrix and its inverse can easily be computed. Subsequently, we introduce constraints. Projection operators α and β are then constructed from the linear velocity relations. These operators split the tangent space of the configuration manifold into two parts. The first part corresponds to the admissible velocities u_α, while the second corresponds to the velocities u_β vanishing in view of the constraint relations. The equations of motion are accordingly also split into two parts: one describing the motion of the constrained system in terms of the variables of the free system, the other yielding the reactive forces. Using the technique of partitioned matrices, the inverse mass matrix of the constrained system is expressed in terms of an inverse matrix of lower order. In favorable cases the system of differential equations can be expressed in explicit form.

We would like to emphasize that although the method is based on the Hamilton or Euler-Hamilton equations, it is in no way restricted to conservative systems. Also it is well suited for dealing with problems of ideally elastic or inelastic collisions, since the basic equations of impulsive motion take their simplest form when expressed in terms of canonical variables [8-11]. Furthermore, it has the advantage that geometric singularities are easily avoided.

2. Classical Equations

We consider first scleronomic, holonomic, conservative mechanical systems. Let us assume that the Hamiltonian is expressed as the sum of the kinetic and potential energy and that the kinetic energy is a homogenous quadratic form in the velocities

$$T(q^k,\dot{q}^k) = \frac{1}{2} M_{ij}(q^k)\, \dot{q}^i \dot{q}^j \ , \qquad (2.1)$$

where $M_{ij}(q^k)$ is the mass matrix of the system. The summation convention is used here, and the indices go from 1 to n, where n denotes the degree of freedom of the system. Introducing canonical momenta

$$p_k = M_{kh} \dot{q}^h \,, \tag{2.2}$$

the Hamiltonian can be written as

$$H(p_k,q^k) = \frac{1}{2} A^{ij}(q^k) \, p_i p_j + V(q^k) \,, \tag{2.3}$$

where $A^{ij}(q^k)$ denotes the inverse mass matrix and $V(q^k)$ is a potential function. The canonical equations are then

$$\dot{p}_k = - \frac{\partial H}{\partial q^k} = - \frac{1}{2} A^{ij}{}_{,k} \, p_i p_j - V_{,k} \,, \tag{2.4'}$$

$$\dot{q}^k = \frac{\partial H}{\partial p_k} = A^{kh} \, p_h \,, \tag{2.4"}$$

where the partial derivative with respect to q^k is denoted by $(\ )_{,k}$. Since A^{ij} is the inverse of M_{ij}, one finds

$$A^{ij}{}_{,k} = - A^{ih} \, M_{h\ell,k} \, A^{\ell j} \,, \tag{2.5}$$

and the canonical equations take the algorithmically simpler form

$$\dot{p}_k = \frac{1}{2} M_{ij,k} \, \dot{q}^i \dot{q}^j - V_{,k} \,, \tag{2.6'}$$

$$\dot{q}^k = A^{kh} \, p_h \,. \tag{2.6"}$$

Here, of course, the $\dot{q}^k$ in the first equation represent just a shorthand way of writing the right hand side of the second equation.

For the type of system under consideration, the Lagrangian is

$$L(q^k,\dot{q}^k) = \frac{1}{2} M_{ij}(q^k) \, \dot{q}^i \dot{q}^j - V(q^k) \tag{2.7}$$

and, since the canonical momenta are given by

$$p_k = \frac{\partial L}{\partial \dot{q}^k} \,, \tag{2.8}$$

it is clear that equation (2.6') is nothing but a restatement of Lagrange's equation

$$\left(\frac{\partial L}{\partial \dot{q}^k}\right)^{\bullet} - \frac{\partial L}{\partial q^k} = 0 \quad . \tag{2.9}$$

The equations of Lagrange and Hamilton are therefore equivalent; however, the introduction of canonical momenta is crucial, if one aims at a simple formulation. The use of the velocities $\dot{q}^k$ as a shorthand notation in (2.6') is also important. If the inverse mass matrix A^{ij} is not available in closed form, equation (2.6") has to be replaced by its implicit form

$$M_{kh}\, \dot{q}^h = p_k \quad . \tag{2.6"'}$$

Only the mass matrix M_{ij} and its derivative $M_{ij,k}$ are then needed to formulate the equations of motion for a system with a given potential function.

For nonconservative systems, (2.6) is replaced by

$$\dot{p}_k = \frac{1}{2}\, M_{ij,k}\, \dot{q}^i \dot{q}^j + Q_k \quad , \tag{2.10'}$$

$$\dot{q}^k = A^{kh}\, p_h \quad , \tag{2.10"}$$

where the generalized forces Q_k are defined by the virtual work relation

$$\delta W = Q_k\, \delta q^k \quad . \tag{2.11}$$

Equation (2.10) is a simple consequence of Lagrange's equations for nonconservative systems. The generalized forces Q_k will be given functions of the position and velocity variables, but they may also explicitly depend on the time.

If the system considered is nonholonomic, or if nonholonomic velocities are used in a holonomic system, the equations of Euler-Lagrange or Euler-Hamilton are applicable. Let the velocities u^k be defined as

$$u^k = b^k_h(q^j)\, \dot{q}^h \quad , \tag{2.12}$$

with a nonsingular coefficient matrix b^k_h, and let the kinetic energy be given as

$$T(q^k,u^k) = \frac{1}{2} M_{ij}(q^k)\, u^i u^j. \tag{2.13}$$

The momenta will now be defined as

$$p_k = \frac{\partial T}{\partial u^k} = M_{kh}\, u^h \tag{2.14}$$

and the generalized forces by

$$\delta W = Q_k\, b^k_h\, \delta q^h \quad . \tag{2.15}$$

The Euler-Hamilton equations can then be stated as

$$\dot{p}_k + C^h_{ki}\, p_h u^i = \frac{\partial T}{\partial \tilde{q}^k} + Q_k \ , \tag{2.16'}$$

$$u^k = A^{kh}\, p_h \ . \tag{2.16"}$$

They have to be supplemented by

$$\dot{q}^k = \tilde{b}^k_h\, u^h \ , \tag{2.17}$$

where $\tilde{b}^k_h$ is the inverse of b^k_h. The coefficient matrices C^h_{ki} in (2.16') are given by

$$C^h_{ki} = \left(b^h_{\ell,j} - b^h_{j,\ell}\right) \tilde{b}^\ell_k\, \tilde{b}^j_i \ , \tag{2.18}$$

and on the right hand side of (2.16') the abbreviation

$$\frac{\partial T}{\partial \tilde{q}^k} = \frac{\partial T}{\partial q^h}\, \tilde{b}^h_k = \frac{1}{2} M_{ij,h}\, \tilde{b}^h_k\, u^i u^j \tag{2.19}$$

is used. The Euler-Hamilton equations were first stated by Pöschl [4]. In the form they are given here, they are equivalent to the equations of Euler-Lagrange [3]. In addition to the mass matrix, the coefficient matrices C^h_{ki} have to be computed, which in some cases may be tedious. The inversion of the matrix b^k_h appearing in (2.12) will generally not give rise to serious difficulties, if care is taken in the introduction of the velocities u^k. Again, if the inverse mass matrix is not available, equation (2,16") will have to be replaced by its implicit form

$$M_{kh}\, u^h = p_k \ . \tag{2.16"'}$$

3. Constraints and Projections

Let us now study the effect of constraints on a given system. By differentiating holonomic constraints of the form

$$f_\mu(q^k) = 0 , \quad \mu = 1,\dots,m \tag{3.1}$$

or by formulating nonholonomic constraint equations, we obtain a set of m linear relations

$$e_{\mu k}(q^j)\, u^k = 0 , \quad \mu = 1,\dots,m \tag{3.2}$$

for the velocities u^k. If the unconstrained system is holonomic, $u^k = \dot{q}^k$, otherwise u^k is given by (2.12). In matrix notation we may write (3.2) as

$$E^T u = 0 , \tag{3.3}$$

where E is an m by n matrix. We may assume that the relations (3.2) are linearly independent. The matrix E is then of rank m and $E^T E$ is nonsingular. Hence,

$$\beta = E\,(E^T E)^{-1}\, E^T \tag{3.4}$$

is well defined. Since

$$\beta^2 = \beta , \tag{3.5}$$

β is a projection operator. A complementary projection α is defined through the relation

$$\alpha = I - \beta , \tag{3.6}$$

where I stands for the unit matrix. The matrices α and β satisfy the relations

$$\alpha\beta = \beta\alpha = 0 \tag{3.7}$$

and

$$\alpha^T = \alpha , \quad \beta^T = \beta . \tag{3.8}$$

The two projections decompose the tangent space of the configuration manifold into two subspaces. The first, α, maps it onto the tangent space of the constrained system. In fact, if we let

$$u_\alpha = \alpha u , \quad u_\beta = \beta u , \tag{3.9}$$

then u_α are the admissible velocities of the constrained system, while u_β vanish in view of the constraint relations (3.2). Similarly,

$$\delta q_\alpha = \alpha \delta q \ , \quad \delta q_\beta = \beta \delta q \tag{3.10}$$

represent the admissible and the inadmissible virtual displacements, respectively. The virtual work equation

$$\delta W = Q^T \ \delta q = Q_\alpha^T \ \delta q_\alpha + Q_\beta^T \ \delta q_\beta \tag{3.11}$$

then yields a corresponding decomposition

$$Q_\alpha = \alpha Q \ , \quad Q_\beta = \beta Q \tag{3.12}$$

of the generalized forces. Q_α is the generalized force of the constrained system, while Q_β contains the internal and external reactive forces due to the constraints.

Introducing the matrices

$$\begin{aligned} M_{\alpha\alpha} &= \alpha M \alpha \ , \quad M_{\alpha\beta} = \alpha M \beta \ , \\ M_{\beta\alpha} &= \beta M \alpha \ , \quad M_{\beta\beta} = \beta M \beta \ , \end{aligned} \tag{3.13}$$

the kinetic energy of the unconstrained system may be written as

$$T = \frac{1}{2} \ u_\alpha^T \ M_{\alpha\alpha} u_\alpha + u_\alpha^T \ M_{\alpha\beta} u_\beta + \frac{1}{2} \ u_\beta^T \ M_{\beta\beta} u_\beta \ , \tag{3.14}$$

and we obtain the momenta

$$\begin{aligned} p_\alpha &= \frac{\partial T}{\partial u_\alpha} = M_{\alpha\alpha} u_\alpha + M_{\alpha\beta} u_\beta \ , \\ p_\beta &= \frac{\partial T}{\partial u_\beta} = M_{\beta\alpha} u_\alpha + M_{\beta\beta} u_\beta \ . \end{aligned} \tag{3.15}$$

Clearly, the relations

$$p_\alpha = \alpha p \ , \quad p_\beta = \beta p \tag{3.16}$$

are valid. The kinetic energy of the constrained system is

$$T = \frac{1}{2} \ u_\alpha^T \ M_{\alpha\alpha} u_\alpha \tag{3.17}$$

and, therefore, $M_{\alpha\alpha}$ represents the mass matrix of the constrained system. Its inverse can be expressed by using the technique of partitioned matrices [11] as

$$M_{\alpha\alpha}^{\div} = A_{\alpha\alpha} - A_{\alpha\beta} \ A_{\beta\beta}^{\div} \ A_{\beta\alpha} \ , \tag{3.18}$$

where

$$A_{\alpha\alpha} = \alpha A\alpha \ , \quad A_{\alpha\beta} = \alpha A\beta \ ,$$
$$A_{\beta\alpha} = \beta A\alpha \ , \quad A_{\beta\beta} = \beta A\beta \tag{3.19}$$

and A is the inverse mass matrix of the unconstrained system. When written in terms of the variables of the unconstrained system, the matrices $M_{\alpha\alpha}$ and $A_{\beta\beta}$ are singular. $M^{\div}_{\alpha\alpha}$ and $A^{\div}_{\beta\beta}$ denote their generalized inverses [12]. The notation used here is the same as that introduced by Davis [13]. If these matrices are expressed in appropriate bases of the respective subspaces, they are nonsingular and their generalized inverses can be obtained by computing ordinary inverses. For $A_{\beta\beta}$ a basis is easily constructed from the rows of the matrix $e_{\mu k}$ of (3.2). If the number of constraints m is small, $A_{\beta\beta}$ can then be inverted and by making use of (3.18) the inverse mass matrix can be computed. Observe, that (3.18) also implies

$$\alpha \, M^{\div}_{\alpha\alpha} = M^{\div}_{\alpha\alpha} \, \alpha = M^{\div}_{\alpha\alpha} \ . \tag{3.20}$$

4. Dynamics of Constrained Systems

We now want to study the dynamics of a system obtained by introducing constraints into a given system in the manner presented above. In applications to multibody systems, the unconstrained system may simply be chosen as a system of free bodies, acted upon by the forces of the constrained system.

Let us first discuss the case where the unconstrained system is holonomic and described by holonomic velocity variables $u^k = \dot{q}^k$. In view of (3.16) the momenta of the constrained system are

$$p_{\alpha k} = \alpha^h_k \, p_h \ . \tag{4.1}$$

Differentiating, we obtain

$$\dot{p}_{\alpha k} = \alpha^h_k \, \dot{p}_h + \alpha^h_{k,j} \, p_h \, \dot{q}^j \ , \tag{4.2}$$

and inserting Hamilton's equations (2.10) in the form

$$\dot{p}_h = \frac{\partial T}{\partial q^h} + Q_h \ , \tag{4.3}$$

we have

$$\dot{p}_{\alpha k} = \frac{\partial T}{\partial q^h} \alpha_k^h + Q_h \alpha_k^h + \alpha_{k,j}^h p_h \dot{q}^j . \tag{4.4}$$

Making use of (3.15') with $u_\beta=0$ we can thus formulate the system of equations

$$\dot{p}_k = \frac{\partial T}{\partial q_\alpha^k} + Q_{\alpha k} + \alpha_{k,j}^h M_{hi} u^i u^j , \tag{4.5'}$$

$$\dot{q}^k = u^k = \left(M_{\alpha\alpha}^{\div}\right)^{kh} p_h . \tag{4.5"}$$

Here, the index α has been dropped in the velocities and the momenta. The velocities u^k in the first equation represent again a shorthand notation for the right hand side of the second equation. Also, the abbreviation

$$\frac{\partial T}{\partial q_\alpha^k} = \frac{\partial T}{\partial q^h} \alpha_k^h = \frac{1}{2} M_{ij,h} u^i u^j \alpha_k^h \tag{4.6}$$

is used, in which the derivatives of the mass matrix of the unconstrained system appear. Observe, that the generalized forces Q_α are simply the generalized forces of the constrained system and do not depend on the reactive forces. We now have the following

Theorem: For initial data satisfying the constraints of the system, the solutions of (4.5) describe motions of the constrained system.

The proof is obvious: First, due to (3.20) $M_{\alpha\alpha}^{\div}$ maps the momenta onto the admissible velocities u_α^k and the second equation together with the initial data for the coordinates implies that $q^k(t)$ satisfies the constraint relations for t>0. Second, if the constraints are satisfied, (4.5') is a correct statement of the dynamical equations of the constrained system.

If the inverse mass matrix $M_{\alpha\alpha}^{\div}$ is not available, equation (4.5") has to be replaced by its implicit form

$$p_k = \left(M_{\alpha\alpha}^{\div}\right)_{kh} u^h \tag{4.7'}$$

supplemented by

$$\beta^h_k \; u^k = 0 \; . \tag{4.7"}$$

It may be important to note that, due to the fact that the constraints (3.2) can be nonholonomic, equation (4.5) is a simple alternative to the equation of Euler-Hamilton.

In order to compute the reactive forces, we start from

$$p_{\beta k} = \beta^h_k \; p_h \; . \tag{4.8}$$

Proceeding as above, we obtain

$$\dot{p}_{\beta k} = \frac{\partial T}{\partial q^k_\beta} + Q_{\beta k} - \alpha^h_{k,j} \; M_{hi} \; u^i_\alpha \; u^j_\alpha \; , \tag{4.9}$$

where

$$\frac{\partial T}{\partial q^k_\beta} = \frac{\partial T}{\partial q^h} \; \beta^h_k = \frac{1}{2} \; M_{ij,h} \; u^i_\alpha \; u^j_\alpha \; \beta^h_k \; . \tag{4.10}$$

Since

$$p_{\beta k} = \left(M_{\beta\alpha}\right)_{kh} \; u^h_\alpha \; , \tag{4.11}$$

the left hand side of (4.9) is known from (4.5) and the Q_β, which contain the reactive forces, can be computed. A considerable simplification is obtained by choosing the projections orthogonal in the mass-metric. In fact, (3.4) can be generalized to

$$\beta = C \; E \; (E^T C E)^{-1} \; E^T \tag{4.12}$$

with an arbitrary positive symmetric matrix C. If the inverse mass matrix A of the unconstrained system is used, the two projections α and β will be mass-orthogonal and

$$M_{\beta\alpha} = 0 \; . \tag{4.13}$$

Then the left hand side of (4.9) vanishes and the computation of Q_β is straightforward. Of course, the labor is just carried over to the computation of $(E^T A E)^{-1}$.

Let us finally state the analog of equation (4.5) for the case that the unconstrained system is nonholonomic or that general velocity variables are used. We obtain

$$\dot{p}_{\alpha k} = \frac{\partial T}{\partial \tilde{q}^k_\alpha} + Q_{\alpha k} + \tilde{C}^h_{kj} \; M_{hi} \; u^i u^j \; , \tag{4.14'}$$

$$\dot{q}^k = u^k = \left[M^{\div}_{\alpha\alpha}\right]^{kh} p_h \, , \tag{4.14"}$$

where

$$\frac{\partial T}{\partial \tilde{q}^k_\alpha} = \frac{\partial T}{\partial q^h} \alpha^h_j \tilde{b}^j_k \tag{4.15}$$

and

$$\tilde{C}^h_{kj} = \alpha^h_{k,i} b^i_j + \alpha^i_k C^h_{ij} \, . \tag{4.16}$$

The theorem stated for the holonomic case carries over to these equations in an obvious fashion.

References

1. Wittenburg, J.: Dynamics of Systems of Rigid Bodies. B.G. Teubner, Stuttgart 1977.
2. Kane, T.R., Levinson, D.A.: Dynamics: Theory and Application. McGraw-Hill inc., 1985.
3. Hamel G.: Die Lagrange-Eulerschen Gleichungen der Mechanik. Teubner, Leipzip, 1903.
4. Pöschl, Th.: Sur les équations canoniques des systèmes non holonomes. C.R.Acad.Sc. Paris, 156, (1913).
5. Weber, R.W.: Eine alternative Herleitung der Euler-Hamilton Gleichungen der Mechanik. Zeitschr.ang.Math.Phys., 31, (1980).
6. Weber, R.W.: Kanonische Theorie nichtholonomer Systeme. Diss. ETH Nr. 6878, Verlag P.Lang, Bern, 1981.
7. Weber, R.W.: Hamiltonian Systems with Constraints and their Meaning in Mechanics. Arch.Rat.Mech.Anal., to appear.
8. Synge, J.L.: Classical Dynamics, Handbuch d. Phys. III/1, Springer Verlag Berlin, 1960.
9. Thomson, W., Tait, P.G.: Handbuch d. theor. Phys., Bd. 1, Vieweg u. Sohn, Braunschweig, 1871.
10. Raher, W.: Zur Theorie des Stosses starrer Körper. Oest. Ing.-Arch. 9, (1955), 55-68.
11. Brauchli, H.: Matrix Formulation of Impulsive Motion. In preparation.
12. Penrose, R.: A Generalized Inverse for Matrices. Proc. Cambr. phil.Soc. 51, (1955), 406-413.
13. Davis, P.J.: Circulant Matrices. J. Wiley & Sons, New York, 1979.

Dynamic Effects of Active Elements in Manipulators and their Influence upon the Controlling Drives

K.Desoyer, P.Lugner, H.Springer

Department of Mechanical Engineering
Technical University of Vienna, Austria

SUMMARY

Approximate equations of motion for tree-structured robots and manipulators can be obtained by neglecting relative rotational inertia effects of the active elements(motors) that drive the links of the system. Starting from these equations the Gibbs-Appell(GA)-calculus is used in this paper in order to determine generalized mass matrices and gyroscopic forces that take care of the dynamics of the active elements in the system.
Additional terms have then to be added to the original equations of motion without further changes. It is shown how the GA-function of a robot is modified when the links are considered as gyrostats carrying axissymmetric rotors. For a three-link RRR-manipulator an example is carried out that studies various effects upon the controlling drives of a robot.

1. INTRODUCTION

For a modern design of control systems for robots and manipulators it is necessary to model a given mechanical robot and derive the corresponding dynamic equations of motion as accurate as possible in order to obtain proper relations between a desired motion (track curve) of a robot and the required driving torques. Considering a chain-structured manipulator, in general each link carries a controlling motor that moves either the link itself or the succeeding link. Because of the moment of inertia of the controlling elements gyroscopic forces are generated in a link which usually may be neglected compared with the dynamic forces of the rigid link itself, [1] . In this way approximate equations of motion for a rigid multibody robot system can be established and may succesfully be used for a heavy-construction design. However, when the rotor speed of the controlling motors is high and a light-construction robot design is performed then the above mentioned forces of the controlling elements may no longer be negligible small and the approximate equations of motion may then no longer yield sufficiently accurate results. Once the investigation is started with a set of approximate euqations of motion,neglecting those rotatory effects, it might

Dynamics of Multibody Systems
IUTAM/IFToMM Symposium Udine 1985
Editors: G. Bianchi and W. Schiehlen

be very advantageous to use given kinematic relations for the links in order to establish correcting terms to the equations of motion that include dynamic characteristics of the driving elements. The application of the GA-calculus for rigid body systems [2,4,5] makes it possible to determine the additional terms in a very straight forward and easy manner. It is shown in this paper that complementary generalized mass matrices and gyroscopic force vectors, that correspond to the moment of inertia of the rotors of the driving motors, can be obtained. After assembling these matrices and force vectors they have simply to be added to the approximate equations of motion.

Nomenclature

${}^{o}\underline{a}_{Ci}$	absolute acceleration of center of mass C_i of body i
${}^{o}\underline{\varepsilon}_{i}$	absolute angular acceleration of body i
${}^{i}\underline{\varepsilon}_{Ri}$	relative angular acceleration of rotor R_i of driving motor attached to link i
${}^{o}G_{i}$	absolute GA-Function of body i
$\underline{I}_{Ci}$	inertial tensor of body i with respect to center of mass C_i
$\underline{I}_{Ri}$	inertial tensor of rotor Ri with respect to its center of mass C_{Ri}
${}^{o}\underline{L}_{Ci}$	absolute translational accel. matrix of Ci
${}^{o}\underline{L}_{\omega i}$	absolute rotational accel. matrix of body i
${}^{i}\underline{L}\omega_{Ri}$	relative rotational accel. matrix of rotor Ri
m_i	mass of body i
${}^{o}\underline{N}_{i}$	generalized mass matrix corresponding to body i
$\nu_{i,j}$	gear ratio
${}^{o}\underline{\omega}_{i}$	absolute angular speed of body i
${}^{o}\underline{\Omega}_{i}$	corresponding tensor to ${}^{o}\underline{\omega}_{i}$
${}^{i}\underline{\omega}_{Ri}$	relative angular speed of rotor Ri with respect to link i
${}^{o}\underline{\Psi}_{i}$	generalized gyroscopic force vector corresponding to body i
$\underline{q}$	generalized coordinates
$\underline{u}$	generalized speeds
$\dot{\underline{u}}$	generalized accelerations
T_i^M	magnetic driving moment of motor M_i
T_i^F	frictional moment in the gearing, reduced to the axis of rotor R_i

2. GENERAL THEORY

2.1 Gibbs-Appell Function for a Rigid Multi-Body System

The GA-Function of a body "i" with respect to an inertial reference frame "O" is a scalar quantity (so-called "acceleration energy") that is defined by

$$ {}^{o}G_{i} = \frac{1}{2}\int_{m_i} {}^{o}\underline{a}^{T}\ {}^{o}\underline{a}\ dm \qquad (1) $$

For a p-body system the resulting GA-Function is

$$ {}^{o}G = \sum_{i=1}^{p} {}^{o}G_i \tag{2} $$

Considering the equations of motion (10), it can be seen that terms, independent of the generalized acceleration $\dot{\underline{u}}$, may be suppressed in the GA-function. Hence, for a rigid body i equation (1) can be carried out and yields a modified GA-Function

$$ {}^{o}\bar{G}_i = \frac{1}{2} m_i \, {}^{o}\underline{a}_{Ci}^{T} \, {}^{o}\underline{a}_{Ci} + \frac{1}{2} \, {}^{o}\underline{\varepsilon}_i^{T} \, \underline{I}_{Ci} \, {}^{o}\underline{\varepsilon}_i + {}^{o}\underline{\varepsilon}_i^{T} \, {}^{o}\underline{\Omega}_i \, \underline{I}_{Ci} \, {}^{o}\underline{\omega}_i \tag{3} $$

When introducing the generalized speeds $\underline{u}$ of a multibody system the acceleration terms of the i'th body can be written in the form, [3]

$$ {}^{o}\underline{a}_{Ci} = {}^{o}\underline{L}_{Ci} \, \dot{\underline{u}} + {}^{o}\underline{a}_{Ci}^{*} \, , \qquad {}^{o}\underline{\varepsilon}_i = {}^{o}\underline{L}_{\omega i} \, \dot{\underline{u}} + {}^{o}\underline{\varepsilon}_i^{*} \tag{4),(5} $$

Inserting (4),(5) into (3) and omitting terms independent of $\dot{\underline{u}}$, we get a further modification of the GA-Function in the form

$$ {}^{o}\bar{\bar{G}}_i = \frac{1}{2} \dot{\underline{u}}^{T} \, {}^{o}\underline{N}_i \, \dot{\underline{u}} + {}^{o}\underline{\Psi}_i^{T} \, \dot{\underline{u}} \tag{6} $$

with

$$ {}^{o}\underline{N}_i = m_i \, {}^{o}\underline{L}_{Ci}^{T} \, {}^{o}\underline{L}_{Ci} + {}^{o}\underline{L}_{\omega i}^{T} \, \underline{I}_{Ci} \, {}^{o}\underline{L}_{\omega i} \tag{7} $$

being a generalized symmetric mass matrix of dimension nxn and

$$ {}^{o}\underline{\Psi}_i = m_i \, {}^{o}\underline{L}_{Ci}^{T} \, {}^{o}\underline{a}_{Ci}^{*} + {}^{o}\underline{L}_{\omega i}^{T} \, \underline{I}_{Ci} \, {}^{o}\underline{\varepsilon}_i^{*} + {}^{o}\underline{L}_{\omega i}^{T} \, {}^{o}\underline{\Omega}_i \, \underline{I}_{Ci} \, {}^{o}\underline{\omega}_i \tag{8} $$

being a generalized gyroscopic force vector of dimension nx1, where n is the number of independent generalized speeds (degrees of freedom) of the system. From (2) for a rigid p-body system follows

$$ {}^{o}\bar{\bar{G}} = \frac{1}{2} \dot{\underline{u}}^{T} \, {}^{o}\underline{N} \, \dot{\underline{u}} + {}^{o}\underline{\Psi}^{T} \, \dot{\underline{u}} \tag{9a} $$

with

$$ {}^{o}\underline{N} = \sum_{i=1}^{p} {}^{o}\underline{N}_i \, , \qquad {}^{o}\underline{\Psi} = \sum_{i=1}^{p} {}^{o}\underline{\Psi}_i \tag{9b} $$

2.2 Equations of Motion

The equations of motion for a rigid body system can be established in different ways, for example, see [2,3,4,6]. Using the scalar GA-Function these equations have the form

$$ \frac{\partial \, {}^{o}G}{\partial \dot{\underline{u}}} \equiv \frac{\partial \, {}^{o}\bar{\bar{G}}}{\partial \dot{\underline{u}}} = \underline{Q}^{T}(\underline{q}, \underline{u}, t) \tag{10} $$

with $\underline{Q}$ being the generalized force vector.

Consider a manipulator with tree or chain structur having n+1 links, i.e. i=0 for the foundation and i=1,2,...n for the movable links with n degrees of freedom $q_1,q_2,\ldots q_n$ and n independent generalized speeds $u_1,u_2,\ldots u_n$. Figure 1, for example, shows a schematic diagram of a RRR-robot with four links and three degrees of freedom. The motion of the moveable links is generated by driving motors and gears that are attached to the links, see figure 2. In this paper it is assumed that the i'th link carries a motor that drives the link i itself or the succeeding link i+1 (called case 1 and case 2, respectively).

If the relative motion of the driving system (rotor and gear) with respect to the links is neglected then each link i can be considered as a fictitious rigid body L_i that includes mass and moments of inertia of the link and the driving system. Equations of motion can then be established by (9) and (10) in the form

$$ {}^{O}\underline{N}_L \; \dot{\underline{u}} + {}^{O}\underline{\Psi}_L = \underline{Q}_D + \underline{Q}_G \qquad (11)$$

where ${}^{O}\underline{N}_L$ and ${}^{O}\underline{\Psi}_L$ are to be calculated from (7),(8), and (9b). In (11)

$$ \underline{Q}_D = \{(T_0^M - T_0^F)\nu_{0,1},\ldots(T_i^M - T_i^F)\nu_{i,j}\ldots\}^T \qquad (12)$$

is the generalized force vector that is generated by the driving system with T_i^M being the magnetic (or hydraulic) driving moment of the motor operating in the i'th link. T_i^F represents a friction moment within the driving system that is reduced to the axis of the motor. The gear ratio of the driving system is defined by

$$ \nu_{i,i} = {}^{i}\omega_{Ri}/u_i \qquad \text{and} \qquad \nu_{i,i+1} = {}^{i}\omega_{Ri}/u_{i+1} \qquad (13)$$

for case 1 and case 2, respectively, with $u_{i+1}=\dot{q}_{i+1}$ being the relative (rotational) speed of the link (i+1) with respect to the link i. In (11) $\underline{Q}_G$ is the generalized force vector due to external weight loads acting upon the manipulator.

2.3 Relative Inertia Effects of the Driving System.

Assume that the driving system that is operating in a link performs a rotational relative motion only without any translational components. Furthermore, if the mass distribution of the rotor of the driving motor and the gearing is symmetric with respect to the axis of relative rotation then each link may be considered as a gyrostat. Let ${}^{O}\bar{G}_{Li}$ be the modified GA-Function of the i'th fictitious rigid link L_i ,including the nonrotating driving system as to be calculated from (3). Then the total GA-Function of the gyrostat can be carried out in the form

$$^{o}\bar{G}_{Tot\ i} = {}^{o}\bar{G}_{Li} + {}^{o}\bar{G}_{Ri} \tag{14}$$

where

$$^{o}\bar{G}_{Ri} = \frac{1}{2}\ {}^{i}\underline{\varepsilon}_{Ri}^{T}\ \underline{I}_{Ri}\ {}^{i}\underline{\varepsilon}_{Ri} + {}^{o}\underline{\varepsilon}_{i}^{T}\ \underline{I}_{Ri}\ {}^{i}\underline{\varepsilon}_{Ri} + {}^{o}\underline{\varepsilon}_{i}^{T}\ {}^{o}\underline{\Omega}_{i}\ \underline{I}_{Ri}\ {}^{i}\underline{\omega}_{Ri} \tag{15}$$

is a complementary internal term representing the relativ rotation of the rotor, with ${}^{i}\underline{\omega}_{Ri}$ and ${}^{i}\underline{\varepsilon}_{Ri}$ being relative angular speed and acceleration of the rotor, respectively, and $\underline{I}_{Ri}$ being the rotor's inertial tensor with respect to its center of mass. Similar to (5), kinematic relations for the angular accelerations

$${}^{i}\underline{\varepsilon}_{Ri} = {}^{i}\underline{L}_{\omega_{Ri}}\ \underline{\dot{u}} + {}^{i}\underline{\varepsilon}_{Ri}^{*} \tag{16}$$

can be established which lead to a complementary generalized mass matrix

$${}^{o}\underline{N}_{Ri} = {}^{i}\underline{L}_{\omega_{Ri}}^{T}\ \underline{I}_{Ri}\ {}^{i}\underline{L}_{\omega_{Ri}} + {}^{o}\underline{L}_{\omega i}^{T}\ \underline{I}_{Ri}\ {}^{i}\underline{L}_{\omega_{Ri}} + {}^{i}\underline{L}_{\omega_{Ri}}^{T}\ \underline{I}_{Ri}\ {}^{o}\underline{L}_{\omega i} \tag{17}$$

and generalized gyroscopic forces

$${}^{o}\underline{\Psi}_{Ri} = ({}^{o}\underline{L}_{\omega\,i}^{T} + {}^{i}\underline{L}_{\omega_{Ri}}^{T})\underline{I}_{Ri}\,{}^{i}\underline{\varepsilon}_{Ri}^{*} + {}^{i}\underline{L}_{\omega_{Ri}}^{T}\underline{I}_{Ri}\,{}^{o}\underline{\varepsilon}_{i}^{*} + {}^{o}\underline{L}_{\omega\,i}^{T}\,{}^{o}\underline{\Omega}_{i}\underline{I}_{Ri}\,{}^{i}\underline{\omega}_{Ri} \tag{18}$$

for the i'th link.

Corresponding to the scalar characteristic of the GA-Function the assembling process is performed by a summation to be taken over all driving systems

$${}^{o}\underline{N}_{R} = \Sigma\ {}^{o}\underline{N}_{Ri}, \qquad {}^{o}\underline{\Psi}_{R} = \Sigma\ {}^{o}\underline{\Psi}_{Ri} \tag{19}$$

The exact equations of motion can be written in the final form

$$({}^{o}\underline{N}_{L} + {}^{o}\underline{N}_{R})\underline{\dot{u}} + ({}^{o}\underline{\Psi}_{L} + {}^{o}\underline{\Psi}_{R}) = \underline{Q}_{D} + \underline{Q}_{G} \tag{20}$$

Equations (17) and (18) are carried out in the following for a single rotor operating in link i and driving the link itself or the succeeding link i+1. It is recommended to describe the relative angular speed and angular acceleration of the driving rotor in a reference frame i' that is rigidly connected to the link with the $x_{i'}$-axis coinciding with the axis of the rotor. In this case in (16) the local acceleration ${}^{i}\underline{\varepsilon}_{Ri}^{*}$ vanishes and the matrix of acceleration becomes the simple form

$${}^{i}\underline{L}_{\omega_{Ri/i'}} = \begin{pmatrix} 0,\ldots,\nu_{i,j},\ldots 0 \\ 0,\ldots,\ 0,\ldots.0 \\ 0,\ldots,\ 0,\ldots.0 \end{pmatrix} \qquad j=i \text{ or } (i+1) \tag{21}$$

If the motor drives the link i then the gear ratio $\nu_{i,i}$ appears in the first row and i-th column of the above matrix while in the second case $\nu_{i,i+1}$ has to be placed in the (i+1)st column. In the reference frame i' the inertial tensor of the axis-symmetric rotor has a diagonal form

$$\underline{I}_{Ri/i'} = \begin{bmatrix} I_{Ri} & 0 & 0 \\ 0 & I_{ai} & 0 \\ 0 & 0 & I_{ai} \end{bmatrix} \tag{22}$$

Let the acceleration matrix ${}^{0}\underline{L}_{\omega i/i}$ in equations (17),(18) be represented in a reference frame i, rigidly connected to link i, as commonly used in robot mechanics. Then a unitary transformation matrix $\underline{A}_{ii'}$ can be employed to transform Equ.(21) and (22) from frame i' into frame i. The elements of $\underline{A}_{ii'}$ to be used for this transformation are composed by the cosines of α_i, β_i, and γ_i, being the angles between $x_{i'}$-axis and the x_i, y_i, z_i-axes, respectively. With ${}^{i}\underline{\varepsilon}^{*}_{Ri} = \underline{0}$ and the abbreviation

$$\underline{B}_{i/i} = \underline{A}_{ii'}\, \underline{I}_{Ri/i'}\, {}^{i}\underline{L}_{\omega Ri/i'}$$

equations (17) and (18) become

$$\begin{aligned} {}^{0}\underline{N}_{Ri} &= {}^{i}\underline{L}^{T}_{\omega Ri/i'}\, \underline{I}_{Ri/i'}\, {}^{i}\underline{L}_{\omega Ri/i'} + ({}^{0}\underline{L}^{T}_{\omega i/i}\, \underline{B}_{i/i} + \underline{B}^{T}_{i/i}\, {}^{0}\underline{L}_{\omega i/i}) \\ {}^{0}\underline{\Psi}_{Ri} &= \underline{B}^{T}_{i/i}\, {}^{0}\underline{\varepsilon}^{*}_{i/i} + {}^{0}\underline{L}^{T}_{\omega i/i}\, \underline{\Omega}_{i/i}\, \underline{B}_{i/i}\, \underline{u} \end{aligned} \tag{23}$$

With $l_{\alpha\beta/i}$ being an element of the acceleration matrix ${}^{0}\underline{L}_{\omega i/i}$ the complementary generalized mass matrix for a single rotor operating in link i can be written in the form

$${}^{0}\underline{N}_{Ri} = \begin{bmatrix} & & \kappa_{1j/i} & & \\ & & \kappa_{2j/i} & & \\ & & \vdots & & \\ \kappa_{j1/i}, & \kappa_{j2/i}, \dots & \kappa_{jj/i}, \dots & \kappa_{jn/i} \\ & & \vdots & & \\ & & \kappa_{nj/i} & & \end{bmatrix} \tag{24}$$

with $$\kappa_{jk/i} = \kappa_{kj/i} = I_{R_i}\nu_{i,j}(l_{1k/i}\cos\alpha_i + l_{2k/i}\cos\beta_i + l_{3k/i}\cos\gamma_i)$$

$$\text{for } j=i \text{ or } (i+1) \qquad j \neq k \tag{25}$$

$$\kappa_{jj/i} = I_{R_i}\nu^{2}_{i,j} + 2I_{R_i}\nu_{i,j}(l_{1j/i}\cos\alpha_i + l_{2j/i}\cos\beta_i + l_{3j/i}\cos\gamma_i)$$

$$l_{\alpha,i+1/i} = 0 \tag{26}$$

The gyroscopic force vector is given by

$$
{}^{o}\underline{\Psi}_{R_i} = \begin{pmatrix} \lambda_{1/i} \\ \lambda_{2/i} \\ \vdots \\ \lambda_{j/i}+\bar{\lambda}_{j/i} \\ \vdots \\ \lambda_{n/i} \end{pmatrix} \qquad j=i \quad \text{or} \quad (i+1) \tag{27}
$$

$$
\lambda_{k/i} = I_{R_i}\nu_{i,j}u_j[(l_{2k/i}\,{}^{o}\omega_{z_i/i}-l_{3k/i}\,{}^{o}\omega_{y_i/i})\cos\alpha_i + (l_{3k/i}\,{}^{o}\omega_{x_i/i}-l_{1k/i}\,{}^{o}\omega_{z_i/i})\cos\beta_i + (l_{1k/i}\,{}^{o}\omega_{y_i/i}-l_{2k/i}\,{}^{o}\omega_{x_i/i})\cos\gamma_i] \tag{28}
$$

$$
\bar{\lambda}_{j/i} = I_{R_i}\nu_{i,j}({}^{o}\varepsilon^{*}_{x_i/i}\cos\alpha_i + {}^{o}\varepsilon^{*}_{y_i/i}\cos\beta_i + {}^{o}\varepsilon^{*}_{z_i/i}\cos\gamma_i) \tag{29}
$$

For a chain-structured robot, for example the RRR-system as shown in figure 1, certain elements in Equ.(24) and (27) will vanish. If the links are consecutively numbered from i=0 (foundation) up to i=n then $\kappa_{jk/i}=0$ and $\lambda_{k/i}=0$ holds for $j,k > i$. In this case the assembling procedure (19) becomes very simple.

3. NUMERICAL EXAMPLE

For a three-link RRR-robot, see figure 1, the dynamical effects of the driving devices will be investigated. In this example case 2 is considered, i.e. a motor that is attached to the i'th link drives link i+1, (i=0,1,2). For i=2,3 it is assumed that the principal axes of inertia with respect to C_i are parallel to the axes x_i, y_i, z_i of reference frame i. Using data from table 1 the approximate set of equations of motion (11) for the generalized speeds $\underline{u}^T = (\dot{\varphi}_1, \dot{\varphi}_2, \dot{\varphi}_3)$ is represented by

$$
{}^{o}\underline{N}_L = \begin{pmatrix} n_{L11} & 0 & 0 \\ 0 & 0.04m_2 + 0.27m_3 & 0.02m_3 \\ 0 & 0.02m_3 & 0.02m_3 \end{pmatrix} \tag{a}
$$

$$
{}^{o}\underline{\Psi}_L = \begin{pmatrix} -(0.058S_2C_2m_2+0.024S_{23}C_{23}m_3)\dot{\varphi}_1\dot{\varphi}_2-0.024S_{23}C_{23}m_3\dot{\varphi}_3\dot{\varphi}_1 \\ (0.029S_2C_2m_2 + 0.012S_{23}C_{23}m_3)\dot{\varphi}_1^2 \\ 0.012S_{23}C_{23}m_3\dot{\varphi}_1^2 \end{pmatrix} \tag{b}
$$

$$
\underline{Q}_D+\underline{Q}_G = \begin{pmatrix} (T_0^M-T_0^F)\nu_{0,1} \\ (T_1^M-T_1^F)\nu_{1,2} \\ (T_2^M-T_2^F)\nu_{2,3} \end{pmatrix} + \begin{pmatrix} 0 \\ -0.5C_2m_3g \\ 0 \end{pmatrix} \tag{c}
$$

The complementary terms which lead to equation (20) are carried out to

$$
{}^{0}\underline{N}_R = \begin{bmatrix} I_{R_0}\nu_{0,1}^2 & I_{R_1}\nu_{1,2}\cos\beta_1 & I_{R_2}\nu_{2,3}(S_2\cos\alpha_2+C_2\cos\beta_2) \\ I_{R_1}\nu_{1,2}\cos\beta_1 & I_{R_1}\nu_{1,2}^2 & I_{R_2}\nu_{2,3}\cos\gamma_2 \\ I_{R_2}\nu_{2,3}(S_2\cos\alpha_2+C_2\cos\beta_2) & I_{R_2}\nu_{2,3}\cos\gamma_2 & I_{R_2}\nu_{2,3}^2 \end{bmatrix} \qquad \text{(d)}
$$

$$
{}^{0}\underline{\Psi}_R = \begin{bmatrix} I_{R_2}\nu_{2,3}(C_2\cos\alpha_2-S_2\cos\beta_2)\dot\varphi_2\dot\varphi_3 \\ -I_{R_2}\nu_{2,3}(C_2\cos\alpha_2-S_2\cos\beta_2)\dot\varphi_1\dot\varphi_3 \\ I_{R_2}\nu_{2,3}(C_2\cos\alpha_2-S_2\cos\beta_2)\dot\varphi_1\dot\varphi_2 \end{bmatrix} \qquad \text{(e)}
$$

In the above expressions the following abbreviations were used

$$n_{L11}=0.09m_1+(0.0036S_2^2+0.032C_2^2+0.113)m_2+[0.0025S_{23}^2+0.0144C_{23}^2+(0.3+0.5C_2)^2]m_3$$

$S_2 = \sin\varphi_2 \qquad C_2 = \cos\varphi_2 \qquad S_{23} = \sin(\varphi_2+\varphi_3) \qquad C_{23} = \cos(\varphi_2+\varphi_3)$

$I_{\alpha i} = m_i \; i_{\alpha i}^2 \qquad \alpha = x,y,z$

Two problems can be investigated. First, the calculation of the driving torques T_i^M for a given motion $\varphi_1(t),\varphi_2(t),\varphi_3(t)$ of the robot. Second, the calculation of the motion for given driving torques.

Figure 3 shows the acting torques $T_i=T_i^M-T_i^F$ as functions of time for a desired motion represented by given angular velocities $\dot\varphi_i(t)$. The full lines represent the exact solutions including rotatory inertia and gyroscopic effects of the driving elements. The dotted lines represent solutions which are obtained by cancelling the matrices (d) and (e) in the equations of motion, i.e. relative rotatory inertia effects of the driving systems are completely neglected. The major differences appear when a link of the robot is accelerated or braked. The required torque T_{i-1} for a driven link i shows that effect, corresponding to the time history of $\dot\varphi_i$. For periodes when all links move with constant angular velocities $\dot\varphi_i$ the torques remain comparatively small because the weight moment of each link i is balanced with respect to its joint (i-1). Only the weight moment of link 3 has to be compensated by torque T_1 in this example. The influences of the gyroscopic terms in (e) and the off-diagonal terms of the matrix (d) remain very small.

These influences can be seen better in figure 4, where the dotted lines represent the system without the off-diagonal and gyroscopic terms. By given initial torques, acting over short periods, the robot is initiated to a motion of all its links. Compared with figure 3, the present value of the torque T_0 is almost twice as much, and therefore the generated angular speed $\dot{\varphi}_1$ is substantially higher and consequently increases the gyroscopic effects. There is practically no gyroscopic influence upon the angular velocity $\dot{\varphi}_1$ within the considered period of one second. The angular speeds $\dot{\varphi}_2$ and $\dot{\varphi}_3$ show small deviations increasing with time. However, gyroscopic effects of this amount can easily be compensated by control circuits.

4. CONCLUSIONS.

If equations of motion for a robot system are available that do not consider dynamic effects of rotating components in the links an appropriate algorithm can be employed to obtain the complete equations. For each driving rotor, operating in a link, only one column and one row have to be added to the given mass matrix and a complementary gyroscopic force vector has to be added to the given equations of motion.

While the effects of rotatory inertia of the driving rotors are essential for the motion of the links, the gyroscopic effects and the off diagonal terms in the complementary mass matrix are quite small for common robot design. However, for future light-construction robot design also small dynamic effects will have increasing importance.

REFERENCES

1. Desoyer,K,; Kopacek,P,; Troch,I., Industrieroboter und Handhabungsgeräte. R.Oldenbourg Verlag München,Wien, 1985.,(pp.118-129).

2. Lugner,P.; Desoyer,K.; Springer,H., Dynamik von Starrkörpersystemen. Vorlesungsmanuskript, TU-Wien, Sommersemester 1985.

3. Kreuzer,E., Symbolische Berechnung der Bewegungsgleichungen von Mehrkörpersystemen. VDI-Fortschrittsbericht VDI-Z. Reihe 11, Nr.32, 1979.

4. Pars,L.A., A Treatise on Analytical Dynamics. London, Heinemann Edition, 1965 ,(pp.201-227).

5. Fischer,U.; Stephan,W., Prinzipien und Methoden der Dynamik. Leipzig, VEB-Verlag, 1972,(pp.78-79).

6. Schwertassek,R,; Roberson,R.E.; Schiehlen,W.O.; Wittenburg,J., Multibody System Dynamics-Computer Aided Generation of System Equations and Simulation. CCG-COURSE V1.07, Oberpfaffenhofen 1984.

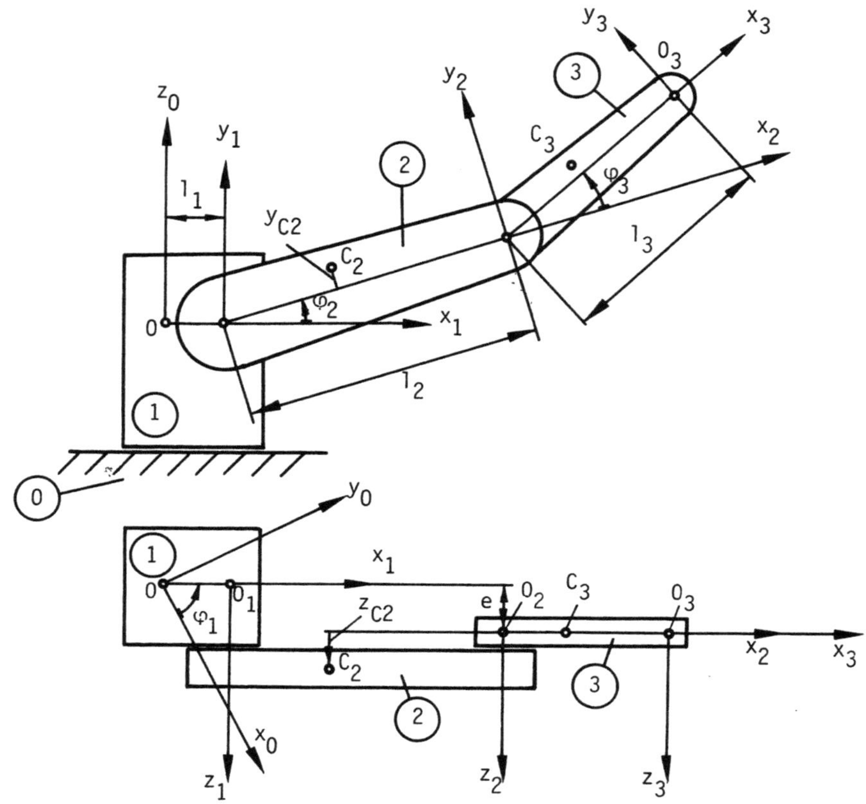

Fig.1. Schematic diagram of a four-link RRR-robot with n=3 degrees of freedom

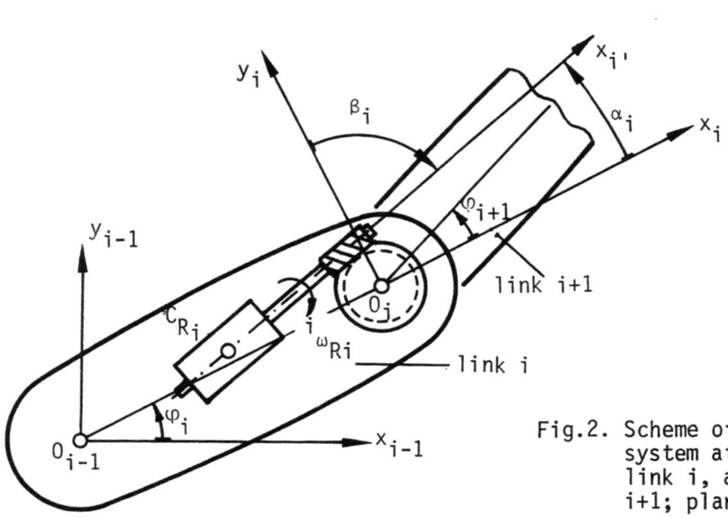

Fig.2. Scheme of the driving system attached to a link i, activating link i+1; plane arrangement

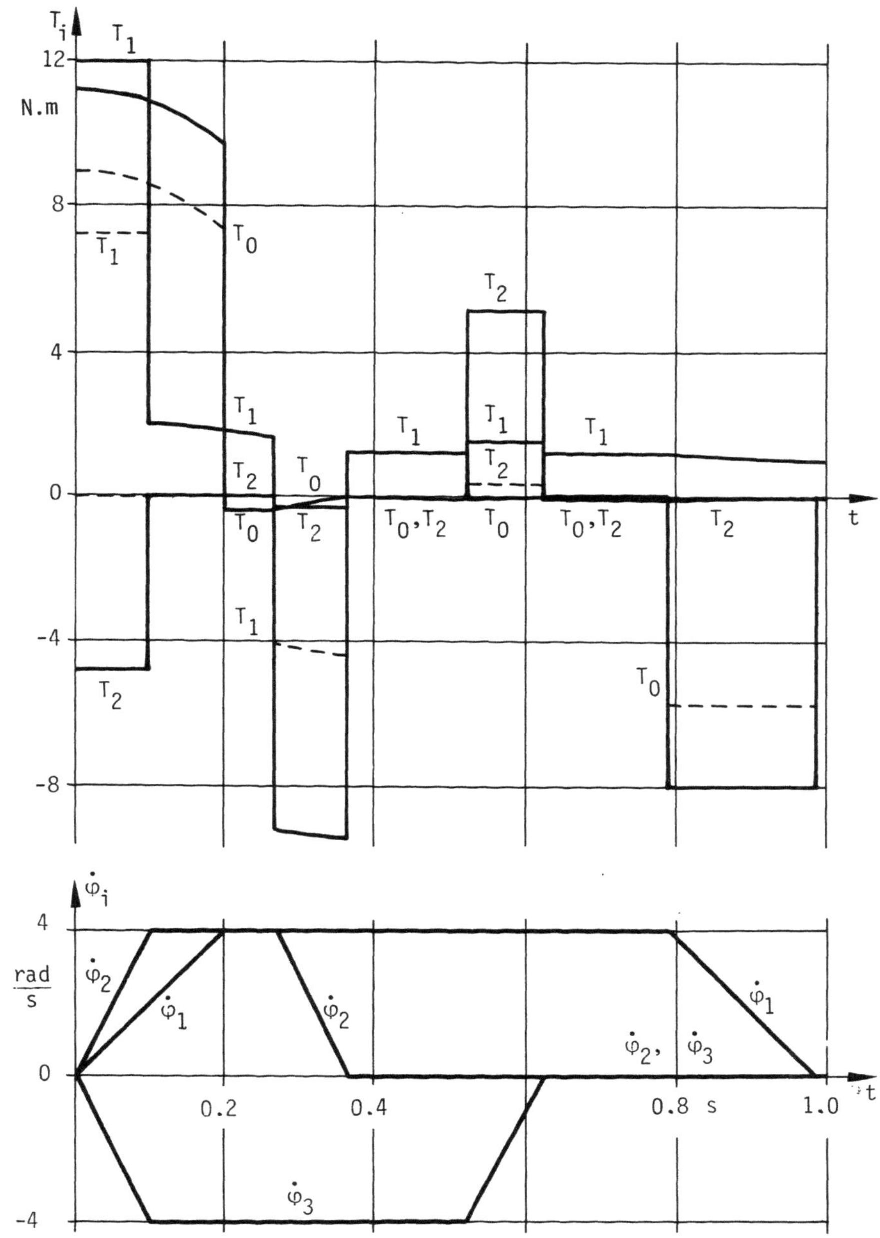

$\varphi_1 .. 0 \to \pi$, $\varphi_2 .. 0 \to \pi/3$, $\varphi_3 .. 0 \to -2\pi/3$

Fig.3. Required torques T_{i-1} for the given angular speeds $\dot{\varphi}_i$
—— with
– – – without rotatory inertia of the rotors.

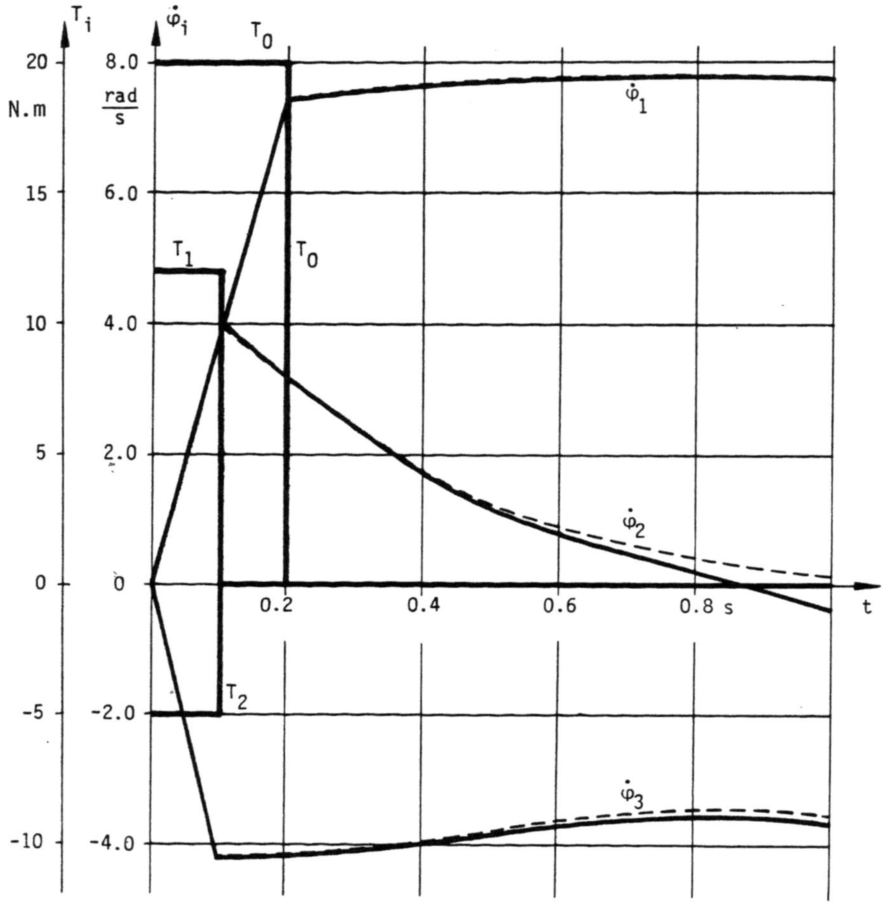

Fig.4. Angular speeds $\dot\varphi_i$ for the given torques T_{i-1}
—— with
– – without
gyroscopic effects of the rotors

	m_i	l_i	$x_{C_i}, y_{C_i}, z_{C_i}$	$i_{x_i}, i_{y_i}, i_{z_i}$	i_{z_0}	$\alpha_i, \beta_i, \gamma_i$
Unit	kg	m	m	m	m	rad
Link 1	70	0.3			0.3	$\pi/2$, 0, $\pi/2$
Link 2	80	0.5	-0.5, 0, 0.15	0.06, 0.18, 0.20		0, $\pi/2$, $\pi/2$
Link 3	40	0.5	-0.5, 0, 0	0.05, 0.12, 0.14		
$e=0$, $I_{R_0}=I_{R_1}=I_{R_2}=0.0012$ kg.m^2 , $\nu_{0,1}=\nu_{1,2}=\nu_{2,3}=100$						

Table 1: Main data for robot as shown in fig.1.

Dynamics of Flexible Machines: A Variational Approach

Edward J. Haug, Shih C. Wu and Sung-Soo Kim

Center for Computer Aided Design and
Department of Mechanical Engineering
The University of Iowa
Iowa City, Iowa

Summary

Recent developments in dynamics of mechanical systems with flexible bodies in spacecraft dynamics [1 and references cited therein] and in flexible machine dynamics [2 and references cited therein] illustrate an emerging capability for coupling geometrically nonlinear gross motion of mechanical systems with elastic deformation. A variational approach for developing the equations of motion of flexible dynamic systems is presented in this paper, using vibration and static correction elastic deformation modes. A slider-crank mechanism with a flexible coupler is used to illustrate the technique and the value of static correction modes in accurate approximation of system dynamic response.

1. Equations of Motion of Flexible Body Systems

Consider a deformable body that occupies volume Ω, with boundary σ, in an inertial x-y-z frame, as shown in Fig. 1. An x'-y'-z' reference frame is chosen to locate and orient the body in its undeformed state. Vector $\mathbf{r}$ from the origin of the x-y-z frame to the origin of the x'-y'-z' body reference frame defines the global position of the body. Linear combinations of deformation modes, prescribed in the x'-y'-z' frame, define the deformation field in the body.

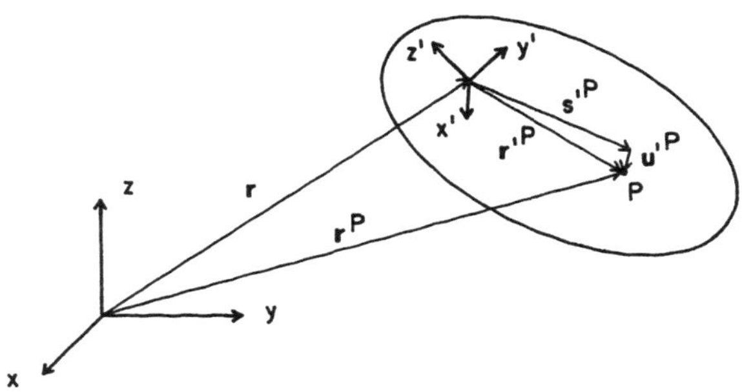

Fig. 1. Reference Generalized Coordinates

Dynamics of Multibody Systems
IUTAM/IFToMM Symposium Udine 1985
Editors: G. Bianchi and W. Schiehlen

The position vector of a typical point P on the body can be written as

$$\mathbf{r}^P = \mathbf{r} + \mathbf{A}(\mathbf{s}'^P + \mathbf{u}'^P) \equiv \mathbf{r} + \mathbf{A}\mathbf{r}'^P \tag{1}$$

where $\mathbf{A}$ is the transformation matrix from the x'-y'-z' frame to the x-y-z frame, $\mathbf{s}'^P$ is the position vector of point P in the x'-y'-z' frame before the body is deformed, and $\mathbf{u}'^P$ is the displacement of point P in the x'-y'-z' frame due to deformation. The displacement $\mathbf{u}'^P$, can be represented in terms of deformation modes as

$$\mathbf{u}'^P = \boldsymbol{\Psi}^P \mathbf{a} \tag{2}$$

where $\boldsymbol{\Psi}^P$ is the translational deformation modal matrix of point P and $\mathbf{a}$ is the modal coordinate vector. Methods of generating and selecting deformation modes are discussed in Section 3.

The variational form of the equations of motion of the body is [1]

$$-\int_\Omega \rho \delta\mathbf{r}^{P^T} \ddot{\mathbf{r}}^P d\Omega + \int_\Omega \delta\mathbf{r}^{P^T} \mathbf{f}^P d\Omega + \int_\sigma \delta\mathbf{r}^{P^T} \mathbf{T}^P d\sigma = \int_\Omega \delta\boldsymbol{\varepsilon}^T \boldsymbol{\tau}\, d\Omega \tag{3}$$

where ρ is material density, $\mathbf{f}^P$ is body force, $\mathbf{T}^P$ is surface traction, $\delta\mathbf{r}^P$ is virtual displacement of point P that is consistent with constraints, $\ddot{\mathbf{r}}^P$ is the acceleration vector of point P, and $\boldsymbol{\tau}$ and $\boldsymbol{\varepsilon}$ are vectors of stress and strain.

A virtual displacement of the origin of the x'-y'-z' reference frame may be denoted $\delta\mathbf{r}$. By Euler's theorem, the change in orientation of the coordinate system may be represented as a virtual rotation $\delta\boldsymbol{\pi}'$ which is a vector along the axis about which the rotation occurs to achieve the orientation variation, in the x'-y'-z' frame. Let $\tilde{\mathbf{a}}\mathbf{b}$ represent the vector product between vectors $\mathbf{a}$ and $\mathbf{b}$; i.e., $\tilde{\mathbf{a}}\mathbf{b}$ is the column vector of components of $\vec{a} \times \vec{b}$, where

$$\tilde{\mathbf{a}} = \begin{bmatrix} 0 & -a_z & a_y \\ a_z & 0 & -a_x \\ -a_y & a_x & 0 \end{bmatrix}$$

Then, the virtual displacement of point P is represented in terms of x'-y'-z' frame variations as [4]

$$\delta \mathbf{r}^P = \delta \mathbf{r} - \mathbf{A}\tilde{\mathbf{r}}'^P \delta \boldsymbol{\pi}' + \mathbf{A}\delta \mathbf{u}'^P \tag{4}$$

From Eq. 2,

$$\delta \mathbf{u}'^P = \boldsymbol{\Psi}^P \delta \mathbf{a} \tag{5}$$

Equation 4 can then be written as

$$\delta \mathbf{r}^P = \delta \mathbf{r} - \mathbf{A}\tilde{\mathbf{r}}'^P \delta \boldsymbol{\pi}' + \mathbf{A}\boldsymbol{\Psi}^P \delta \mathbf{a} \tag{6}$$

The acceleration vector of a typical point can be obtained by taking the derivative with respect to time of Eq. 1 twice, which gives [4]

$$\ddot{\mathbf{r}}^P = \ddot{\mathbf{r}} + \mathbf{A}\tilde{\boldsymbol{\omega}}'\tilde{\boldsymbol{\omega}}'\mathbf{r}'^P - \mathbf{A}\mathbf{r}'^P\dot{\boldsymbol{\omega}}' + 2\mathbf{A}\tilde{\boldsymbol{\omega}}'\boldsymbol{\Psi}^P\dot{\mathbf{a}} + \mathbf{A}\boldsymbol{\Psi}^P\dot{\mathbf{a}} + \mathbf{A}\boldsymbol{\Psi}^P\ddot{\mathbf{a}} \tag{7}$$

Substituting Eqs. 6 and 7 into Eq. 3, the first term is

$$-\int_\Omega \rho \delta \mathbf{r}^{P^T} \ddot{\mathbf{r}}^P d\Omega = -\int_\Omega \rho \left(\delta \mathbf{r}^T + \delta \boldsymbol{\pi}'^T \tilde{\mathbf{r}}'^P \mathbf{A}^T + \delta \mathbf{a}^T \boldsymbol{\Psi}^{P^T} \mathbf{A}^T \right)$$

$$\times \left(\ddot{\mathbf{r}} + \mathbf{A}\tilde{\boldsymbol{\omega}}'\tilde{\boldsymbol{\omega}}'\mathbf{r}'^P - \mathbf{A}\tilde{\mathbf{r}}'^P\dot{\boldsymbol{\omega}}' + 2\mathbf{A}\tilde{\boldsymbol{\omega}}'\boldsymbol{\Psi}^P\dot{\mathbf{a}} + \mathbf{A}\boldsymbol{\Psi}^P\ddot{\mathbf{a}} \right) d\Omega$$

$$= -[\delta \mathbf{r}^T, \delta \boldsymbol{\pi}'^T, \delta \mathbf{a}^T] \begin{bmatrix} \int_\Omega \rho \, d\Omega \, \mathbf{I}, & -\mathbf{A}\int_\Omega \rho \tilde{\mathbf{r}}'^P d\Omega, & \mathbf{A}\int_\Omega \rho \boldsymbol{\Psi}^P d\Omega \\ [-\mathbf{A}\int_\Omega \rho \tilde{\mathbf{r}}'^P d\Omega]^T, & -\int_\Omega \rho \tilde{\mathbf{r}}'^P \tilde{\mathbf{r}}'^P d\Omega, & \int_\Omega \rho \tilde{\mathbf{r}}'^P \boldsymbol{\Psi}^P d\Omega \\ [\mathbf{A}\int_\Omega \rho \boldsymbol{\Psi}^P d\Omega]^T, & [\int_\Omega \rho \tilde{\mathbf{r}}'^P \boldsymbol{\Psi}^P d\Omega]^T, & \int_\Omega \rho \boldsymbol{\Psi}^{P^T} \boldsymbol{\Psi}^P d\Omega \end{bmatrix} \begin{bmatrix} \ddot{\mathbf{r}} \\ \dot{\boldsymbol{\omega}}' \\ \ddot{\mathbf{a}} \end{bmatrix}$$

$$+ \begin{bmatrix} \mathbf{A}\int_\Omega \rho \tilde{\boldsymbol{\omega}}'\tilde{\boldsymbol{\omega}}'\mathbf{r}'^P d\Omega + 2\mathbf{A}\tilde{\boldsymbol{\omega}}' \int_\Omega \rho \boldsymbol{\Psi}^P d\Omega \, \dot{\mathbf{a}} \\ \int_\Omega \rho \tilde{\mathbf{r}}'^P \tilde{\boldsymbol{\omega}}'\tilde{\boldsymbol{\omega}}'\mathbf{r}'^P d\Omega + 2\int_\Omega \rho \tilde{\mathbf{r}}'^P \tilde{\boldsymbol{\omega}}'\boldsymbol{\Psi}^P d\Omega \, \dot{\mathbf{a}} \\ \int_\Omega \rho \boldsymbol{\Psi}^{P^T} \tilde{\boldsymbol{\omega}}'\tilde{\boldsymbol{\omega}}'\mathbf{r}'^P d\Omega + 2\int_\Omega \rho \boldsymbol{\Psi}^{P^T} \tilde{\boldsymbol{\omega}}'\boldsymbol{\Psi}^P d\Omega \, \dot{\mathbf{a}} \end{bmatrix}$$

$$\equiv -[\delta \mathbf{r}^T, \delta \boldsymbol{\pi}'^T, \delta \mathbf{a}^T] \, \mathbf{M} \begin{bmatrix} \ddot{\mathbf{r}} \\ \dot{\boldsymbol{\omega}}' \\ \ddot{\mathbf{a}} \end{bmatrix} + \mathbf{S}(\boldsymbol{\omega}', \dot{\mathbf{a}}) \tag{8}$$

Similarly, the second and third terms in Eq. 3 become

$$\int_\Omega \delta\mathbf{r}^{P^T}\mathbf{f}^P d\Omega + \int_\sigma \delta\mathbf{r}^{P^T}\mathbf{T}^P d\sigma$$

$$= [\delta\mathbf{r}^T,\ \delta\pi'^T,\ \delta\mathbf{a}^T] \begin{bmatrix} \int_\Omega \mathbf{f}^P d\Omega + \int_\sigma \mathbf{T}^P d\sigma \\ \int_\Omega \tilde{\mathbf{r}}'^P \mathbf{A}^T\mathbf{f}^P d\Omega + \int_\sigma \tilde{\mathbf{r}}'^P \mathbf{A}^T\mathbf{T}^P d\sigma \\ \int_\Omega \boldsymbol{\Psi}^{P^T}\mathbf{A}^T\mathbf{f}^P d\Omega + \int_\sigma \Psi^{P^T}\mathbf{A}^T\mathbf{T}^P d\sigma \end{bmatrix}$$

$$\equiv [\delta\mathbf{r}^T,\ \delta\pi'^T,\ \delta\mathbf{a}^T]\ \mathbf{Q} \tag{9}$$

If deformations are small, then in terms of a modal strain matrix $\mathbf{B}$,

$$\boldsymbol{\varepsilon} = \mathbf{B}\mathbf{a} \tag{10}$$

and for linear elasticity,

$$\boldsymbol{\tau} = \mathbf{D}\boldsymbol{\varepsilon} \tag{11}$$

where $\mathbf{D}$ is the material properties matrix. Substituting Eqs. 10 and 11 into the right side of Eq. 3,

$$\int_\Omega \delta\boldsymbol{\varepsilon}^T\boldsymbol{\tau} d\Omega = \delta\mathbf{a}^T\int_\Omega \mathbf{B}^T\mathbf{D}\mathbf{B} d\Omega\ \mathbf{a}$$

$$= [\delta\mathbf{r}^T,\ \delta\pi'^T,\ \delta\mathbf{a}^T] \begin{bmatrix} 0 \\ 0 \\ \int_\Omega \mathbf{B}^T\mathbf{D}\mathbf{B} d\Omega\ \mathbf{a} \end{bmatrix}$$

$$\equiv [\delta\mathbf{r}^T,\ \delta\pi'^T,\ \delta\mathbf{a}^T]\ \mathbf{U}(\mathbf{a}) \tag{12}$$

Using, Eqs. 8-12, Eq. 3 can be written as

$$[\delta\mathbf{r}^T,\ \delta\pi'^T,\ \delta\mathbf{a}^T]\left\{\mathbf{M}\begin{bmatrix} \ddot{\mathbf{r}} \\ \dot{\boldsymbol{\omega}}' \\ \ddot{\mathbf{a}} \end{bmatrix} + \mathbf{S}(\dot{\mathbf{a}},\boldsymbol{\omega}') + \mathbf{U}(\mathbf{a}) - \mathbf{Q}\right\} = 0$$

for all virtual displacements $\delta\mathbf{r}$, virtual rotations $\delta\pi'$, and variations $\delta\mathbf{a}$ in modal coordinates that are consistent with holonomeic constraints; i.e., which satisfy $\boldsymbol{\Phi}_\mathbf{q}\delta\mathbf{q} = \mathbf{0}$, where $\delta\mathbf{q} = [\delta\mathbf{r}^T,\ \delta\pi'^T,\ \delta\mathbf{a}^T]^T$ and $\boldsymbol{\Phi}(\mathbf{q}) = \mathbf{0}$.

By Farkas' lemma, there exists a Lagrange multiplier vector $\boldsymbol{\lambda}$ such that

$$\delta q^T \left\{ \mathbf{M} \begin{bmatrix} \ddot{\mathbf{r}} \\ \dot{\boldsymbol{\omega}}' \\ \ddot{\mathbf{a}} \end{bmatrix} + \mathbf{s}(\dot{\mathbf{a}},\boldsymbol{\omega}') + \mathbf{U}(\mathbf{a}) + \boldsymbol{\Phi}_{\mathbf{q}}^T\lambda - Q \right\} = 0 \tag{13}$$

for arbitrary δq. Therefore, the coefficient of $\delta \mathbf{q}$ in Eq. 13 must be $\mathbf{0}$, yielding the equation of motion of a flexible body in the form

$$\mathbf{M} \begin{bmatrix} \ddot{\mathbf{r}} \\ \dot{\boldsymbol{\omega}}' \\ \ddot{\mathbf{a}} \end{bmatrix} + \mathbf{S}(\dot{\mathbf{a}},\boldsymbol{\omega}') + \mathbf{U}(\mathbf{a}) + \boldsymbol{\Phi}_{\mathbf{q}}^T\lambda - \mathbf{Q} = \mathbf{0} \tag{14}$$

In order to evaluate integral terms in Eqs. 8, 9, and 12, a lumped mass material distribution approximation from a finite element code can be used that models a flexible body as a collection of many mass points. The vector $\mathbf{r}_i'^P$ defines the position of the lumped mass at node point P_i in the x'-y'-z' reference frame. The lumped mass of each point, denoted as m_i, is obtained using finite element analysis. The forces $\mathbf{f}_i^P$ and tractions $\mathbf{T}_i^P$ are defined as nodal body forces on each lumped mass and external force on lumped masses located at the boundary of the flexible body, respectively. The modal matrix $\boldsymbol{\Psi}_i^P$ is defined at point P_i. For simplification of notation, $\mathbf{r}_i'$, $\mathbf{f}_i$, $\mathbf{T}_i$, and $\boldsymbol{\Psi}_i$ represent $\mathbf{r}_i'^P$, $\mathbf{f}_i^P$, $\mathbf{T}_i^P$, and $\boldsymbol{\Psi}_i^P$, respectively.

Integral terms of the mass matrix $\mathbf{M}$ in Eq. 8 can be evaluated approximately as

$$\int_\Omega \rho d\Omega \cong \sum_{i=1}^{N} m_i$$

where N is the number of lumped masses in the flexible body model and

$$\int_\Omega \rho \mathbf{r}'^P d\Omega \cong \sum_{i=1}^{N} m_i \mathbf{r}_i' ; \quad \int_\Omega \rho \boldsymbol{\Psi}^P d\Omega \cong \sum_{i=1}^{N} m_i \boldsymbol{\Psi}_i$$

$$\int_\Omega \rho \tilde{\mathbf{r}}'^P \tilde{\mathbf{r}}'^P d\Omega \cong \sum_{i=1}^{N} m_i \tilde{\mathbf{r}}_i' \tilde{\mathbf{r}}_i' ; \quad \int_\Omega \rho \mathbf{r}'^P \boldsymbol{\Psi}^P d\Omega \cong \sum_{i=1}^{N} m_i \mathbf{r}_i' \boldsymbol{\Psi}_i$$

$$\int_\Omega \rho \boldsymbol{\Psi}^{P^T} \boldsymbol{\Psi}^P d\Omega \cong \sum_{i=1}^{N} m_i \boldsymbol{\Psi}_i^T \boldsymbol{\Psi}_i ; \quad \int_\Omega \rho \tilde{\boldsymbol{\omega}}' \tilde{\boldsymbol{\omega}}' \mathbf{r}'^P d\Omega \cong \sum_{i=1}^{N} m_i \tilde{\boldsymbol{\omega}}' \tilde{\boldsymbol{\omega}}' \mathbf{r}_i'$$

$$\int_\Omega \rho \Psi^P d\Omega \cong \sum_{i=1}^{N} m_i \boldsymbol{\Psi}_i^P ; \int_\Omega \rho \tilde{\mathbf{r}}'^P \tilde{\boldsymbol{\omega}}' \tilde{\boldsymbol{\omega}}' \mathbf{r}'^P d\Omega \cong \sum_{i=1}^{N} m_i \tilde{\mathbf{r}}_i' \tilde{\boldsymbol{\omega}}' \tilde{\boldsymbol{\omega}}' \mathbf{r}_i'$$

$$\int_\Omega \rho \boldsymbol{\Psi}^{P^T} \tilde{\boldsymbol{\omega}}' \tilde{\boldsymbol{\omega}}' \mathbf{r}'^P d\Omega \cong \sum_{i=1}^{N} m_i \boldsymbol{\Psi}_i^T \tilde{\boldsymbol{\omega}}' \tilde{\boldsymbol{\omega}}' \mathbf{r}_i'$$

$$\int_\Omega \rho \tilde{\mathbf{r}}'^P \tilde{\boldsymbol{\omega}}' \boldsymbol{\Psi}^P d\Omega \cong \sum_{i=1}^{N} m_i \tilde{\mathbf{r}}_i' \tilde{\boldsymbol{\omega}}' \boldsymbol{\Psi}_i ; \int_\Omega \rho \boldsymbol{\Psi}^{P^T} \boldsymbol{\omega}' \boldsymbol{\Psi}^P d\Omega \cong \sum_{i=1}^{N} m_i \boldsymbol{\Psi}_i^T \boldsymbol{\omega}' \boldsymbol{\Psi}_i$$

Integral terms in the generalized force expression **Q** of Eq. 9 are

$$\int_\Omega \mathbf{f}^P d\Omega = \sum_{i=1}^{N} \mathbf{f}_i ; \int_\sigma \mathbf{T}^P d\sigma = \sum_{i=1}^{M} \mathbf{T}_i$$

where M is number of nodes on the boundary of the flexible body. Finally,

$$\int_\Omega \tilde{\mathbf{r}}'^P \mathbf{A}^T \mathbf{f}^P d\Omega = \sum_{i=1}^{N} \tilde{\mathbf{r}}_i' \mathbf{A}^T \mathbf{f}_i ; \int_\sigma \tilde{\mathbf{r}}'^P \mathbf{A}^T \mathbf{T}^P d\sigma = \sum_{i=1}^{M} \tilde{\mathbf{r}}_i' \mathbf{A}^T \mathbf{T}_i$$

$$\int_\Omega \boldsymbol{\Psi}^{P^T} \mathbf{A}^T \mathbf{f}^P d\Omega = \sum_{i=1}^{N} \boldsymbol{\Psi}_i^T \mathbf{A}^T \mathbf{f}_i ; \int_\sigma \boldsymbol{\Psi}^{P^T} \mathbf{A}^T \mathbf{T}^P d\sigma = \sum_{i=1}^{M} \boldsymbol{\Psi}_i^T \mathbf{A}^T \mathbf{T}_i$$

Integral terms in the strain energy expression **U** of Eq. 12 are $\int_\Omega \mathbf{B}^T \mathbf{D} \mathbf{B} d\Omega = \boldsymbol{\Psi}^T \mathbf{K} \boldsymbol{\Psi}$, where $\boldsymbol{\Psi}$ is the 6N × k modal matrix, k is the number of modal coordinates, and **K** is the finite element stiffness matrix obtained by finite element analysis.

2. Kinematic Constraint Formulation

When flexible mechanical systems are modelled, generalized coordinates for each body are not generally independent. They are related by constraint equations that restrict relative motion of the bodies. In order to express geometrical conditions as equations of constraint, joint triads are defined in Fig. 2. Unit vectors $\mathbf{f}_i$, $\mathbf{g}_i$, and $\mathbf{h}_i$ are defined along x_i'', y_i'', and z_i'' axes of the joint triad at point Pi in body i. Geometrical definition of joints requires combinations of following conditions:

(1) the origins of joint triads coincide
(2) Pairs of vectors in each of the bodies are perpendicular
(3) pairs of vectors in each of the bodies are parallel.

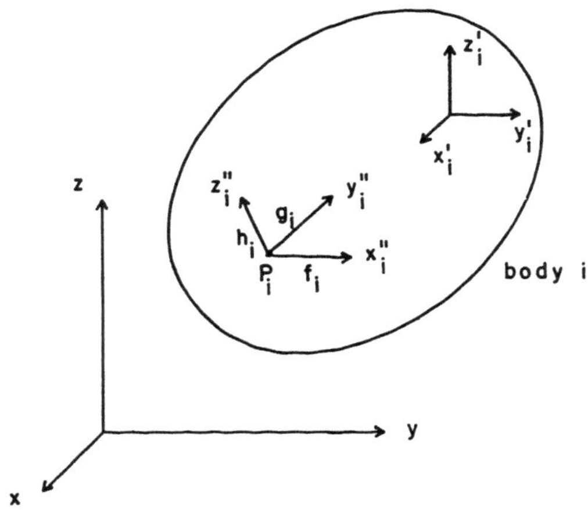

Fig. 2. Joint Triad

For condition (1), three constraint equations are defined, as follows:

$$\Phi_1 = r_j + r_j^{\,p} - r_i - r_i^{\,p}$$

$$= r_j + A_j\left(s_j'^{\,p} + u_j'(a_j)\right) - r_i - A_i\left(s_i'^{\,p} + u_i'(a_i)\right) = 0 \qquad (15)$$

where A_i and A_j are direction cosine orientation matrices of body i and j reference frames, respectively, and a_i and a_j are modal coordinate vectors for body i and body j.

For condition (2), one may require that vectors g_i and f_j be perpendicular. This requires one constraint equation, as follows:

$$\Phi_2 = g_i(a_i)^T f_j(a_j) = g_i'(a_i)^T A_i^{\,T} A_j f_j'(a_j) = 0 \qquad (16)$$

where g_i' and f_j' are unit vectors in terms of body fixed coordinates.

For condition (3), one may require that vectors h_i and h_j be parallel by demanding that h_i be orthogonal to both f_j and g_j; i.e.,

$$\Phi_3 = \begin{bmatrix} h_i^{\,T} f_j \\ h_i^{\,T} g_j \end{bmatrix} = \begin{bmatrix} h_i'(a_i)^T A_i^{\,T} A_j f_j'(a_j) \\ h_i'(a_i)^T A_i^{\,T} A_j g_j'(a_j) \end{bmatrix} = 0 \qquad (17)$$

There are several standard types of kinematic joints. Due to limitation of space in this paper, only spherical and revolute joint constraint equations are presented. For the spherical joint of Fig. 3 with centers at points P_i and P_j on bodies i and j, the constraint equation is Eq. 15. For the revolute joint of Fig. 4, the constraint equations are the combination of Eqs. 15 and 17. The translational joint will be shown for an example in the last section. Its constraint equations are presented in Ref. 5.

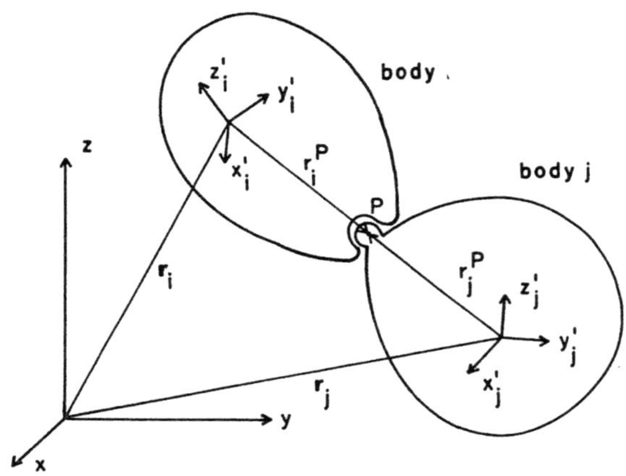

Fig. 3. Spherical Joint

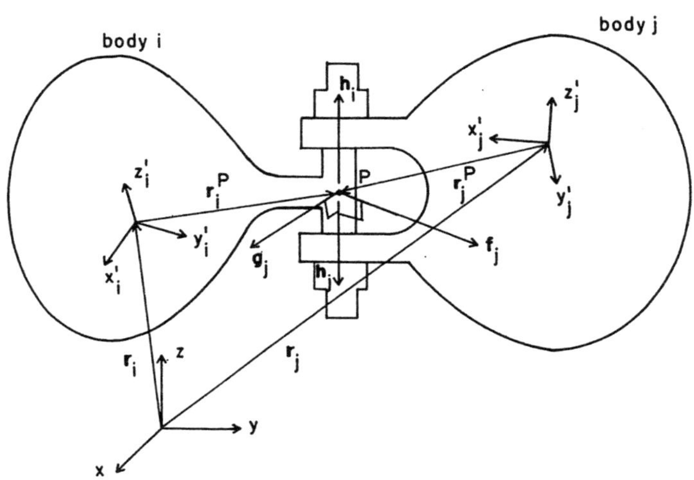

Fig. 4. Revolute Joint

3. Deformation Mode Selection

The conventional approach to selecting elastic deformation modes for dynamic analysis of flexible machines is to use normal vibration modes of individual components to represent elastic deformation. Recent developments in structural mechanics suggest that static correction modes [2] can be used to represent deformation due to forces and torques transmitted by kinematic couplings between substructures. In using static correction modes, one defines a set of nonredundant boundary conditions associated with kinematic constraints that act on each component, for purposes of finite elements analysis. Reaction forces and torques that may occur due to imposing kinematic constraints that were deleted for purposes of deformation analysis are applied to each component and associated static deformation modes are determined. These static correction modes are then employed with normal vibration modes to represent deformation of each component.

Experience to date with combinations of static correction and normal vibration modes [2] indicates that excellent results can be obtained using the minimal essential set of static correction modes and a few normal vibration modes. An example presented in the following section clearly illustrates the value of static correction modes.

4. A Numerical Example

To demonstrate use of equations of motion and selection of deformation modes, a spatial slider-crank mechanism with a flexible coupler is analyzed. The mechanism shown in Fig. 5 is composed of a rigid crank, a flexible circular cross section coupler, and a massless slider. Ground is assumed to be rigid. Revolute joints A and B are located between the ground and crank and the crank and flexible coupler, respectively. Spherical joint C is positioned between the coupler and slider. Translational joint D connects the slider and ground.

The rotational axis of revolute joint A is misaligned two degrees from the global z-axis, in the x-z plane. The rotational axis of revolute joint B is parallel to the global z-axis when the crank arm is in the vertical position along the positive global y-axis. However, the angle between the rotational axis of revolute joint B and the global z-axis varies during rotation of the crank, due to misalignment of revolute joint A. Therefore, deformation of the coupler increases until the crank is vertical downward.

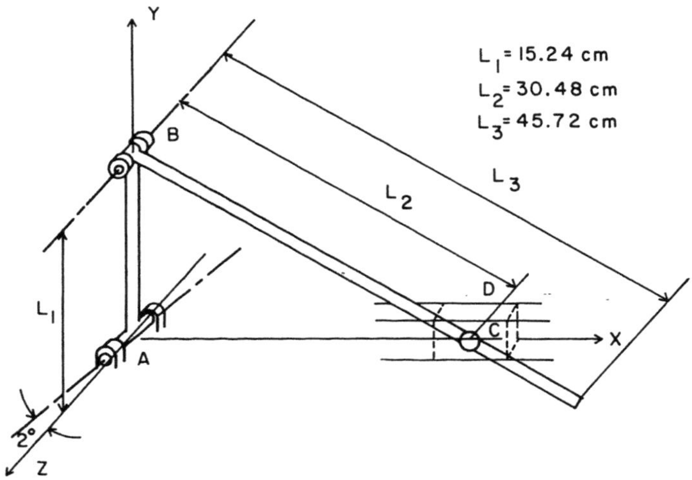

Fig. 5. Spatial Slider - Crank Mechanism

Material properties of the mechanism are as follows: modulus of elasticity $E = 2.0 \times 10^{12}$ dyne/cm, radius of coupler $r = 0.3$ cm, and mass density $\rho = 7.83$ g/cm^3. Initial velocity of slider $\nu_0 = -189.4714$ cm/sec. The coupler is initially undeformed, in the configuration shown in Fig. 5.

A finite element model of the flexible coupler is shown in Fig. 6, discretized into 18 equal length beam elements. Boundary conditions for vibration and static correction mode analysis are chosen such that the x, y, and z displacements and rotation about the x-axis at node 1 are zero and the y and z coordinates at node 13 are zero, all in the coordinate system of Fig. 6. The first vibration mode represents deformation in the y direction (simply supported beam bending mode). The second vibration mode represents deformation in the z-direction (simply supported beam bending mode). Deformation of the coupler due to a unit torque about the y axis at node 1 is defined as a static correction mode, because it is expected that misalignment of revolute joint A will lead to such a reaction torque in the revolute joint at B.

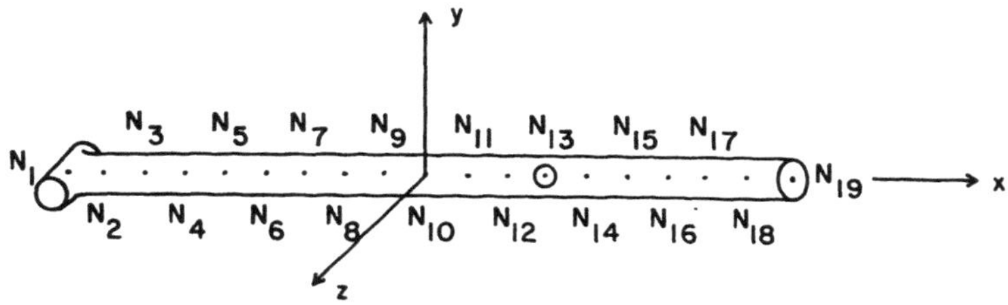

Fig. 6. Finite Element Model of the Coupler

Transient dynamic response is obtained using the Dynamic Analysis and)esign System (DADS) program. Simulations with one static correction mode 1S), with one vibration normal mode and one static correction mode (1N1S), rith two, four, six, eight, and ten vibration normal modes, and with six 'ibration normal modes and one static correction mode (6N1S), are carried ›ut. As shown in Fig. 7, angular velocity of the crank decreases ubstantially, due to increasing strain energy shown in Fig. 8, compared rith the motion of a planar rigid model; i.e., with no misalignment at A. ngular velocities of the crank in simulations 1S, 1N1S, and 6N1S are ssentially the same. As the number of vibration normal modes is ncreased, vibration normal mode results approach that of 1S. This means hat the static correction mode is dominent and the approximate solution ith one static correction mode and a few vibration normal modes is closer o the exact solution than the approximate solution with ten normal odes. The z coordinates of deformation of node 19 from 1S, 1N1S, 6N1S, nd 10N solutions are presentation in Fig. 9. Simulation times in CPU inutes per real time seconds on a PRIME 750 computer are as follows; 1S) ~ 5.87, (1N1S) ~ 27.23, and (10N) ~ 626.98.

Thus, using a combination of static correction modes and vibration ormal modes is more efficient than using purely vibration normal modes.

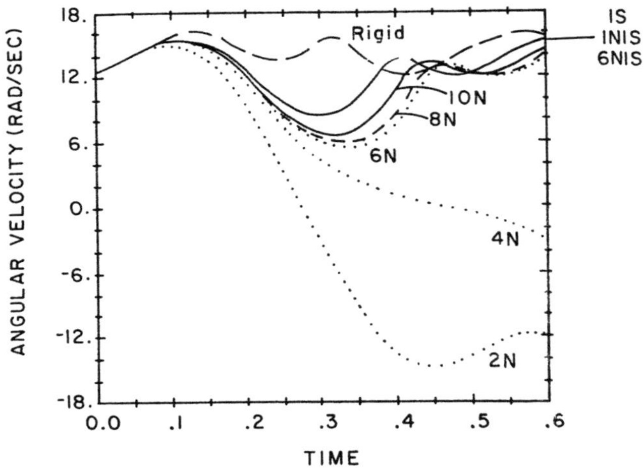

Fig. 7. Angular Velocity of Crank

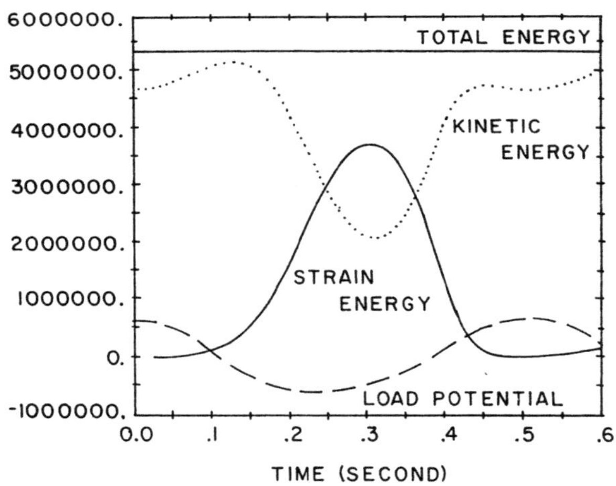

Fig. 8. Strain Energy of Slider-Crank Mechanism

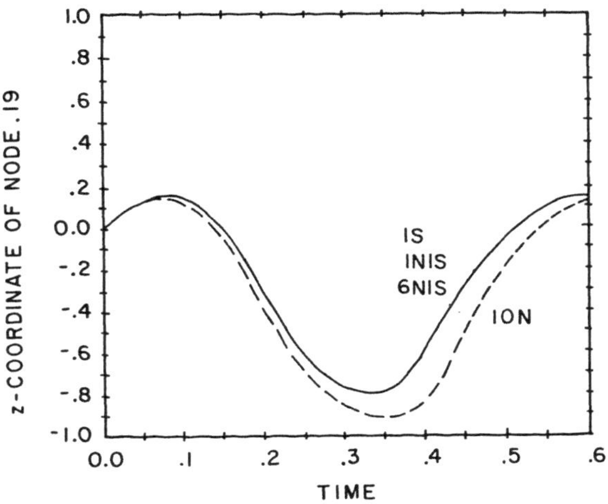

Fig. 9. z Coordinate of Node 19

References

1. Ho, J. 'Developement of Dynamics and Control Simulation Simulation of Large flexible Space Systems', J. Guidance, Vol. 8, No. 3, 1985, pp. 374-383.
2. Yoo, W. S., and Haug, E. J., 'Dynamics of Articulated Structures, Part I: Theory and Part II: Computer Implementation and Applications', J. Str. Mech., to appear.
3. Sokolnikof, I. S., Mathematical Theory of Elasticity, 2nd Ed. McGraw Hill 1978.
4. Haug, E. J., and McCullough, M., 'A Variational-Vector Calculus Approach to Machine Dynamics' J. Mechanism, Transmissions, and Automation in Design, to appear.
5. Haug, E. J., Computer Aided Kinematics and Dynamics of Mechanical Systems, Vol I: Basic Methods, Allyn and Bacon, Newton, MA., to appear.

Useful Procedures in Multibody Dynamics

R. L. HUSTON

Department of Mechanical and Industrial Engineering
University of Cincinnati
Cincinnati, Ohio 45221-0072 USA

Summary

This paper discusses four concepts expected to lead to advances in numerical simulation of multibody system dynamics. They are: 1) Body connection arrays; 2) Euler parameters; 3) Kane's equations; and 4) Orthogonal complement arrays.

Introduction

The interest in multibody dynamics continues to grow. The past decade has seen numerous advances in both modelling and analysis. These advances are documented in the references.

In the writer's opinion four of these advances stand out as major contributing factors in the development of multibody system dynamics. They are: 1) the use of body connection arrays in the automated development of the governing equations; 2) the use of Kane's Equations to obtain efficient, tractable formulations of the dynamics equations; 3) the use of Euler parameters to define the system configuration and to avoid singularities; and 4) the use of the Zero Eigenvalue Theorem and orthogonal complement arrays to obtain the governing equations of "constrained" multibody systems - that is, systems possessing closed loops or systems having specified motion.

The paper is devoted to a brief discussion of each of these concepts with the exception of Euler parameters. Euler parameters are omitted because of space limitations and because they are already well-documented in the literature. The paper employs the notation and formulation of References [4-12].

Dynamics of Multibody Systems
IUTAM/IFToMM Symposium Udine 1985
Editors: G. Bianchi and W. Schiehlen

The reader may wish to consult these references for additional details.

Body Connection Arrays

Body connection arrays provide a basis for developing algorithms for automated (numerical) analyses of multibody systems. To illustrate their use consider a typical multibody system as in Fig. 1. Let the system consist of spherically pinned bodies, numbered as

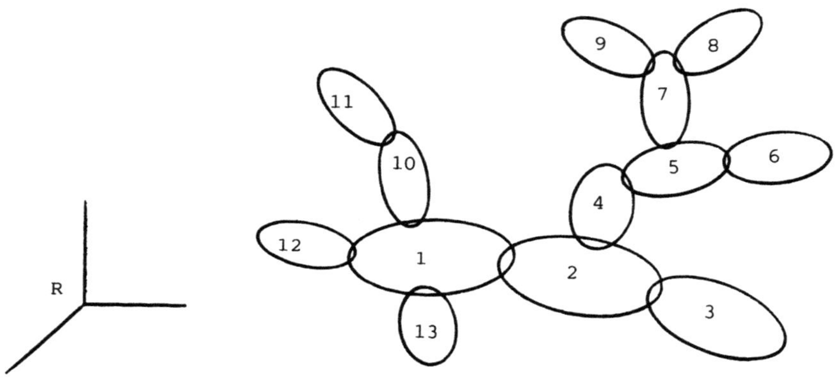

Fig. 1. A Multibody System

shown. Although the numbering system is not unique, it does follow a few simple rules: That is, a typical body is selected and numbered as Body 1 (or B_1). Then the remaining bodies are numbered in ascending progression through the branches.

With this numbering system, each body is connected to a unique lower-numbered body. The array of body numbers of the lower-numbered bodies is called the "body-connection array" or "lower-body array." Let (K) be the array of body numers and let L(K) be the array of lower-numbered bodies. Then for the system of Fig. 1. (K) and L(K) are

$$(K) = (1,2,3,4,5,6,7,8,9,10,11,12) \tag{1}$$

and

$$L(K) = (0,1,2,2,4,5,4,7,7,1,10,1) \tag{2}$$

where the "lower-body" of B_1 is taken as a reference frame R with number 0.

Observe that L(K) may be considered as an operator on (K). Indeed, from Eq. (2) we can obtain the lower-body array of the lower-body array: that is, L(L(K)), denoted by: $L^2(K)$. Then for the system of Fig. 1. $L^2(K)$ is

$$L^2(K) = (0,0,1,1,2,4,2,4,4,0,1,0) \tag{3}$$

where L(0) is taken as 0. In this context, let (K) be identified as $L^0(K)$, L(K) as $L^1(K)$, and $L^n(K)$ as $L(L^{n-1}(K))$.

A listing of $L^0(K)$, ..., $L^5(K)$ for the system of Fig. 1. is given in Table I.

$L^0(K)$	1	2	3	4	5	6	7	8	9	10	11	12
$L^1(K)$	0	1	2	2	4	5	4	7	7	1	10	1
$L^2(K)$	0	0	1	1	2	4	2	4	4	0	1	0
$L^3(K)$	0	0	0	0	1	2	1	2	2	0	0	0
$L^4(K)$	0	0	0	0	0	1	0	1	1	0	0	0
$L^5(K)$	0	0	0	0	0	0	0	0	0	0	0	0

Table I. Listing of $L^0(K)$, ..., $L^5(K)$ for the System of Fig. 1.

Finally, to illustrate one of the many uses of the body-connection array, consider the angular velocity $\underset{\sim}{\omega}_9$ of B_9. From the addition theorem for angular velocities $\underset{\sim}{\omega}_9$ is

$$\hat{\underset{\sim}{\omega}}_9 = \hat{\underset{\sim}{\omega}}_9 + \hat{\underset{\sim}{\omega}}_7 + \hat{\underset{\sim}{\omega}}_4 + \hat{\underset{\sim}{\omega}}_2 + \hat{\underset{\sim}{\omega}}_1 \tag{4}$$

where the "hats" designate relative angular velocity with respect to the adjacent lower-numbered body. The subscripts in Eq. (4) are simply the numbers in the 9th column of Table I.

Kane's Equations

In 1961, T. R. Kane [13] developed a form of dynamical equations which were shown to be especially efficient when used with nonholonomic systems. The equations (referred to at that time as "Lagrange's form of d'Alembert's principle") were developed from d'Alembert's principle. They possess the features of Lagrange's equations in that nonworking constraint forces are automatically eliminated. The equations were also shown to be applicable with h o l o n o m i c systems, but for the systems the advantages over other methods were not as clear.

It was not until several years later that the equations (now referred to as "Kane's equations") were shown to have distinct advantages over other methods in formulating dynamical equations for large multibody systems [14]. For example, for large systems Lagrange's equations are quickly seen to be cumbersome and unweildy as one attempts to differentiate the energy functions. Moreover, with the use of vector differentiation algorithms Kane's equations are ideally suited for automated (numerical) development of the governing equations.

To briefly illustrate the development and use of the equations, consider a set S of N particles P_i with masses m_i. Let these particles be acted upon by a system of applied forces $\underset{\sim}{F}_i$ (i=1,...,N). Then d'Alembert's principle asserts that

$$\underset{\sim}{F}_i - m\underset{\sim}{a}^{P_i} = 0 \qquad i=1,\ldots,N \tag{5}$$

where $\underset{\sim}{a}^{P_i}$ is the acceleration of P_i in an inertial reference frame R. If S has n degrees-of-freedom its behavior may be characterized by n generalized coordinates x_r. Then the velocity of P_i in R may be expressed in the form [15]

$$\underset{\sim}{v}^{P_i} = \sum_{r=1}^{n} \underset{\sim}{v}_{x_r}^{P_i}\, \dot{x}_r + \underset{\sim}{v}_t^{P_i} \tag{6}$$

where $\underset{\sim}{v}_{x_r}^{P_i}$ and $\underset{\sim}{v}_t^{P_i}$ are called "partial velocities" of P_i with respect to x_r and time t. Generalized active and inertia forces may then be defined as

$$F_r \overset{D}{=} \sum_{i=1}^{N} \underset{\sim}{v}_{\dot{x}_r}^{P_i} \cdot \underset{\sim}{F}_i \quad \text{and} \quad F_r^* \overset{D}{=} -\sum_{i=1}^{N} m_i \underset{\sim}{v}_{\dot{x}_r}^{P_i} \cdot \underset{\sim}{a}^{P_i} \quad (r=1,\ldots,n) \tag{7}$$

Hence, by multiplying Eqs. (5) by $\underset{\sim}{v}_{\dot{x}_r}^{P_i}$ and by adding, leads to Kane's equations

$$F_r + F_r^* = 0 \qquad (r=1,\ldots,n) \tag{8}$$

The utility of these equations is based upon a series of theorems which state that a broad class of applied forces do not contribute to the generalized active forces F_r and thus these applied forces may be ignored in dynamical analyses. These forces include constraint forces, certain forces of interaction between particles, and forces exerted over smooth surfaces. (See Reference [15] for details.) Also, it is shown in Reference [15] that for a rigid body B if the system of applied forces is replaced by an equivalent system consisting of a single force $\underset{\sim}{F}$ passing through say the mass center G, together with a couple with torque $\underset{\sim}{T}$, then F_r may be expressed in the simple form

$$F_r = \underset{\sim}{F} \cdot \underset{\sim}{v}^G_{\dot{x}_r} + \underset{\sim}{T} \cdot \underset{\sim}{\omega}^B_{\dot{x}_r} \qquad (r=1,\ldots,n) \tag{9}$$

where $\underset{\sim}{\omega}^B_{\dot{x}_r}$ is the "partial angular velocity of B" - analogous to the partial velocities of Eq. (6). (A similar expression may be written for the generalized inertia force F^*_r.)

Suppose B_k is a typical body of a multibody system S. Let G_k be the mass center of B_k. Then using Eq. (6), Huston et. al. [4,5,6] have shown that the velocity of G_k and the angular velocity of B_k may be expressed as

$$\underset{\sim}{v}_k = v_{k\ell m}\dot{x}_\ell \underset{\sim}{n}_m \qquad \text{and} \qquad \underset{\sim}{\omega}_k = \omega_{k\ell m}\dot{x}_\ell \underset{\sim}{n}_m \tag{10}$$

where $\underset{\sim}{n}_m$ (m=1,2,3) are mutually perpendicular unit vectors fixed in R and where the coefficients $v_{k\ell m}$ and $\omega_{k\ell m}$ are functions of the generalized coordinates x_r $(r=1,\ldots,n)$. These coefficients may be automatically evaluated using body-connection arrays. Also, simple multiplicative algorithms have been developed for computing their derivatives [4,5,6]. Hence, the acceleration of G_k and the angular acceleration of B_k may be expressed as

$$\underset{\sim}{a}_k = (\dot{v}_{k\ell m}\dot{x}_\ell + v_{k\ell m}\ddot{x}_\ell)\underset{\sim}{n}_m \quad \text{and} \quad \underset{\sim}{\alpha}_k = (\dot{\omega}_{k\ell m}\dot{x}_\ell + \omega_{k\ell m}\ddot{x}_\ell)\underset{\sim}{n}_m \tag{11}$$

where repeated indices designate a sum over the range of the index. Then, by using Eqs. (8) and (9) the governing dynamical equations of S may be written as

$$a_{\ell p}\ddot{x}_p = f_\ell \qquad (\ell=1,\ldots,n) \tag{12}$$

where $a_{\ell p}$ and f_ℓ are [4,5,6].

$$a_{\ell p} = m_k v_{kpm} v_{k\ell m} + I_{kmn}\omega_{kpm}\omega_{k\ell n} \tag{13}$$

and

$$f_{\ell} = F_{\ell} - (m_k v_{k\ell m} \dot{v}_{kqm} \dot{x}_q + I_{kmn} \omega_{k\ell m} \dot{\omega}_{kqn} \dot{x}_q + e_{nmh} I_{kmr} \omega_{kqn} \omega_{ksr} \omega_{k\ell h} \dot{x}_q \dot{x}_s) \tag{14}$$

where I_{kmn} are $\underset{\sim}{n}_m$ and $\underset{\sim}{n}_n$ components of the inertia dyadic of B_k for G_k and where e_{nmk} is the permutation symbol.

Orthogonal Complement Arrays

If a multibody system is subjected to constraints there will occur, in addition to Eqs. (12), a set of constraint equations of the form

$$b_{rp} \dot{x}_p = g_r \qquad r=1,\ldots,m \tag{15}$$

where the b_{rp} and g_r are functions of the x_p and time and where $m<n$. Eqs. (12) and (15) thus form a set of m+n equations for the n x_p. Since m of these are constraint equations the number of degrees of freedom of the system is reduced from n to n-m. Hence, the question arises: What are the best methods for efficiently reducing Eqs. (12) and (15) to n consistent equations - that is, to n-m dynamical equations and to m constraint equations?

To answer this question, Let Eqs. (12) and (14) be written in matrix form as

$$A\ddot{x} = f \quad \text{and} \quad B\dot{x} = g \tag{16}$$

where A is an nxn array, B is an mxn array, and x, f, and g are nx1 column arrays. Let C be an nx(n-m) array such that

$$BC = 0 \quad \text{or} \quad C^T B^T = 0 \tag{17}$$

The columns of C may then be viewed as being vectors in n-dimensional space of x_{ℓ}, which are "orthogonal" to the rows of B. Hence, C is called an "orthogonal complement" array. In this context, if the rows of B are considered as being "restraining vectors" in the n-dimensional space then the columns of C represent directions for which the system is unrestrained. Therefore, the system is free to move in the directions of the columns of C. Hence, the reduced set of dynamical equations is

[11,12,21]

$$C^T A\ddot{x} = C^T f \tag{18}$$

Eqs. (18) together with the constraint equations of (16) thus form the consistent set of n equations for the n x_ℓ.

The remaining question is how to find C. Although C is not unique, one version of C may be obtained as follows: Noting that the nxn matrix $B^T B$ is symmetrical, let y be an eigenvector of $B^T B$ associated with a z e r o eigenvalue of $B^T B$. Since the rank of $B^T B$ is at most m, there are at least n-m such vectors y. Let C be the array whose columns are these vectors. Then

$$B^T BC = 0 \quad \text{or} \quad C^T B^T BC = 0 \quad \text{or} \quad BC = 0 \tag{19}$$

Discussion

The foregoing procedures were developed independently, at different times, for different purposes. However, when they are applied with the analysis of multibody systems they become interactive and their individual usefulness is increased. For example, the automated generation of the governing equations through body-connection arrays is made possible through using Kane's equations. Also, if a numerical solution of the equations is sought singularities must be eliminated. The use of Euler parameters [5] accomplishes this task. Moreover, as discussed in Reference [5] when Euler parameters are used it is convenient to use relative angular velocity components as generalized coordinate derivatives. Kane's equations are ideally suited for such a variable definition [22]. Finally, if constraints are imposed upon the system the reduced governing equations are readily obtained with orthogonal complement arrays. Again Kane's equations are ideally suited for this reduction procedure.

Acknowledgement

Support for research for this paper has been partially provided by the National Science Foundation under Grant MEA-8313439 and is gratefully acknowledged.

References

1. Bayazitoglu, Y. O. and Chace, M. A.: Methods of Automated Dynamic Analysis of Discrete Mechanical Systems. J. Appl. Mech. 40 (1973) 809-819.

2. Hollerback, J. M.: A Recursive Lagrangian Formulation of Manipulator Dynamics and a Comparative Study of Dynamics of Formulation Complexity. IEEE Trans. Sys. Man. Cyber. SMC-10 (1980) 730-736.

3. Hooker, W. W. and Margulies, G.: The Dynamical Attitude Equations for an n-Body Satellite. J. Astr. Sci. 12 (1965) 123-128.

4. Huston, R. L. and Passerello, C. E.: On the Dynamics of Chain Systems. ASME Paper 74-WA/Aut 11 (1974).

5. Huston, R. L., Passerello, C. E., and Harlow, M. W.: Dynamics of Multi-Rigid-Body Systems. J. Appl. Mech. 45 (1978) 889-894.

6. Huston, R. L. and Passerello, C. E.: On Multi-Rigid-Body System Dynamics. Comp. and Struct. 19 (1979) 439-446.

7. Huston, R. L. and Passerello, C. E.: On Lagrange's Form of d'Alembert's Principle. Mat. and Ten. Quart. 23 (1973) 109-112.

8. Huston, R. L. and Passerello, C. E.: Eliminating Singularities in Governing Equations of Mechanical Systems. Mech. Res. Comm. (1976) 361-365.

9. Huston, R. L.: Multibody Dynamics Including the Effects of Flexibility and Compliance. Comp. and Struct. 14 (1981) 443-451.

10. Huston, R. L. and Passerello, C. E.: Multibody Structural Dynamics Including Translation Between the Bodies. Comp. and Struct. 11 (1980) 715-720.

11. Huston, R. L. and Passerello, C. E.: On Constraint Equations - A New Approach. J. Appl. Mech. 41 (1974) 1130-1131.

12. Kamman, J. W. and Huston, R. L.: Constrained Multibody System Dynamics. Comp. and Struct. 18 (1984) 999-1003.

13. Kane, T. R.: Dynamics of Nonholonomic Systems. J. Appl. Mech. 28 (1961) 574-578.

14. Kane, T. R. and Levinson, D. A.: Formulation of Equations of Motion for Complex Spacecraft. J. Guid. and Cont. 3 (1980) 99-112.

15. Kane, T. R. and Levinson, D. A.: Dynamics: Theory and Application. New York. McGraw Hill. 1985.

16. Lilov, L. K. and Chirikov, V. A.: On the Dynamics of Systems of Interconnected Bodies. Prikl. Matem. Mekhan. 45 (1981) 525-534.

17. Nikravesh, P. E. and Chung, I. S.: Application of Euler Parameters to the Dynamic Analysis of Three-Dimensional Constrained Mechanical Systems. J. Mech. Des. 104 (1982) 785-791.

18. Scheihlen, W. O. and Kreuzer, E. J.: Symbolic Computerized Derivation of Equations of Motion. Dynamics of Multibody Systems (K. Magnus, Ed.) Berlin. Springer 1977.

19. Wittenburg, J.: Nonlinear Equations of Motion for Arbitrary Systems of Interconnected Rigid Bodies. Dynamics of Multibody Systems (K. Magnus, Ed.) Berlin. Springer 1977.

20. Hemami, H. and Weimer, F. C.: Modelling of Nonholonomic Dynamic Systems with Applications. J. Appl. Mech. 48 (1981) 177-182.

21. Walton, W. C., Jr. and Steeves, E. C.: A New Matrix Theorem and Its Application for Establishing Independent Coordinates for Complex Dynamical Systems with Constraints. NASA Tech. Rep. TR-R-326 (1969).

22. Kane, T. R. and Wang, C. F.: On the Derivation of Equations of Motion. J. Soc. Ind. and Appl. Math. 13 (1965) 487-492.

On Steady-State Motions of an Axisymmetric Rigid Body on a »String« Suspension

A.Iu.ISCHLINSKY, V.A.STOROJENKO, M.E.TEMCHENKO

Institute for Problems of Mechanics USSR Academy of Sciences, Moscow;
Institute of Mathematics Ukrainian Academy of Sciences, Kiev

Summary

Studies of rotating bodies, suspended on a string or fiber, began, in fact, in the days of the Great Patriotic War with experiments of M.A.Lavrentiev and S.V.Malaschenko concerned at first with effect of rotation on a cumulation phenomenon and later with stability of liquid-filled bodies. In 1943 Malaschenko proposed quite a simple string suspension, originally designed as a thin steel string with one of its ends attached to a vertical drive rotor and another one-to the body to be studied. In this design high speeds of rotation (up to 3000 s^{-1} and higher) could be achieved, and a lot of stability studies had been made with such a suspension, particularly for bodies, partly filled with liquid. In these studies the influence of string properties on rotating body behavieur was investigated. The string first was assumed as being infinitely flexible and completely inextensible (i.e. providing a simple geometrical constraint). However later the string was found to be inadequate for rotating large-size bodies. So the existing design of the "string" suspension had been made as a set of rather heavy axisymmetric bodies (rods), coupled through Hooke's joints. An interesting apparent paradox was observed in fifties, namely, the coexistence of several varions stable steady-state configurations of the axisymmetric rotating body with a string suspension for certain identical values of angular speed ω . The analysis of behaviour of bodies with fiber or "string" suspension is the principal aim of this report. Analysis of a suspension mass effect has shown, that substituting heavy rod for an ideal string doesn't lead to qualitative changes of rotation patterns, the bifurcational values of rotation speed being located between similar values for the string case.
Generally "string" suspension is assumed to be composed of n rods. Stability of the basic equilibrium state with symmetry axes of all rods being aligned along a single line is proved. The existence of various deffected equilibrium states is established, their number being dependent on the number of rods within a suspension and relative values of inertia moments of the body. For a certain specific suspension with four rods equilibrium configurations have been obtained numerically.

Dynamics of Multibody Systems
IUTAM/IFToMM Symposium Udine 1985
Editors: G. Bianchi and W. Schiehlen

1°. Consider an axisymmetric rigid body of the mass m, suspended on a perfectly flexible weightless fiber or string with the length l. Consider an equilibrium state relative to frame $\xi\eta\zeta$, which rotates around ζ with a constant angular velocity ω (Fig.1). It is clear, that ζ - axis, string and symmetry axis of the body should lie within a single plane [1,2], the inclination angles of string and symmetry axis relative to vertical being denoted as α and θ respectively.

Equations of forces and moments balance in the above frame $\xi\eta\zeta$ may be written as

$$T\cos\alpha - mg = 0$$

$$-T\sin\alpha + m\omega^2(l\sin\alpha + a\sin\theta) = 0 \qquad (1)$$

$$\omega^2[-(A-C)\sin\theta\cos\theta - m(l\sin\alpha + a\sin\theta)(l\cos\alpha + a\cos\theta)] + mg(l\sin\alpha + a\sin\theta) = 0,$$

with T being tension of the string, a - distance between mass center G of the body and string attachment point, A, C - equatorial and axial inertia moments respectively.

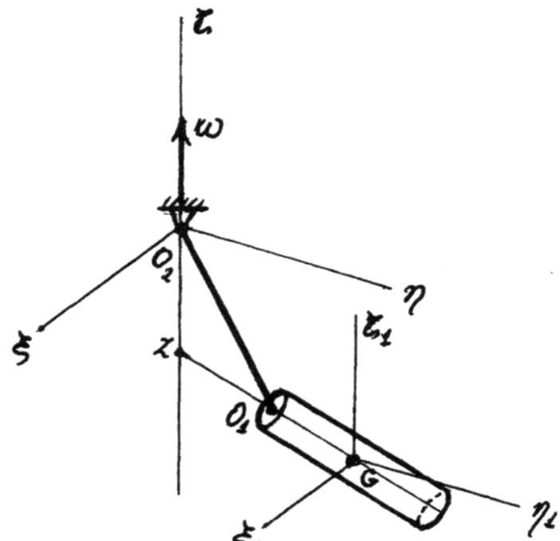

Fig. 1

Excluding tension from the first two equilibrium conditions (1) yields

$$\omega^2 l\sin\alpha\cos\alpha + \omega^2 a\sin\theta\cos\alpha - g\sin\alpha = 0$$

$$\omega^2[(A-C)\sin\theta\cos\theta + ma^2\sin\theta\cos\theta + mal\sin\alpha\cos\theta] - mga\sin\theta = 0 \qquad (2)$$

Equations (2) have a trivial solution $\alpha \equiv \theta \equiv 0$, corresponding to vertical dynamic equilibrium configuration (Fig.2a), which is stable for any ω [3]. Other solutions may be found by linearizing (2) for small α, θ. Such nonzero solutions exist provided the corresponding characteristic equation is satisfied, namely

$$\omega^4 - g/\ell \left[1 + \frac{ma(\ell + a)}{A - C}\right]\omega^2 + \frac{mga}{A - C}\, g/\ell = 0 \qquad (3)$$

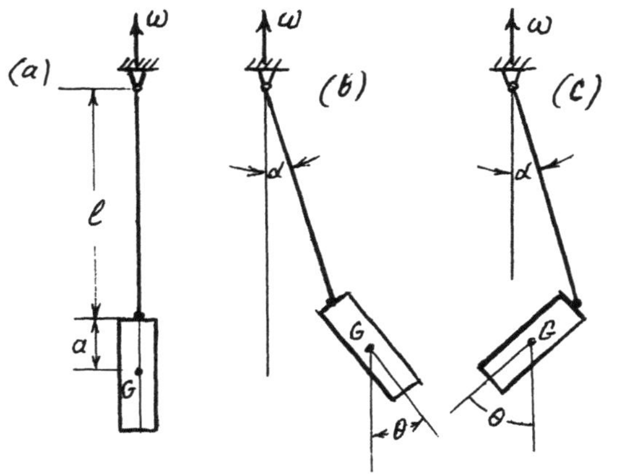

Fig.2

For an oblong axisymmetric body, such that $A = B > C$ this quadratic equation for ω^2 has two distinct roots ω_1^2, and ω_2^2. They define bifurcational values of angular velocity, at which new dynamic equilibrium configurations do appear. The configuration, illustrated in Fig.2b, corresponds to the lowest root of (5) and is stable [4], whereas the configuration, shown in Fig.2c, corresponds to the highest root of (5); its stability has been studied in [5,6].

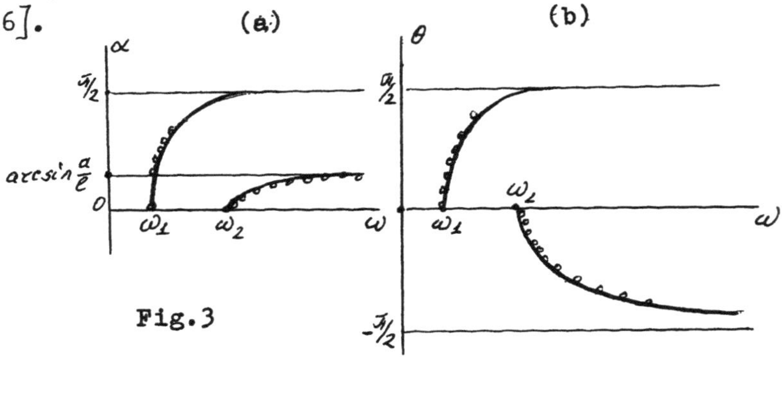

Fig.3

From the original equations (2) α and θ may be found as functions of ω . Fig.3 illustrates these calculated functions, which agree well with experimental data, shown by circles.

2°. Consider now the case of a so-called "string suspension" which in fact is an axisymmetric rigid rod with a gimbal at the fixed point O . The other end of the rod is attached to the axisymmetric rotating body through a spherical joint (Fig.4). For the rod properties and sizes the same notations will be used as defined above for those of the body, with subscript 1 added for the rod (see also Fig.4).

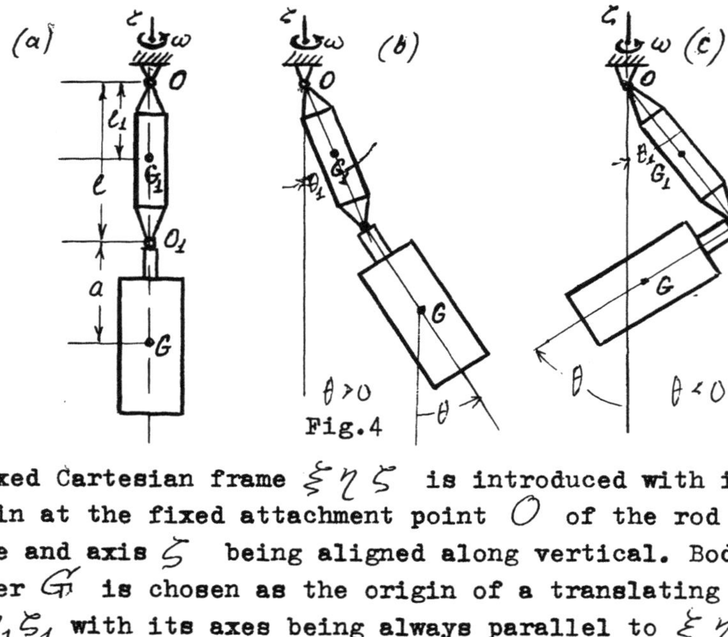

Fig.4

A fixed Cartesian frame $\xi\eta\zeta$ is introduced with its origin at the fixed attachment point O of the rod to drive and axis ζ being aligned along vertical. Body mass center G is chosen as the origin of a translating frame $\xi_1\eta_1\zeta_1$ with its axes being always parallel to $\xi\eta\zeta$ and also as the origin of a body-fixed Cartesian frame xyz with z - axis oriented along body symmetry axis. Similar frames $\xi'\eta'\zeta'$ and $x_1y_1z_1$ for rod are introduced as well.

Orientation of body-fixed frame xyz relative to frame $\xi'\eta'\zeta'$ may be defined by three Krylov angles α,β,γ (Fig.5). The latter may be used as generalized coordinates together with similar angles $\alpha_1, \beta_1, \gamma_1$ for rod. Then the Lagrange equations of motions of the second kind for this six-degree-of-freedom system may be written as [8]

$$A^*\cos\beta_1\ddot{\alpha}_1 + 2A^*\sin\beta_1\dot{\alpha}_1\dot{\beta}_1 + C_1(\dot{\alpha}_1\sin\beta_1 + \dot{\gamma}_1)\dot{\beta}_1 + ma\ell\{(\cos\beta\ddot{\alpha} - 2\sin\beta\dot{\alpha}\dot{\beta})\cos(\alpha-\alpha_1) -$$
$$-[(\dot{\alpha}^2+\dot{\beta}^2)\cos\beta + \sin\beta\ddot{\beta}]\sin(\alpha-\alpha_1)\} + (m_1\ell_1 + m\ell)g\sin\alpha_1 = 0$$

$$A^*(\ddot{\beta} + \dot{\alpha}_1^2\sin\beta_1\cos\beta_1) - C_1\dot{\alpha}_1\cos\beta_1(\dot{\alpha}_1\sin\beta_1 + \dot{\gamma}_1) + ma\ell(\cos\beta\ddot{\beta} - \dot{\beta}^2\sin\beta)\cos\beta_1 +$$
$$+ ma\ell\{[(\dot{\alpha}^2+\dot{\beta}^2)\cos\beta + \ddot{\beta}\sin\beta]\cos(\alpha-\alpha_1) + (\cos\beta\ddot{\alpha} - 2\sin\beta\dot{\alpha}\dot{\beta})\sin(\alpha-\alpha_1)\}\sin\beta_1 +$$
$$+ g(m_1\ell_1 + m\ell)\sin\beta_1\cos\alpha_1 = 0$$

$$A^0(\ddot{\alpha}\cos\beta - 2\sin\beta\dot{\alpha}\dot{\beta}) + C\dot{\beta}(\dot{\alpha}\sin\beta + \dot{\gamma}) + ma\ell\{[(\dot{\alpha}_1^2 + \dot{\beta}_1^2)\cos\beta_1 + \ddot{\beta}_1\sin\beta_1]\sin(\alpha-\alpha_1) +$$
$$+ (\ddot{\alpha}_1\cos\beta_1 - 2\dot{\alpha}_1\dot{\beta}_1\sin\beta_1)\cos(\alpha-\alpha_1)\} + mga\sin\alpha\cos\beta = 0$$

$$A^0(\ddot{\beta} + \dot{\alpha}^2\cos\beta\sin\beta) - C(\dot{\alpha}\sin\beta + \dot{\gamma})\dot{\alpha}\cos\beta + ma\ell(\ddot{\beta}_1\cos\beta_1 - \dot{\beta}_1^2\sin\beta_1)\cos\beta + ma\ell\{[(\ddot{\beta}_1\sin\beta_1 +$$
$$+ (\dot{\alpha}_1^2 + \dot{\beta}_1^2)\cos\beta_1]\cos(\alpha-\alpha_1) - (\ddot{\alpha}_1\cos\beta_1 - 2\dot{\alpha}_1\dot{\beta}_1\sin(\alpha-\alpha_1)\}\sin\beta + mga\cos\alpha\sin\beta = 0$$

$$C_1\frac{d}{dt}(\dot{\alpha}_1\sin\beta_1 + \dot{\gamma}_1) = 0; \quad C\frac{d}{dt}(\dot{\alpha}\sin\beta + \dot{\gamma}) = 0 \tag{4}$$

where

$$A^0 = A + ma^2; \qquad A^* = A_1 + m_1\ell_1^2 + m\ell^2 \tag{5}$$

3°. The set of equations (4) has a particular solution

$$\alpha = \beta = \alpha_1 = \beta_1 = 0; \qquad \dot{\gamma} = \dot{\gamma}_1 = \omega = \text{const} \tag{6}$$

which corresponds to a steady-state motion with vertical symmetry axes of body and rod (Fig.4). Stability of this motion in the first-order approximation may be studied by linearizing (4) for small $\alpha, \beta, \alpha_1, \beta_1$ and $\dot{\gamma} = \dot{\gamma}_1 = \omega$. The characteristic equation for this linearized set

$$[A^0A^* - (mal)^2]\lambda^4 - \omega[CA^* + C_1A^0]\lambda^3 + (-mgaA^* - C_1C\omega^2 - KgA^0)\lambda^2 + (C_1ma + CK)g\omega\lambda + Kg^2ma = 0 \qquad (7)$$

always has real roots [8], implying that steadystate rotations with vertical orientation of both body and rod symmetry axes are always stable.

4°. For the analysis of other possble steadystate configurations of the coupled system body-rod classical Euler angles seem to be more siutable. Therefore, angles ψ, θ, φ and $\psi_1, \theta_1, \varphi_1$ will be used to define orientations of body and rod respectively relative to translating frames ξ_1, η_1, ζ_1 and $\xi'\eta'\zeta'$. Considering steady-state rotation around vertical, we may without any loss of generality put $\varphi = \varphi_1 = 0$. Then the following set of equations for ψ, θ, ψ_1, θ_1 is obtained [8]

$$\begin{aligned}
&\sin(\psi - \psi_1) = 0 \\
&(A^* - C_1)\omega^2 \sin\theta_1 \cos\theta_1 + mal \sin\theta \cos\theta_1 \cos(\psi - \psi_1)\omega^2 = Kg\sin\theta_1 \\
&(A^0 - C)\omega^2 \sin\theta\cos\theta + mal\sin\theta_1 \cos\theta\cos(\psi - \psi_1)\omega^2 = mga\sin\theta
\end{aligned} \qquad (8)$$

This set can be analyzed by the approach as in sect. 1° for small deviations of symmetry axes from vertical (small θ and θ_1). Then the last two of equations (8) may be reduced to

$$\begin{aligned}
&[(A^* - C_1)\omega^2 - Kg]\theta_1 + mal\omega^2\theta = 0 \\
&mal\omega^2\theta_1 + [\omega^2(A^0 - C) - mga]\theta = 0
\end{aligned} \qquad (9)$$

The last set has a nonzero solution provided

$$f_1(\omega) = [(A^* - C_1)(A^0 - C) - (mal)^2]\omega^4 - g[ma(A^* - C_1) + K(A^0 - C)]\omega^2 + Kmag^2 = 0 \qquad (10)$$

(this condition may be obtained directly from (7) as well, by putting $\lambda = \omega$). For $A = B > C$ and $A_1 = B_1 > C_1$ eg. (10) always has two positive roots $(\omega_1^*)^2$ and $(\omega_2^*)^2$. At each of these bifurcational values of angular speed two other distinct steady-state configurations appear with inclined symmetry axes, one with positive, the other one - with negative value of θ_1/θ (see Fig.4).

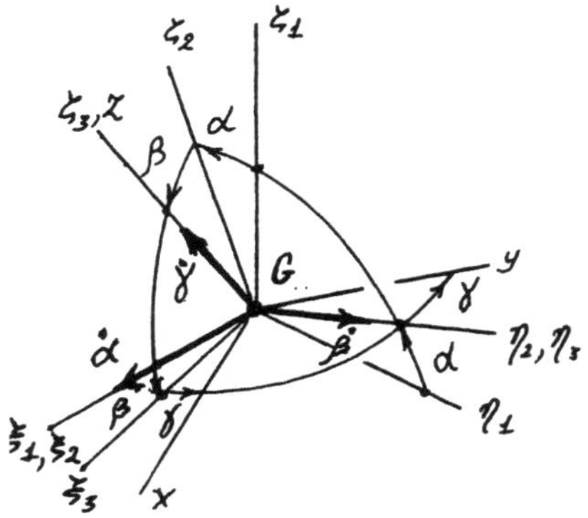

Fig.5

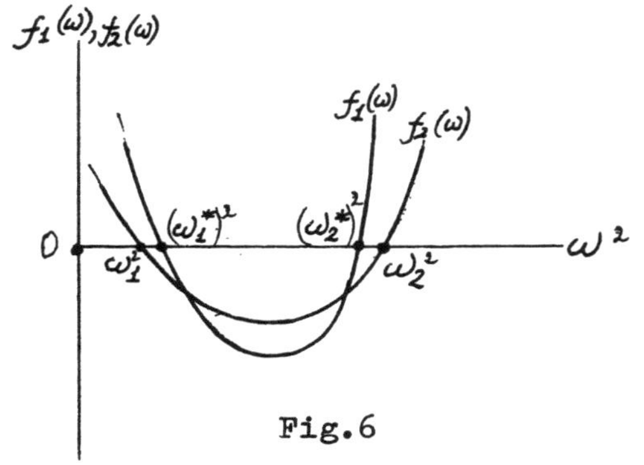

Fig.6

In Fig.6 a graph of $f_1(\omega)$ as defined by (10) is presented together with that of the function

$$f_2(\omega) = ml^2(A^0 - C - ma^2)\omega^4 - mgl(A^0 - C + mal)\omega^2 + m^2 a l g^2 = 0 \tag{11}$$

which is obtained from (10) for the case, when the rod is substituted by a string, regarded as a simple geometrical constraint. It can be clearly seen, that bifurcational values $\omega_1^{*2}, \omega_2^{*2}$ always lie between corresponding ones ω_1^2, ω_2^2 for the system with heavy rod substituted by an ideal string.

5°. Consider now a suspension composed of n rods, connected in series with the Hooke's joints (Fig.7). Each of these joints, as well as of those connecting the suspension with body and drive, provides two degrees of freedom, so that the system, shown in Fig.7, has $2(n+1)$ DOF. All rods are identical cylinders of the height l and radius r . The center of each Hooke's joint lie on a symmetry axis of the corresponding rod (or body).

Let a be the distance of body center mass from the nearest Hooke's joint, other notations for properties being the same as in the single-rod case. Using inclination angles α_j, β_j with respect to vertical of body and rods symmetry axes (Fig.8) as generalized coordinates, leads to the following first-order equations of motion in the vicinity of vertical equilibrium state

$$(\rho_1-\varepsilon)(\ddot z_j+2i\dot z_j-z_j)+[1+4\varepsilon(n-j+\tfrac{1}{2})]\Big(\nu z_j+\sum_{k=1}^{n}(\ddot z_k+2i\dot z_k-z_k)\Big)+$$

$$+\sigma_1(-i\dot z_j+z_j)+\varkappa(\ddot z_{n+1}+2i\dot z_{n+1}-z_{n+1})-4\varepsilon\sum_{p=1}^{n-j}p(\ddot z_{j+p}+2i\dot z_{j+p}-z_{j+p})=0$$

$$(j=1,2,\ldots,n)$$

$$\varkappa(1+\rho)(\ddot z_{n+1}+2i\dot z_{n+1}-z_{n+1})+\varkappa\sigma(-i\dot z_{n+1}+z_{n+1})+\sum_{k=1}^{n}(\ddot z_k+2i\dot z_k-z_k)+\nu z_{n+1}=0 \qquad (12)$$

Here

$$\varepsilon=\frac{m_1}{4m};\quad \nu=\frac{g}{\omega^2 l};\quad \varkappa=\frac{a}{l};\quad \rho=\frac{A}{ma^2};\quad \sigma=\frac{C}{ma^2};$$

$$\rho_1=\frac{A_1}{ml^2};\quad \sigma_1=\frac{C_1}{ml^2} \qquad (13)$$

and $z_j=\alpha_j+i\beta_j$ are complex state variables as introduced by a compression procedure [7], whereas $\tau=\omega t$ is a dimensionless time.

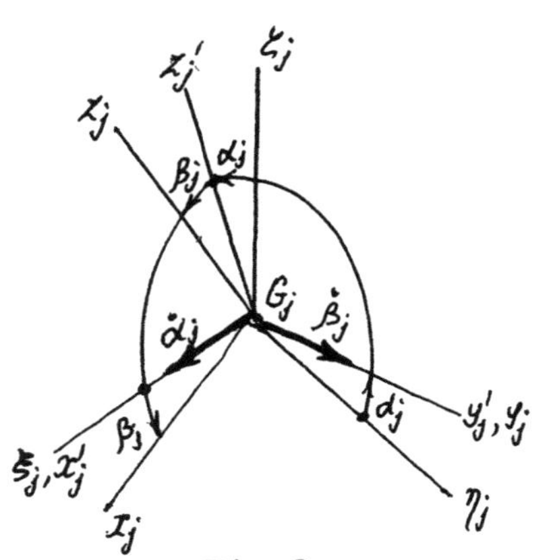

Fig.8

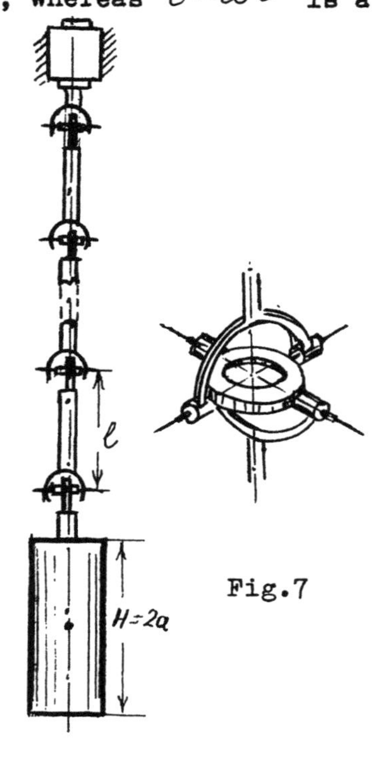

Fig.7

6°. Assuming the solution to (12) as being of the form $z_j = z_{jm} \exp(-\lambda \tau i)$, yields the following characteristic equation

$$\det \left\| -d_{jk}(1+\lambda)^2 + g_{jk}(1+\lambda) + g_{jk} \right\| = 0$$

$$(j, k = 1, 2, \ldots, n+1), \tag{14}$$

where

$$d_{jk} = 1 + 4\varepsilon(n + \tfrac{1}{2} - \max(j,k)) - \delta_{jk}(-\rho + \varepsilon)$$

$$g_{jk} = \delta_{jk}\sigma_1; \quad g_{jk} = \delta_{jk}[1 + 4\varepsilon(n + \tfrac{1}{2} - j)); \quad d_{j,n+1} = d_{n+1,j} = 1;$$
$$g_{j,n+1} = g_{n+1,j} = g_{j,n+1} = g_{n+1,j} = 0; \quad d_{n+1,n+1} = 1 + \rho;$$
$$g_{n+1,n+1} = \sigma; \quad g_{n+1,n+1} = \frac{1}{\chi} \qquad (j = 1, 2, \ldots, n);$$

(15)

and δ_{ik} is Kronecker delta. The trivial solution to (12) would be stable provided all roots $u_\mu = 1 + \mu\lambda$ of (14) are real. This would be the casė, as shown in [11], if all principal minors of the symmetric matrix $D = \| d_{ik} \|$ or (in view of the inertia law for quadratic forms [12]) those of the transformed matrix $D_1 = S'DS$, where S is any nonsingular $(n+1) \times (n+1)$ matrix are positive. Using

$$S = \begin{Vmatrix} 1 & 0 & 0 & \cdots & 0 & 0 \\ -1 & 1 & 0 & \cdots & 0 & 0 \\ \cdot & \cdot & \cdot & \cdots & \cdot & \cdot \\ 0 & 0 & 0 & \cdots & 1 & 0 \\ 0 & 0 & 0 & \cdots & -1 & 1 \end{Vmatrix} \tag{16}$$

leads to D_1 (and therefore D) with all principal minors being positive, implying the first-order stability of a vertical dynamic equilibrium configuration for all rotation velocities.

7°. Other, nonvertical equilibrium configurations of the system would be possible, provided eg.(12) have nonzero constant solutions $z_j = z_j^0$ = const. Such solutions would exist provided the determinant of (12) with time derivative terms being dropped vanishes, that is, if

$$\det \| d_{jk} - g_{jk} - q_{jk} \| = 0 \tag{17}$$

where d_{jk}, g_{jk}, q_{jk} are defined by (15). This may be regarded as an equation for ν, and positive real roots $\nu_\mu = g/\omega^2_\mu l$ correspond to real bifurcational values of angular velocity ω, at which other stable equilibrium configurations do appear.

Since ν appears only in main diagonal terms, (17) may be rewritten as

$$\det \| c_{jk} - \nu \delta_{jk} \| \quad ; \quad c_{jk} = (d_{jk} - g_{jk}) / \sqrt{g_{jj}, g_{kk}} \tag{18}$$

implying that roots ν_μ of (18) may be regarded as eigenvalues of the matrix $C = \| c_{jk} \|$. All of these are real since C is symmetric. Moreover, in view of the Jacobi theorem [12] the number of positive eigenvalues is equal to that of constant sign subintervals within the series 1, Δ_1, $\Delta_2, \ldots, \Delta_{n+1}$. Here Δ_j are principal minors of C or of $C_1 = P'CP$ where P is an arbitrary nonsingular matrix. The latter matrix can be chosen in such a way as to obtain C_1 identical to the matrix $D_1 = S'DS$ with $\xi = \rho_1 - \sigma_1$ and $\eta = \rho - \sigma$ substituted for ρ_1 and ρ respectively (S is defined by (16)). Thus closed-form expressions for principal minors of C_1 may be obtained. Since each rod of the suspension is an oblong cylinder, i.e. $\xi = (A_1 - C_1)/ml^2 > 0$, first n principal minors of C_1 are positive, whereas sign of Δ_{n+1} depends on η. If the suspended body is also oblong $(A > C)$, then $\eta > 0$, so that $n+1$ bifurcational values of angular velocity do exist, whereas only n such values exist when $C > A + ml^2$ $(\eta < -1)$ and therefore $\Delta_{n+1} < 0$. For the case $A < C < A + ml^2$ the sign of Δ_{n+1} should be established by a direct calculation.

8°. Calculations of bifurcated equilibrium configurations have been made by numerical solution of equations (12) with time derivative terms being dropped for the case of a cylindrical body with m =2500 g, R =27.5 sm, H =2a=10sm and a suspension with four rods with m_1 =50g, ℓ =8sm, r =0.5sm. The results are presented in the following table

γ_f	ω_i, c^{-1}	x_1	x_2	x_3	x_4
4,295	$1,511\ 10^{-2}$	3,375	3,376	2,227	3,377
$6,153\ 10^{-4}$	$1,263\ 10^{2}$	$-8,668\ 10^{3}$	$-3,593\ 10^{3}$	$3,591\ 10^{3}$	$8,675\ 10^{3}$
$1,329\ 10^{-4}$	$2,716\ 10^{2}$	$2,959\ 10^{4}$	$-2,958\ 10^{4}$	$-2,960\ 10^{4}$	$2,962\ 10^{4}$
$5,021\ 10^{-5}$	$4,420\ 10^{2}$	$-5,055\ 10^{4}$	$1,220\ 10^{5}$	$-1,219\ 10^{5}$	$5,046\ 10^{4}$

In the first column the results of numerical solution to (17) are presented, whereas ω_i are the corresponding bifurcational values of angular speed. Four such values are found, as it should be expected from the above results of analytical study for the case of a flattened cylinder. Subsequent columns contain calculated values of $x_j = \alpha_j^0 / \alpha_{n+1}^0$. The results of these calculations, illustrating four various dynamic equilibrium configurations, are presented in Fig.9.

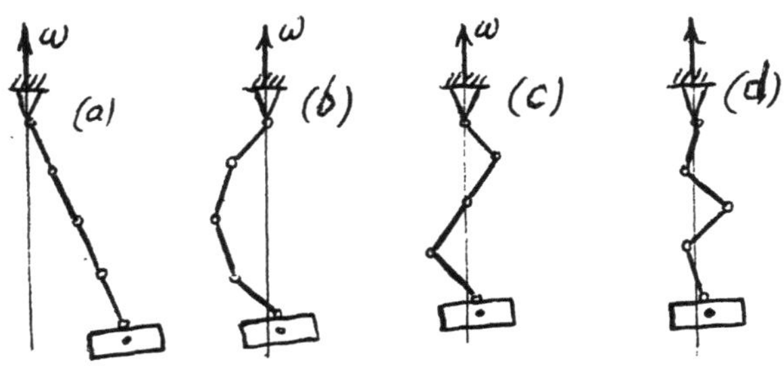

Fig.9

References

I. Ишлинский А.Ю. Пример бифуркации, не приводящей к появлению неустойчивых форм стационарного движения. - Докл.АН СССР, 1957, т.117, № 1, с.47-49.

2. Ишлинский А.Ю., Малашенко С.В., Темченко М.Е. О разветвлении устойчивых положений динамического равновесия одной механической системы. - Изв.АН СССР, ОТН, 1958, №8, с. 53-61.

3. Морозова Е.П. Об устойчивости вращения твердого тела, подвешенного на струне. - ПММ, 1956, т.20, вып.5, с.621-626.

4. Темченко М.Е. Об устойчивости одного из положений динамического равновесия одной механической системы. - Докл. АН СССР, 1957, т.117, №1, с.50-52.

5. Скимель В.Н. О движении гиростата, подвешенного на струне. - Тр. межвуз.конф. по прикладной теории устойчивости движения и аналитической механике. Казань: Изд-во Казан. авиац. ин-та, 1964, с.118-122.

6. Стороженко В.А. Об устойчивости одной из форм стационарного движения продолговатого тела вращения, подвешенного на струне. - Изв.АН СССР, МТТ, 1983, № 4, с.32-40.

7. Ишлинский А.Ю. Ориентация, гироскопы и инерциальная навигация. М., Наука, 1976, 670 с.

8. Темченко М.Е. О стационарных движениях двух связанных тел. В сб. Динамика и устойчивость сложных систем.Киев. Ин-т математики АН УССР, 1981, с.95-105.

9. Ишлинский А.Ю., Темченко М.Е. Об устойчивости вращения на струне твердого тела с эллипсоидальной полостью, целиком наполненной идеальной несжимаемой жидкостью.-Прикл. мат. и мех., 1966, т.30, с.30-41.

10.Темченко М.Е. Об исследовании критериев устойчивости движения подвешенного на струне твердого тела и волчка при наличии у них эллипсоидальной полости, наполненной жидкостью. Изв. АН СССР, МТТ, 1969, №1, с.26-31.

11.Стороженко В.А. Об устойчивости вращения тела, подвешенного посредством системы последовательно соединенных стержней. - Изв. АН СССР, МТТ, 1985, №1, с.45-52.

12.Гантмахер Ф.Р. Теория матриц. М.:Наука, 1967, 575 с.

Startup Torsional Vibration of Rotating Machine Driven by Synchronous Motor

Takuzo IWATSUBO, Yoshito YAMAMOTO, Ryoji KAWAI

Faculty of Engineering, Kobe University,
Rokko, Nada, Kobe, Japan

Abstracts

In this paper, electric-mechanical transient process by starting synchronous motors in the gear trained rotor system is investigated. The mathematical model is represented and it is numerically simulated to investigate the torsional vibration due to interaction between electrical and mechanical system. Then the characteristics of torsional vibration due to the interaction of electric and mechanical system, effect of backlash in the gear trained rotors and so on are studied. As the results, it is found that the interaction between electric and mechanical system is very big in the transient vibration of startup, nonlinearity of the system is strong in both the electric circuit and gear teeth meshing, the amplitude of transient vibration depends on the interaction of the electro-mechanical system and high frequency vibration is incurred due to backlash of the gears.

1. Introduction

Synchronous-motors are commonly used for driving rotating machinery for the benefit of high efficiency of energy transposition. In the startup of the rotating machinery, torque fluctuation occurs with the variation of magnetic flux in the motor. In the compressor system, for example, rotating speed is increased by gears. Thus, in a certain speed range of the startup, big torsional vibration occurs and sometimes the gear teeth may be broken. The cause of this vibration is due to the resonance of torsional vibration and the torque fluctuation of the motor. In the common analysis of this phenomena, torque fluctuation has been fixed with rotating speed and the torsional resonance of the rotor has been only considered, but, in practice, this vibration is due to the interaction of the electric system and the mechanical system in the energy transposition.

Dynamics of Multibody Systems
IUTAM/IFToMM Symposium Udine 1985
Editors: G. Bianchi and W. Schiehlen

The startup torsional vibration of rotating machine coupled by gears has been investigated by Inoue et.al.[1] for the case that the compressor is driven by synchronous motor. In that analysis, they assumed that the torque changes sinusoidally and torque variation and electric system has no interaction. But in practice these interact very much. Bendl and Schreier [2] have studied the electric-mechanical transient process by starting induction motors with given initial currents in the rotor. But in his analysis, the effect of gear coupling is not considered.

In this paper, the relation between electric circuit and torque and the relation between torque and torsional vibration are represented in the nonlinear mathematical model with multi degrees of freedom and by using the special translation of the coordinate, the mathematical model is numerically simulated to investigate the torsional vibration due to the interaction between electric and mechanical system. Then the stability of the system, characteristics of torsional vibration due to the interaction of electric and mechanical system, effect of backlash in the gear train and so on are discussed.

2.Nomencleature

- B : A half of the backlash clearance of the gear
- $\mathbb{C}$: Damping matrix
- C_m : damping coefficient due to loss of wind and bearing of rotor of electric motor
- $\mathbb{D}_i$: Transfer matrix of coordinate
- $\mathbb{F}$: External force vector
- $\mathbb{G}$: Back e.m.f. matrix
- $\mathbb{i}$: Current vector
- i_i : Current of coil of the i-th axis
- $\mathbb{J}$: Rotary inertia matrix
- $\mathbb{K}$: Stiffness matrix
- $\mathbb{L}$: Inductance matrix
- L_r : Inductance of rotor coil
- L_s : Inductance of stator coil
- M : Mutual inductance between coils

$\mathbb{R}$: Resistance matrix
R_{bi} : Raius of basic circle of the i-th gear
R_r : Resistance of rotor coil
R_s : Resistance of stator coil
T_e : Transient torque of electric motor
T_L : Transient torque of driven rotor
$\mathbf{v}$: Voltage vector
v_i : Voltage of coil of the i-th axis
$\mathbf{X}$: Angular vector
λ : Number of polar couple
ω_e : Angular velocity of power supply
ω_m : Angular velocity of electric motor
ζ : Damping ratio of shaft for critical damping

3.Definition of model and equation of motion

3.1 Model of electric motor and equations of motion

Fig.1 shows a motor model, in which electric motor consists of

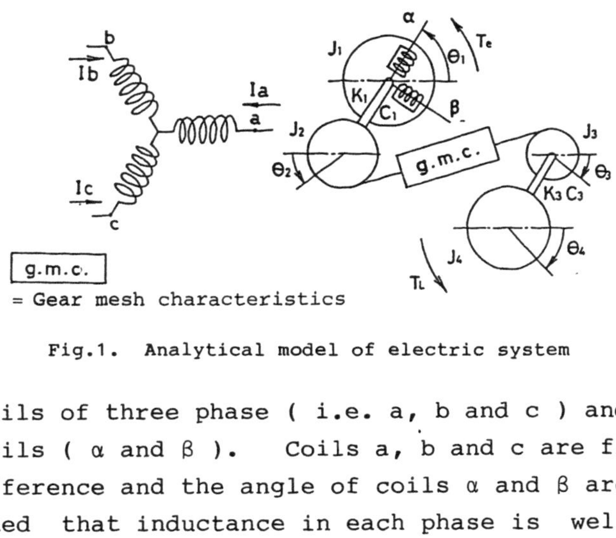

Fig.1. Analytical model of electric system

stator coils of three phase (i.e. a, b and c) and rotor coil of two coils (α and β). Coils a, b and c are fixed in 120° phase difference and the angle of coils α and β are 90°. It is assumed that inductance in each phase is well balanced. Then the voltage equation for the coil of the rotor is represented as,

$$\begin{pmatrix} v_\alpha \\ v_\beta \end{pmatrix} = \begin{pmatrix} R_\alpha + \frac{d}{dt} L_\alpha & \frac{d}{dt} M_{\alpha\beta} \\ \frac{d}{dt} M_{\beta\alpha} & R_\beta + \frac{d}{dt} L_\beta \end{pmatrix} \begin{pmatrix} i_\alpha \\ i_\beta \end{pmatrix} + \begin{pmatrix} \frac{d}{dt} M_{\alpha a} & \frac{d}{dt} M_{\alpha b} & \frac{d}{dt} M_{\alpha c} \\ \frac{d}{dt} M_{\beta a} & \frac{d}{dt} M_{\beta b} & \frac{d}{dt} M_{\beta c} \end{pmatrix} \begin{pmatrix} i_a \\ i_b \\ i_c \end{pmatrix} \quad (1)$$

The voltage equation for the coil of the stator is

$$\begin{pmatrix} V_a \\ V_b \\ V_c \end{pmatrix} = \begin{pmatrix} \frac{d}{dt}M_{a\alpha} & \frac{d}{dt}M_{a\beta} \\ \frac{d}{dt}M_{b\alpha} & \frac{d}{dt}M_{b\beta} \\ \frac{d}{dt}M_{c\alpha} & \frac{d}{dt}M_{c\beta} \end{pmatrix} \begin{pmatrix} i_\alpha \\ i_\beta \end{pmatrix} + \begin{pmatrix} R_a+\frac{d}{dt}L_a & \frac{d}{dt}M_{ab} & \frac{d}{dt}M_{ac} \\ \frac{d}{dt}M_{ba} & R_b+\frac{d}{dt}L_b & \frac{d}{dt}M_{bc} \\ \frac{d}{dt}M_{ca} & \frac{d}{dt}M_{cb} & R_c+\frac{d}{dt}L_c \end{pmatrix} \begin{pmatrix} i_a \\ i_b \\ i_c \end{pmatrix} \tag{2}$$

Then the coordinates system of the stator (axis a, b and c) and the rotor (axis α and β) are translated into axis (D, Q,0) and axis (d, q), where transfer matrices $\mathbb{D}_1$ and $\mathbb{D}_2$ are written as ;

$$\mathbb{D}_1 = \sqrt{3/2} \begin{pmatrix} 1/\sqrt{2} & 1 & 0 \\ 1/\sqrt{2} & -1/2 & \sqrt{3}/2 \\ 1/\sqrt{2} & 1/2 & \sqrt{3}/2 \end{pmatrix} \tag{3}$$

$$\mathbb{D}_2 = \begin{pmatrix} \cos\theta & \sin\theta \\ \sin\theta & -\cos\theta \end{pmatrix} \tag{4}$$

$$\mathbf{v} = \mathbb{R}\mathbf{i} + \mathbb{L}\dot{\mathbf{i}} + \lambda\omega_m \mathbb{G}\mathbf{i} \tag{5}$$

$$v_0 = R_0 i_0 \tag{6}$$

$$\mathbb{R} = \begin{pmatrix} R_r & & & 0 \\ & R_r & & \\ & & R_r & \\ 0 & & & R_r \end{pmatrix} \qquad \mathbb{L} = \begin{pmatrix} L_r & 0 & M & 0 \\ 0 & L_r & 0 & M \\ M & 0 & L_s & 0 \\ 0 & M & 0 & L_s \end{pmatrix}$$

$$\mathbb{G} = \begin{pmatrix} 0 & L_r & 0 & M \\ -L_r & 0 & -M & 0 \\ 0 & 0 & 0 & 0 \\ 0 & 0 & 0 & 0 \end{pmatrix} \qquad \mathbf{i} = \begin{pmatrix} i_d \\ i_q \\ i_D \\ i_Q \end{pmatrix} \qquad \mathbf{v} = \begin{pmatrix} v_d \\ v_q \\ v_D \\ v_Q \end{pmatrix}$$

where R_r and R_s are resistance of the rotor coil and the stator coil, respectively, L_s mutual inductance between (D, Q, 0)axis, L_r mutual inductance between (d, q)axis, M mutual inductance between (D, Q, 0)axis and (d, q)axis. If the three phase voltage of power supply is balanced,

Eq.(3.4) can be neglected. Then the electric torque at the rotor of the induction motor is represented as,

$$T_e = \lambda \dot{\mathbf{i}}^T G \dot{\mathbf{i}} \tag{7}$$

3.2 Model of rotor system and equation of motion

As shown in Fig.2 rotor system consists of the rotor of electric motor, driving gear, driven gear and driven rotor,

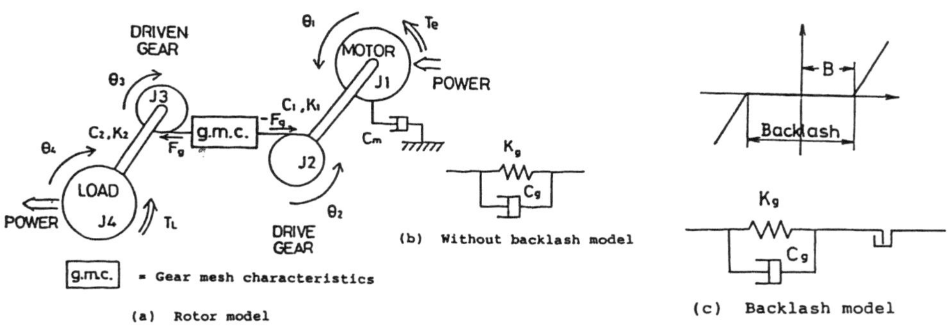

Fig.2. Analytical model of mechanical system

whose moments of inertia are denoted by J_1, J_2, J_3 and J_4, respectively.

The followings are assumed to simplify the equation,

1) The bending stiffness of the shaft is higher enough to neglect the bending of shaft.
2) There is no unbalance in each shaft, so unbalance vibration is neglectable.

Then the equation of motion is written as ;

$$\mathbb{J}\,\ddot{\mathbf{x}} + \mathbb{C}\,\dot{\mathbf{x}} + \mathbb{K}\,\mathbf{x} = \mathbb{F} \tag{8}$$

where $\mathbb{J} = \begin{pmatrix} J_1 & & & 0 \\ & J_2 & & \\ & & J_3 & \\ 0 & & & J_4 \end{pmatrix} \quad \mathbb{X} = \begin{pmatrix} \theta_1 \\ \theta_2 \\ \theta_3 \\ \theta_4 \end{pmatrix}$

if the gear is meshing ; $|R_{b1}\theta_1 - R_{b2}\theta_3| > B/2$

$$\mathbb{C} = \begin{pmatrix} C_1 & -C_1 & 0 & 0 \\ -C_1 & C_1+C_gR_{b1}{}^2 & -C_gR_{b1}R_{b2} & 0 \\ 0 & -C_gR_{b1}R_{b2} & C_2+C_gR_{b2}{}^2 & -C_2 \\ 0 & 0 & -C_2 & C_2 \end{pmatrix}$$

$$\mathbb{K} = \begin{pmatrix} K_1 & -K_1 & 0 & 0 \\ -K_1 & K_1+K_gR_{b1}{}^2 & -K_gR_{b1}R_{b2} & 0 \\ 0 & -K_gR_{b1}R_{b2} & K_2+K_gR_{b2}{}^2 & -K_2 \\ 0 & 0 & -K_2 & K_2 \end{pmatrix}$$

$$\mathbb{F} = \begin{pmatrix} T_e \\ -(1/2)\operatorname{sgn}(R_{b1}\theta_2-R_{b2}\theta_3)BR_{b1}K_g \\ (1/2)\operatorname{sgn}(R_{b1}\theta_2-R_{b2}\theta_3)BR_{b2}K_g \\ T_L \end{pmatrix}$$

and if the gear is not meshing ; $|R_{b1}\theta_2 - R_{b2}\theta_3| < B/2$

$$\mathbb{C} = \begin{pmatrix} C_1+C_m & -C_1 & & \\ -C_1 & C_1 & & \\ & & C_2 & -C_2 \\ & & -C_2 & C_2 \end{pmatrix} \quad (\text{off-diagonal blocks } 0)$$

$$\mathbb{K} = \begin{pmatrix} K_1 & -K_1 & & \\ -K_1 & K_1 & & \\ & & K_2 & -K_2 \\ & & -K_2 & K_2 \end{pmatrix} \quad (\text{off-diagonal blocks } 0) \qquad \mathbb{F} = \begin{pmatrix} T_e \\ 0 \\ 0 \\ T_L \end{pmatrix}$$

4.Numerical calculation

The equation coupled by electric system and mechanical system is heavy nonlinear, so it is impossible to obtain a closed solution. Therefore a direct numerical integration method is used for obtaining the physical behavior of the vibration of geared rotor system. That is, Newmark's β method is applied for the direct numerical integration (digital simulation). In this paper time step Δt is usually set to 0.5msec and β is selected $\beta = 1/4$.

For the calculation considering the backlash of gear meshing, the time which corresponds to the critical point of teeth touch is iteratively calculated and the stiffnesses of gear teeth on meshing and without meshing are chosen. In this calculation high accurary is demanded, so double precision calculation is applied.

Data used in the numerical calculation are listed in Tables 1 and 2. Another special data for calculation are presented in each section.

Table 1 Data of mechanical system

J_1	1.25×10^{-2} (kgm^2)	K_1	503 (Nm/rad)
J_2	6.37×10^{-2} (kgm^2)	K_g	7.36×10^{7} (N/m)
J_3	1.16×10^{-3} (kgm^2)	K_2	509 (Nm/rad)
J_4	9.80×10^{-2} (kgm^2)	R_{b1}/R_{b2}	37.5/75.1 (mm/mm)

Table 2 Data of electric system

Rr	1.64	(Ω)
Rs	2.18	(Ω)
Lr	0.4283	(H)
Ls	0.4283	(H)
M	0.4171	(H)

5.Effect of interaction between electric and mechanical system

5-1 Simulation result of the electric-mechanical transient system

In this calculation, the effect of backlash is neglected, i.e. B in Eq.(9) is set to be zero. The data of electric system and mechanical system is listed in Table 1 and 2, respectively and the model used in this chapter is the one that the backlash in Fig.2 is eliminated.

Fig.3 shows the simulation results when the load of the rotor is zero. As the instant power is applied to the motor, large power is consumed and electric torque is fluctuated. This period is only about 0.6sec. This period is called the vibration of the first period in this paper. The frequency of the first period is 58Hz, which corresponds to the frequency of slip of rotor. The vibration of the second period occurs at the critical speed of the shaft torsional vibration. In the vibration of the second period the electric torque is synchronously changed with the torsional vibration of the shaft ; this means the resonance oscillation. As the rotating speed increases, the shaft vibration decreases and shaft torque, electric torque and consuming electric power decrease.

Fig.4 shows the relation between rotating speed and output torque of the electric motor for different damping ratios of

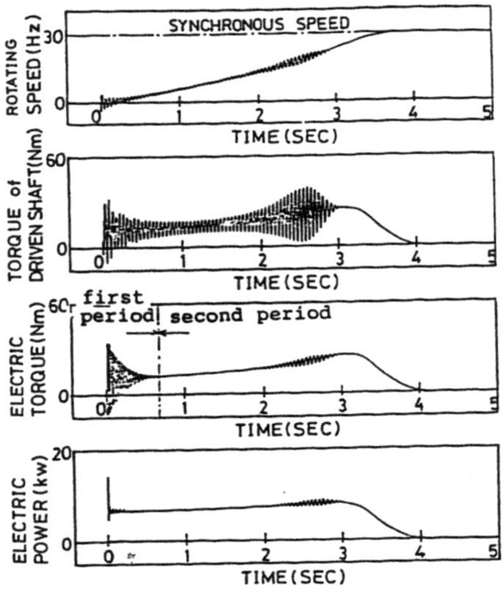

Fig.3. Simulated results of rotating speed, torque of driven shaft, electric torque and electric power

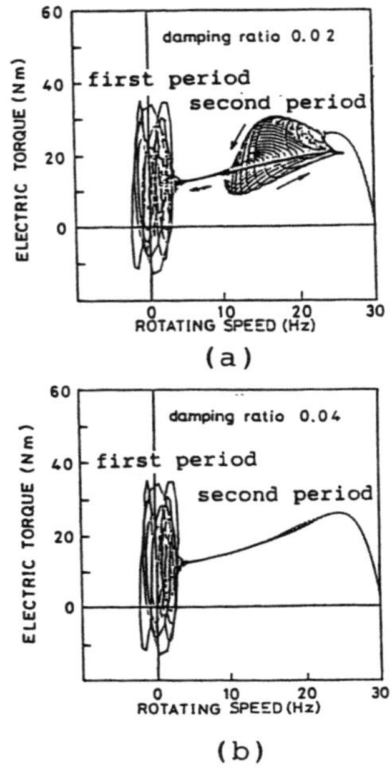

Fig.4. Transient electric torque for the damping ratio of shaft

the shaft. From these figures it is found that the vibration of the first period occurs independently from the mechanical system, but the vibration of the second period occurs by a close interaction between the electric system and the mechanical system, which is known from the deference of electric torque for different damping ratios of the shaft.

5-2 Comparison of interaction model with non-interaction model

The equation of motion derived in section 3 is an coupled model because the electric system and the mechanical system is coupled each other. Here the non-interacting model is defined. The non-interacting model means that the electric system is independent from the mechanical system and the output torque is as same as static characteristics shown in Fig.5 and driving torque fluctuation is given by the slip. Figs.6 (a) and (b) show the calculated result of tansient drive shaft torque for interacting models and non-interacting

Table 3 Data of rotor models

		Model No. 1	2	3
Stiffness	K1 (Nm/rad)	893.3	446.5	1787
	K_g (N/m)	3.68×10^7		
	K_2 (Nm/rad)	2233	178.6	714.6
Eigen-frequency	1st (Hz)	6.53	3.34	6.69
	2nd (Hz)	16.2	6.37	12.7
	3rd (Hz)	121.6	121.2	121.5

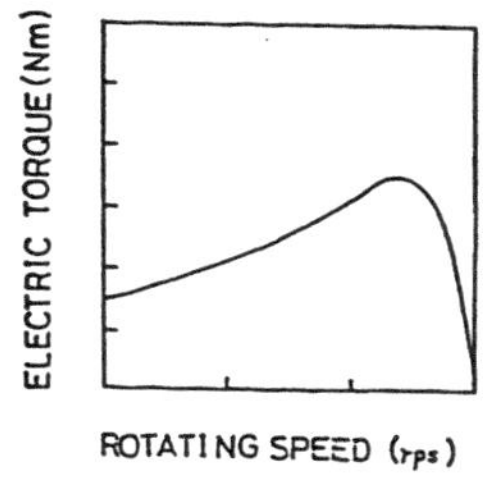

Fig.5. Static characteristics of electric torque for rotating speed

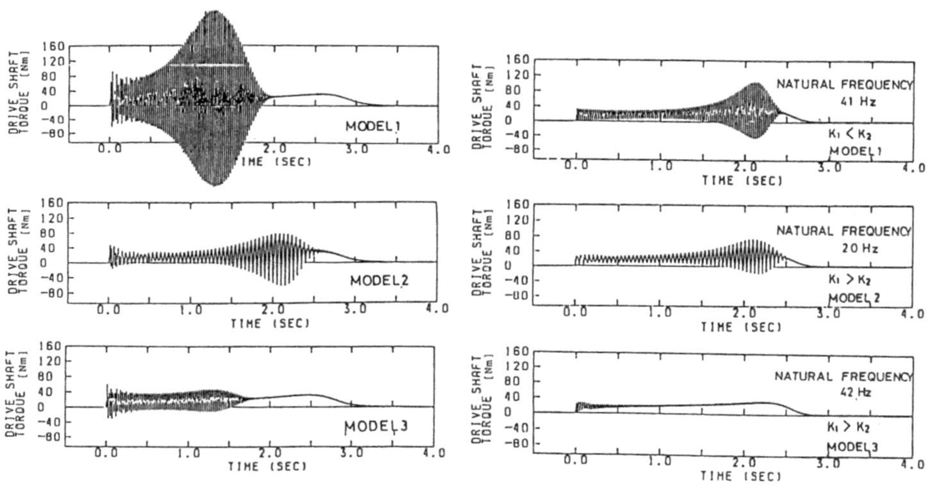

(a) Interaction model, without backlash

(b) Non interaction model, without backlash

Fig.6. Drive shaft torque for different combinations of shaft torsional stiffness,

models, respectively. For the non-interacting models, the amplitude of the shaft vibration becomes maximum at the rotating speed of the maximum incline of the torque chart shown in Fig.5. But for the interacting model the rotating speed where the amplitude of shaft vibration becomes maximum depends on the shaft torsional stiffness or the torsional natural frequency of the shaft. The rotating speed of this maximum amplitude occurs lower than that of non-interacting model when the natural frequency of the shaft system is high then reverse of this is also true. This phenomena occurs only by the interaction between the electric system and the mechanical system. The magnitude of the vibration amplitude, with the interacting system is larger than with the non-interacting system. So, the interacting model should be considered in order to design a safe rotor system from the viewpoint the startup torsional vibration of rotating machinery coupled by a gear.

6.Effect of backlash of the gear

In this section the effect of backlash on the startup torsional vibration is investigated. The backlash is reduced as shown in Fig.2(c).

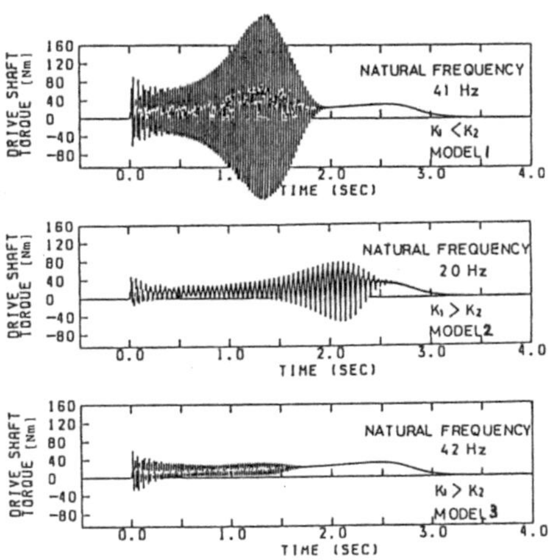

Fig.7. Drive shaft torque for different combinations of shaft torsional stiffness (interaction model, with backlash)

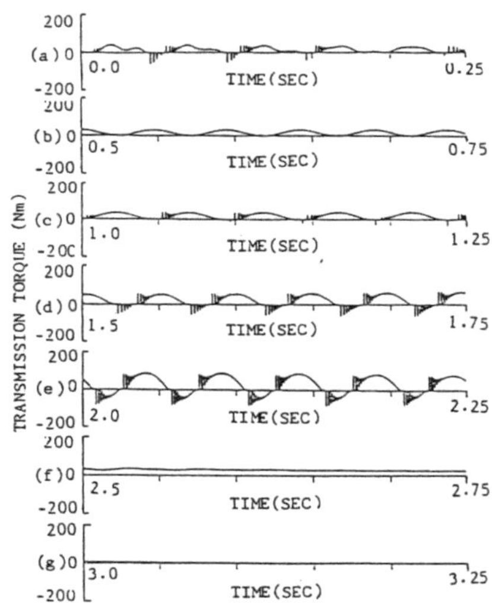

Fig.9. Transmission torque of backlash model for each period of starting up process

Fig.7 shows the calculated results of the transient drive shaft torque considering the backlash of the gear. The data of these models 1 to 3 are shown in Table 3 and the results correspond to the ones without backlash in Fig.6 for comparing the results for the cases with and without backlash. From the results, it is known that the effect of the backlash is not so big as the effect of interaction of the electric system and mechanical system and the amplitude of the vibration becomes smaller than the case neglected the backlash.

Fig.8 (a),(b) shows the calculated results of the transient electric torque, transmission torque and drive shaft torque for the cases considering and not considering the backlash, respectively. From these figures, there is not so much difference for the torsional vibration, but if the time scale is enlarged for the transmission torque, we can find that the high frequency vibration is occured by the backlash. Fig.9 shows this effect, that is, Figs (a)-(g) correspond to the time periods (a)-(g) of transmission torque in Fig.8(b). In Fig.9, collision of the meshing gear due to backlash occurs and its frequency correspond to the third eigenvalue of the torsional vibration, which is close to the eigen-frequency of the teeth.

The absolute value of the peak of the teeth transmission force is almost the same, this means that the collision force of the face and the back of the teeth is almost the same. Therefore, if the collision due to backlash has happened, the amplitude of transmission force of the teeth becomes very large and its

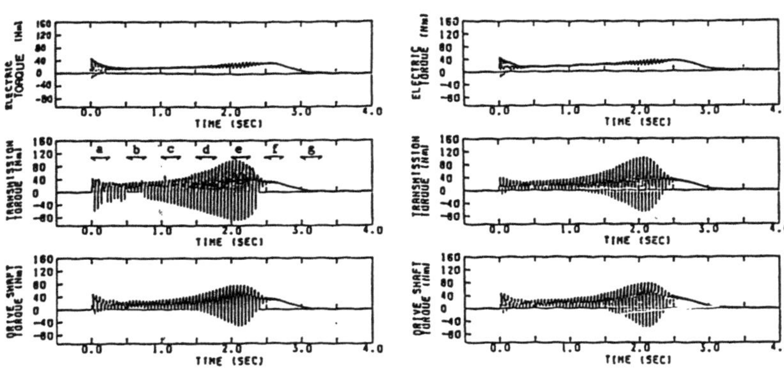

(a) With backlash

(b) Without backlash

Fig.8. Electric torque, transmission torque and drive shaft torque,

frequency is also very large. Thus, it is not good from the view-point of the failure of gear teeth. But this high frequency vibration does not interact with the electric system, so it is not necesary to consider the effect of backlash when we interested in analyzing the interaction between the electric system and the mechanical system for the overall system design. But for the design of teeth strength, the effect of backlash is necessary. Because the amplitude and period of the second vibration heavily depends on the interaction of the electric system and the mechanical system.

7.Conclusion

As a result, the following conclusions are obtained :

1) The interaction between electric system and mechanical system is very large in the transient vibration of the startup of rotating machinery. Thus the torsional vibration cannot be evaluated without considering dynamic behaviour of the electric system.

2) Nonlinearlity of the system is strong in the electric circuit and gear teeth meshing.

3) The amplitude of the vibration in startup is dependent on the relation of the natural frequences of motor shaft and machinery shaft and also the damping coefficients of motor and driven rotor.

4) In the resonance in the startup, high frequency vibration occurs due to backlash of the gears. This frequency, however, corresponds to the natural frequency of the gear tooth then it is not interacting with another system. So, when the torsional vibration in the startup is considered, this effect can be neglected and the gear teeth can be considered as a linear spring and damping model. But for the gear design from the viewpoint of strength, this effect should be taken into account.

References

1. Y.Inoue, T.Shibata, T.Fujikawa, M.Takemura, " Non-linear torsional vibration in synchronous motor driven systems, (First report: analysis, experiment, and field test) ", Bulletin of JSME vol.24, no.197, Nov. 1981, 203.

2. Bendl, J. and Schreier, L., " Electro-Mechanical Transient Process by Starting Induction Motors with Given Initial Currents in the Rotor ", ACTA, Technica CSAV, No.4 (1983) 429.

3. Hancock, N. N., " Matrix Analysis of Electrical Machinery (2nd Edition) ", Pergamon Press, 1974

Dynamics of Active Mechanisms with Flexible Links

J.B. JONKER

Laboratory of Engineering Mechanics
Department of Mechanical Engineering
Delft University of Technology
The Netherlands

Summary

A finite element procedure is presented for dynamic analysis of mechanisms with flexible links. In this analysis the geometrically nonlinear relations for the element deformations in terms of nodal position- and orientation coordinates play a central role. The equations of motion are formulated in terms of mixed sets of generalized coordinates of the mechanism with rigid links and deformation parameters that characterize deformation of the elements. In case of deformable elements constitutive equations have to be supplied. They may express simply linear-elastic behaviour, but by these constitutive equations also active elements in a mechanism can be modeled. A dynamic simulation of a hydraulically actuated servo mechanism with flexible links is presented.

1. Introduction

Historically the theory of kinematics and dynamics of mechanisms has been developed for rigid links. Contact between these rigid links is expressed by nonlinear constraint equations. Characteristic for the finite element method is the decomposition of the mechanism into finite elements with well defined mechanical properties. Permanent contact between the elements is obtained by letting them have nodal points in common and in these nodal points they share some or all the coordinates of these points. Instead of constraint equations in the points of contact of the rigid links, in the finite element method we have to impose conditions on the deformation modes of the elements. The number of deformation modes is equal to the number of nodal coordinates diminished by the number of degrees of freedom of the element as a rigid body. The rigid body condition for a finite element is obtained by putting all deformation modes equal to zero. Because of the fact that we start out from expressions for the deformation modes as nonlinear functions of the nodal coordinates, it is easy to extend the analysis to the case of mechanisms with deformable links. Then constitutive equations for the deformations have to be supplied. They may express simply

Dynamics of Multibody Systems
IUTAM/IFToMM Symposium Udine 1985
Editors: G. Bianchi and W. Schiehlen

linear elastic behaviour, but by these constitutive equations we can also model active elements in a mechanism.
The finite element theory for kinematics and dynamics of mechanisms has been utilized by Van der Werff [1, 2]. In these publications methods are presented for the calculation of the transfer functions of multi-degree of freedom mechanisms. These functions characterize a configuration manifold in the free configuration- and deformation spaces of the mechanism. As degrees of freedom both generalized coordinates of the mechanism with rigid links and possible deformation parameters, describing the deformation modes of the element, can be defined. The equations of motion are derived by the principle of virtual power. They form a nonlinear system of ordinary differential equations of second order and describe the general case of coupled rigid link motion and small elastic deformation. A general purpose computer programme, called SPACAR, based on the presented theory has been developped.

2. Finite element representation of mechanisms

Fig. 1 shows a manipulator mechanism with six degrees of freedom which is modeled by three different types of elements.

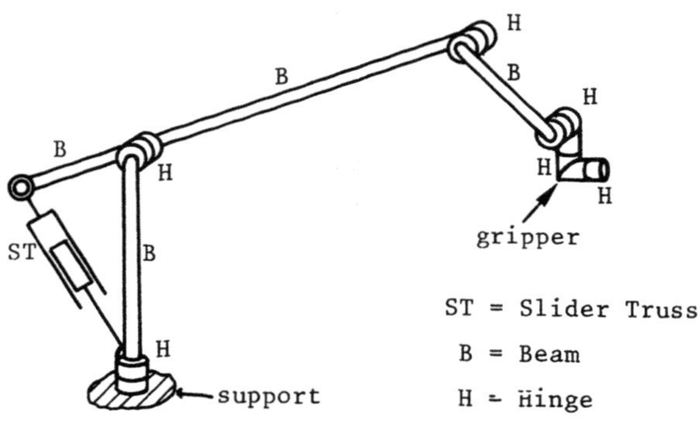

Fig. 1. Manipulator with 6 d.o.f.

The location of each element is specified by a set of reference coordinates $x^k \in X^k$, some of which may be cartesian coordinates of the end nodes, others describe the orientation at the nodal points. For each element we can define a fixed number of deformation parameters $e^k \in E^k$, some of which are associated with large relative displacements and rotations between the element nodes, others describe small elastic deformations of the element and will be denoted by ε^k. We call E^k the deformation space of the element k. The number of deformation parameters is equal to the number of nodal coordinates diminished by the number of degrees of freedom of the element as a rigid body. In this way we can define for each element a map

$$\mathcal{D}^k : X^k \rightarrow E^k, \quad \text{or} \quad e^k = \mathcal{D}^k(x^k). \tag{1}$$

This map represents the geometrically nonlinear relations for the deformation parameters $e^k \in E^k$ as functions of the reference coordinates $x^k \in X^k$.

TRUSS

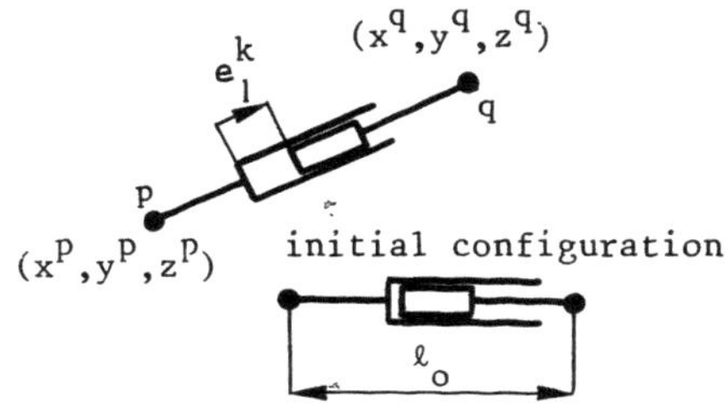

$x^k_{truss} = (x^{pT},x^{qT})^T$,

$\Delta\underline{x} = (x^q-x^p,y^q-y^p,z^q-z^p)^T$,

elongation:

$e^k_1 = (\Delta\underline{x},\Delta\underline{x})^{\frac{1}{2}} - \ell_o$.

BEAM

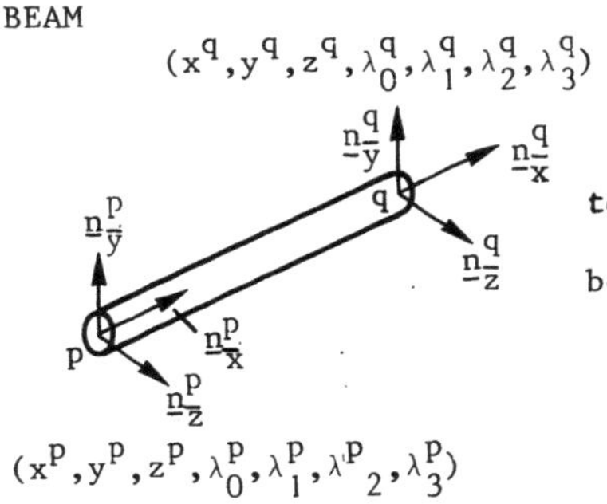

$x^k_{beam} = (x^{pT},x^{qT},\lambda^{pT},\lambda^{qT})^T$,

torsion: $\varepsilon^k_2 = \{(R^p\underline{n}_{\bar{z}},R^q\underline{n}_{\bar{y}})-(R^p\underline{n}_{\bar{y}},R^q\underline{n}_{\bar{z}})\}\ell_o/2$,

bending: $\varepsilon^k_3 = -(R^p\underline{n}_{\bar{z}},\Delta\underline{x})$, $\varepsilon^k_4 = (R^q\underline{n}_{\bar{z}},\Delta\underline{x})$,

$\varepsilon^k_5 = (R^p\underline{n}_{\bar{y}},\Delta\underline{x})$, $\varepsilon^k_6 = -(R^q\underline{n}_{\bar{y}},\Delta\underline{x})$.

HINGE

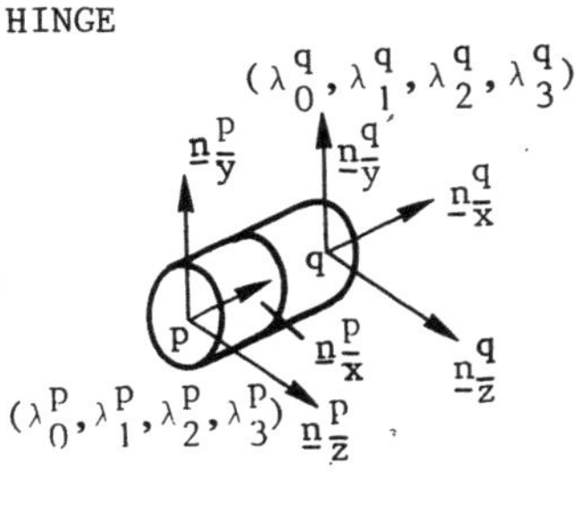

$x^k_{hinge} = (\lambda^{pT},\lambda^{qT})^T$,

relative rotation:

$e^k_1 = \arctan\{-(R^p\underline{n}_{\bar{y}},R^q\underline{n}_{\bar{z}})/(R^p\underline{n}_{\bar{y}},R^q\underline{n}_{\bar{y}})\}$,

bending: $\varepsilon^k_2 = (R^p\underline{n}_{\bar{y}},R^q\underline{n}_{\bar{x}})$,

$\varepsilon^k_3 = (R^p\underline{n}_{\bar{z}},R^q\underline{n}_{\bar{x}})$.

Table 1. Deformation parameters e^k_i for the Truss-, Beam- and Hinge element. In these expressions (,) stands for the inner product of two vectors.

Table 1 lists the expressions for the deformation parameters of a number of elements published in [2]. Bending and torsion deformations are defined with respect to orthogonal transformations which determine the orientation of the orthogonal traids $(\underline{n}_{\bar{x}},\underline{n}_{\bar{y}},\underline{n}_{\bar{z}})$ rigidly attached at the nodal points. In the undeformed state the triads coincide with the element-axis and the principle axes of its cross-section. For numerical determination of angular orientation we found Euler parameters most suitable [3]. In this case the transformation matrix can be written in terms of four Euler parameters $\lambda^T = (\lambda_0,\lambda_1,\lambda_2,\lambda_3)$, as

$$R = \begin{vmatrix} \lambda_0^2+\lambda_1^2-\lambda_2^2-\lambda_3^2 & 2(\lambda_1\lambda_2-\lambda_0\lambda_3) & 2(\lambda_1\lambda_3+\lambda_0\lambda_2) \\ 2(\lambda_1\lambda_2+\lambda_0\lambda_3) & \lambda_0^2-\lambda_1^2+\lambda_2^2-\lambda_3^2 & 2(\lambda_2\lambda_3-\lambda_0\lambda_1) \\ 2(\lambda_1\lambda_3-\lambda_0\lambda_2) & 2(\lambda_2\lambda_3+\lambda_0\lambda_1) & \lambda_0^2-\lambda_1^2-\lambda_2^2+\lambda_3^2 \end{vmatrix} . \tag{2}$$

By definition the Eulerparameters satisfy the identity

$$\lambda_0^2+\lambda_1^2+\lambda_2^2+\lambda_3^2-1 = 0. \tag{3}$$

This condition is of similar form as the undeformability condition $(e^\lambda = 0)$ and is therefore introduced in the description under the name λ-element.

3. Kinematic analysis

The configuration spaces X^k, of the individual finite elements are subspaces of the mechanism configuration space X. We call X the free configuration space of the mechanism. The space of deformations E, for the entire mechanism is introduced as the direct sum of the E^k's. The spaces X and E can both be splitted into subspaces according to the function of their components,

$$X = X^o \oplus X^c \oplus X^m \quad \text{and} \quad E = E^o \oplus E^m \oplus E^c, \tag{4}$$

where the subscripts o, c and m denote the space of invariant, dependent and independent coordinates, respectively.

3.1. Transfer functions

The problem formulation for the kinematical analysis is the evaluation of the reference coordinates and the deformation parameters for given discrete values of the global degrees of freedom x^m and the relative (elastic) degrees of freedom e^m. Hence determine the maps

$$\mathcal{F}^x: X^m \times E^m \to X, \quad x = \mathcal{F}^x(x^m, e^m), \tag{5}$$

$$\mathcal{F}^e: X^m \times E^m \to E, \quad e = \mathcal{F}^e(x^m, e^m), \tag{6}$$

which map the space of degrees of freedom $X^m \times E^m$ to the free configuration space X and the space of mechanism deformation parameters E. The maps $\mathcal{F}^x$ and $\mathcal{F}^e$ are called the transfer functions of the mechanism; they characterize a time independent configuration manifold M(X) and a deformation manifold M(E) in the spaces X and E, Fig. 2.

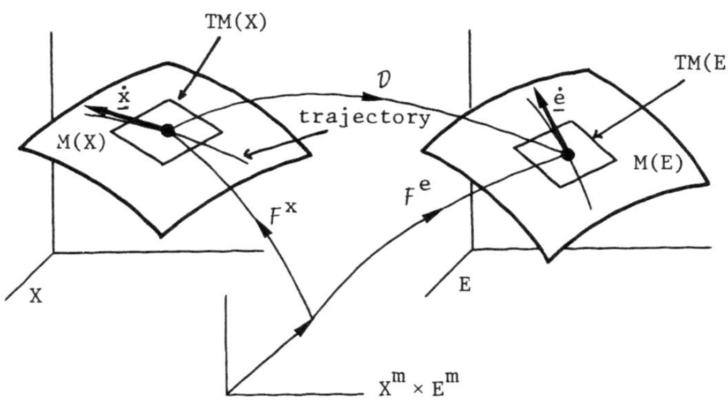

Fig. 2. Visualization of the kinematic relations.

The maps $\mathcal{D}^k$ of the elements, defined in (1), can be taken together in a map for the entire mechanism. We write symbolically

$$\mathcal{D} = \sum_k \mathcal{D}^k, \tag{7}$$

where $\mathcal{D}$ may be viewed as a mapping between the two manifolds M(X) and M(E),

$$\mathcal{D} : M(X) \to M(E), \quad e = \mathcal{D}(x), \quad x \in M(X), \quad e \in M(E).$$

The configuration of the mechanism at any instant is represented by a point on the manifolds. The time history, as the mechanism evolves from state to state, is then represented by a connected set of points on the configuration manifolds and we refer to these sets of points as the trajectories of the mechanism. At each point along the trajectories, there is a tangent vector that corresponds to the velocities of the mechanism. The tangent vectors are elements of the tangent spaces TM(X) and TM(E) at (x^m, e^m). The velocity vectors $\dot{\underline{x}}$ and $\dot{\underline{e}}$ can be calculated from eqs. (5) and (6) as

$$\dot{\underline{x}} = D\mathcal{F}^x.(\dot{\underline{x}}^m, \dot{\underline{e}}^m), \quad \dot{\underline{x}} \in TM(X), \tag{8}$$

$$\dot{\underline{e}} = D\mathcal{F}^e.(\dot{\underline{x}}^m, \dot{\underline{e}}^m), \quad \dot{\underline{e}} \in TM(E), \tag{9}$$

where $D\mathcal{F}^x$ and $D\mathcal{F}^e$ are derivative maps of $\mathcal{F}^x$ and $\mathcal{F}^e$ at (x^m, e^m).
Consider now the composite mapping $\mathcal{D} \circ \mathcal{F}^x$

$$X^m \times E^m \xrightarrow{\mathcal{F}^x} M(X) \xrightarrow{\mathcal{D}} M(E),$$

and the mapping $\mathcal{F}^e$

$$X^m \times E^m \xrightarrow{\qquad \mathcal{F}^e \qquad} M(E).$$

Clearly $\mathcal{D} \circ \mathcal{F}^x$ and $\mathcal{F}^e$ are identically, hence

$$\mathcal{F}^e = \mathcal{D} \circ \mathcal{F}^x, \text{ for all } (x^m, e^m). \tag{10}$$

Due to the non-linear character of the equations the unknown transfer functions cannot be calculated directly from equation (10). It will be shown however that expressions for the derivatives $D\mathcal{F}$ and $D^2\mathcal{F}$ can be obtained from the derivative map $D\mathcal{D}$. $\mathcal{F}$ itself is then found by integration and subsequent iteration. Differentiation of eq. (10) with respect to (x^m, e^m) yields with the chain rule

$$D\mathcal{F}^e = D\mathcal{D} \cdot D\mathcal{F}^x, \text{ for all } (x^m, e^m), \tag{11}$$

where the derivative map $D\mathcal{D}$ may be viewed as a linear mapping between the tangent spaces,

$$D\mathcal{D} : TM(X) \rightarrow TM(E), \quad \text{or} \quad \dot{\underline{e}} = D\mathcal{D}(x) \cdot \dot{\underline{x}}.$$

In accordance with the separations in eq. (4), the system can be partioned as

$$\begin{vmatrix} DF^{eo} \\ DF^{em} \\ DF^{ec} \end{vmatrix} = \begin{vmatrix} D^o\mathcal{D}^o & \boxed{D^c\mathcal{D}^o} & D^m\mathcal{D}^o \\ D^o\mathcal{D}^m & \boxed{D^c\mathcal{D}^m} & D^m\mathcal{D}^m \\ D^o\mathcal{D}^c & D^c\mathcal{D}^c & D^m\mathcal{D}^c \end{vmatrix} \cdot \begin{vmatrix} DF^{xo} \\ DF^{xc} \\ DF^{xm} \end{vmatrix}. \tag{12}$$

The only unknowns in this equations are the maps DF^{xc} and DF^{ec}. For the other partial maps we have

$$DF^{eo} = \left(\frac{\partial e^o}{\partial x^m}, \frac{\partial e^o}{\partial e^m}\right) = (0,\ 0), \qquad DF^{xo} = \left(\frac{\partial x^o}{\partial x^m}, \frac{\partial x^o}{\partial e^m}\right) = (0,\ 0),$$

$$DF^{em} = \left(\frac{\partial e^m}{\partial x^m}, \frac{\partial e^m}{\partial e^m}\right) = (0,\ I), \qquad DF^{xm} = \left(\frac{\partial x^m}{\partial x^m}, \frac{\partial x^m}{\partial e^m}\right) = (I,\ 0).$$

where 0 and I are zero- and unit maps respectively.
If the mechanism is not in a singular position, then the partial map $\begin{pmatrix} D^c\mathcal{D}^o \\ D^c\mathcal{D}^m \end{pmatrix}$ is nonsingular and DF^{xc} can be calculated by

$$DF^{xc} = \begin{pmatrix} D^c\mathcal{D}^o \\ D^c\mathcal{D}^m \end{pmatrix}^{-1} \cdot \left[\begin{pmatrix} DF^{eo} \\ DF^{em} \end{pmatrix} - \begin{pmatrix} D^m\mathcal{D}^o \\ D^m\mathcal{D}^m \end{pmatrix} \cdot DF^{xm}\right] \tag{13}$$

and

$$DF^{ec} = D^c\mathcal{D}^c \cdot DF^{xc} + D^m\mathcal{D}^c \cdot DF^{xm}. \tag{14}$$

Expressions for the second derivatives D^2F^x and D^2F^e can be obtained from the derivative map $D\mathcal{D}$ in a similar way. Differentiation of eq. (11) with respect to $(x^m,\ e^m)$ yields

$$D^2F^e = (D^2\mathcal{D} \cdot DF^x) \cdot DF^x + D\mathcal{D} \cdot D^2F^x, \tag{15}$$

where $D^2\mathcal{D}$ is a bilinear mapping defined as

$$D^2\mathcal{D}\colon\ TM(X) \times TM(X) \to TM(E).$$

In accordance with the coordinate splitting in eq. (4), the system can be written in the form

$$\begin{vmatrix} D^2F^{eo} \\ D^2F^{em} \\ D^2F^{ec} \end{vmatrix} = \begin{vmatrix} (D^2\mathcal{D}^o \cdot DF^x) \cdot DF^x \\ (D^2\mathcal{D}^m \cdot DF^x) \cdot DF^x \\ (D^2\mathcal{D}^c \cdot DF^x) \cdot DF^x \end{vmatrix} + \begin{vmatrix} D^o\mathcal{D}^o & D^c\mathcal{D}^o & D^m\mathcal{D}^o \\ D^o\mathcal{D}^m & D^c\mathcal{D}^m & D^m\mathcal{D}^m \\ D^o\mathcal{D}^c & D^c\mathcal{D}^c & D^m\mathcal{D}^c \end{vmatrix} \cdot \begin{vmatrix} D^2F^{xo} \\ D^2F^{xc} \\ D^2F^{xm} \end{vmatrix}, \tag{16}$$

where

$$D^2F^{eo} = D^2F^{xo} = 0 \quad \text{and} \quad D^2F^{em} = D^2F^{xm} = 0.$$

For the unknown second derivatives D^2F^{xc} and D^2F^{ec}, we then obtain

$$D^2F^{xc} = -\begin{pmatrix} D^c\mathcal{D}^o \\ D^c\mathcal{D}^m \end{pmatrix}^{-1} \cdot \begin{pmatrix} (D^2\mathcal{D}^o \cdot DF^x) \cdot DF^x \\ (D^2\mathcal{D}^m \cdot DF^x) \cdot DF^x \end{pmatrix} \tag{17}$$

and

$$D^2F^{ec} = (D^2\mathcal{D}^c \cdot DF^x) \cdot DF^x + D^c\mathcal{D}^c \cdot D^2F^{xc}. \tag{18}$$

For a mechanism with given kinematical dimensions, starting from a given position we are able to calculate subsequent positions for a sequence of values of the degrees of freedom (x^m, e^m). When a starting position (x_1, e_1) is known, a new position (x_2, e_2) can be calculated by integration,

$$\begin{aligned}(x_2, e_2) &= F(x_1^m, e_1^m) + DF(x_1^m, e_1^m) \cdot (x_2^m - x_1^m, e_2^m - e_1^m) + \\ &+ \frac{1}{2}\,(D^2F(x_1^m, e_1^m) \cdot (x_2^m - x_1^m, e_2^m - e_1^m)) \cdot (x_2^m - x_1^m, e_2^m - e_1^m).\end{aligned} \tag{19}$$

An iteration process is then applied in order to quarantee that ultimately $\mathcal{D}^m(x_2) = e_2^m$ and $\mathcal{D}^o(x_2) = 0$.

4. Equations of motion

The basic laws of motion and the principle of virtual power are used to obtain the equations of motion. The inertia properties of the concentrated and distributed mass of the links are described with the aid of mass matrices. Two formulations are treated in this paper.

4.1. Lumped formulation

In this formulation rigid bodies with equivalent mass and rotational inertia are attached to the end nodes of the elements. The inertia properties of the bodies are calculated by assuming that the elements behave like a rigid body. This assumption therefore excludes dynamic coupling between the translational and rotational motions.

Consider a lumped mass in the form of a rigid body attached to element node p described by the coordinates

$$(x^{pT},\ \lambda^{pT})^T = (x^p, y^p, z^p, \lambda_0^p, \lambda_1^p, \lambda_2^p, \lambda_3^p)^T. \tag{20}$$

The seven coordinates define the location of the $(\bar{x}^p, \bar{y}^p, \bar{z}^p)$ cartesian coordinate system fixed in the body, relative to the fixed (inertial) coordinate system (x^p, y^p, z^p). Fig. 3.

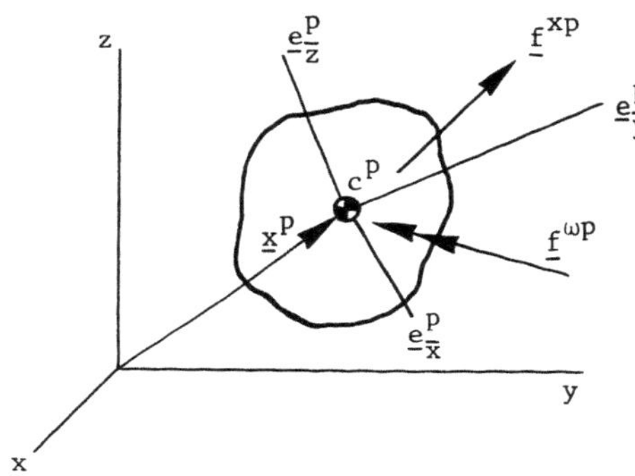

Fig. 3. Position and orientation of a lumped rigid body.

Let $\underline{\omega}^p$ be the absolute angular velocity vector of the body with components relative to the fixed coordinate system (x^p, y^p, z^p). The relationship between these components and the time derivative of the Euler parameters is determined by the transformation [3]

$$\underline{\omega}^p = 2\Lambda^p \dot{\underline{\lambda}}^p, \quad \text{where } \Lambda^p = \begin{vmatrix} -\lambda_1^p & \lambda_0^p & -\lambda_3^p & \lambda_2^p \\ -\lambda_2^p & \lambda_3^p & \lambda_0^p & -\lambda_1^p \\ -\lambda_3^p & -\lambda_2^p & \lambda_1^p & \lambda_0^p \end{vmatrix}. \tag{21}$$

Let $\underline{f}^{xp}$ and $\underline{f}^{\omega p}$ be the node force vector and the node torque vector with respect to the center of mass c^p. Now we can formulate the principle of virtual power for the rigid body by

$$<(\underline{f}^{xp} - M^p \ddot{\underline{x}}^p),\ \dot{\underline{x}}^p> + <(\underline{f}^{\omega p} - \frac{d}{dt}(R^p \bar{J}_c^p R^{pT} \underline{\omega}^p)),\ \underline{\omega}^p> = 0, \tag{22}$$

for all $\dot{\underline{x}}^p$ and for all $\underline{\omega}^p = 2\Lambda^p \dot{\underline{\lambda}}^p$.

Here M^p and $\bar{J}_c^p$ are the mass matrix and the rotational inertia matrix with components related to the body fixed coordinate system $(\bar{x}^p, \bar{y}^p, \bar{z}^p)$. R^p is the orthogonal transformation matrix, defined in section 2, which can be expressed as the result of two successive linear transformations as

$$R^p = \Lambda^p \bar{\Lambda}^{pT}, \text{ where } \bar{\Lambda}^p = \begin{vmatrix} -\lambda_1^p & \lambda_0^p & \lambda_3^p & -\lambda_2^p \\ -\lambda_2^p & -\lambda_3^p & \lambda_0^p & \lambda_1^p \\ -\lambda_3^p & \lambda_2^p & -\lambda_1^p & \lambda_0^p \end{vmatrix}. \tag{23}$$

Substituting of eq. (21) and eq. (23) into eq. (22) yields with the transpose of the transformation Λ^p

$$<(\underline{f}^{xp}-M^p\ddot{\underline{x}}^p),\ \dot{\underline{x}}^p> + <(\underline{f}^{\lambda p}-4\Lambda^{pT}\frac{d}{dt}(\Lambda^p\bar{\Lambda}^{pT}\bar{J}_c^p\bar{\Lambda}^p\Lambda^{pT}\Lambda^p\dot{\underline{\lambda}}^p)),\ \dot{\underline{\lambda}}^p> = 0, \tag{24}$$

for all $\dot{\underline{x}}^p$ and for all $\dot{\underline{\lambda}}^p$.
The components of $\underline{f}^{\lambda p}$ are defined as

$$\underline{f}^{\lambda p} = 2\Lambda^{pT}\underline{f}^{\omega p}. \tag{25}$$

These are the four moment components associated with the Euler parameters λ^p. For an unconstrained lumped rigid body the velocity vectors $\dot{\underline{x}}^p$ and $\dot{\underline{\lambda}}^p$ are independent. From this follows that

$$\underline{f}^{xp} - M^p\ddot{\underline{x}}^p = 0, \tag{26}$$

and

$$\underline{f}^{\lambda p} - 4\Lambda^{pT}\frac{d}{dt}(\Lambda^p\bar{\Lambda}^{pT}\bar{J}_c^p\bar{\Lambda}^p\Lambda^{pT}\Lambda^p\dot{\underline{\lambda}}^p) = 0. \tag{27}$$

Carrying out the differentiation with respect to time by using the identities for the Euler parameters, we obtain [3]

$$\underline{f}^{\lambda p} - 4\bar{\Lambda}^{pT}\bar{J}_c^p\bar{\Lambda}^p\ddot{\underline{\lambda}}^p - 8\dot{\bar{\Lambda}}^{pT}\bar{J}_c^p\bar{\Lambda}^p\dot{\underline{\lambda}}^p = 0. \tag{28}$$

These equations describe the inertia forces associated with the Euler parameters. The lumped mass matrix of element k with nodes p and q may thus be written as

$$M_\ell^k = \begin{vmatrix} (M^x)_\ell^k & 0 \\ 0 & 4\bar{\Lambda}^{kT}\bar{J}_\ell^k\bar{\Lambda}^k \end{vmatrix}, \text{ with } (M^x)_\ell^k = \begin{vmatrix} M^p & 0 \\ 0 & M^q \end{vmatrix}^k,\ \bar{J}_\ell^k = \begin{vmatrix} \bar{J}_c^p & 0 \\ 0 & \bar{J}_c^q \end{vmatrix}^k, \text{ and } \bar{\Lambda}^k = \begin{vmatrix} \bar{\Lambda}^p & 0 \\ 0 & \bar{\Lambda}^q \end{vmatrix}^k. \tag{29}$$

4.2. Consistent formulation

In finite element theory the consistent massmatrix is formulated from the same interpolation functions that are employed in the evaluation of the element stiffness coefficients. Using this formulation we obtain [4]

$$M_c^k = \begin{vmatrix} (M^{xx})_c^k & (M^{x\lambda})_c^k \\ (M^{\lambda x})_c^k & (M^{\lambda\lambda})_c^k \end{vmatrix} = \left|\begin{array}{c:c} R^{kT}(\bar{M}^{xx})_c^k R^k & 2R^{kT}(\bar{M}^{x\omega})_c^k \bar{\Lambda}^k \\ 2\bar{\Lambda}^{kT}(\bar{M}^{\omega x})_c^k R^k & 4\bar{\Lambda}^{kT}(\bar{M}^{\omega\omega})_c^k \bar{\Lambda}^k \end{array}\right|, \tag{30}$$

$$\text{with } R^k = \begin{vmatrix} R^p & 0 \\ 0 & R^q \end{vmatrix}^k, \quad \text{and} \quad \bar{\Lambda}^k = \begin{vmatrix} \bar{\Lambda}^p & 0 \\ 0 & \bar{\Lambda}^q \end{vmatrix}^k. \tag{31}$$

The partioned massmatrices $(\bar{M}^{xx})_c^k$, $(\bar{M}^{x\omega})_c^k$, $(\bar{M}^{\omega x})_c^k$ and $(\bar{M}^{\omega\omega})_c^k$ are the consistent massmatrices of the element k with components related to a element fixed coordinate system $(\bar{x},\bar{y},\bar{z})$ that initially coincides with the fixed coordinate system.

4.3. Equations of motion of mechanisms

In view of the different treatment of the translational and angular velocities in the derivation of the equations of motion, it is useful to split the space X, of mechanism reference coordinates into subspaces according to

$$X = X^x \oplus X^\lambda, \quad x \in X^x, \quad \lambda \in X^\lambda, \tag{32}$$

where X^x is the space of position coordinates x and X^λ is the space of Euler parameters λ. The corresponding transfer functions are defined as

$$F^x: X^m \times E^m \to X^x, \qquad x = F^x(x^m, e^m), \tag{33}$$

$$F^\lambda: X^m \times E^m \to X^\lambda, \qquad \lambda = F^\lambda(x^m, e^m). \tag{34}$$

The velocity vectors $\dot{\underline{x}}$ and $\dot{\underline{\lambda}}$ can be calculated from these equations as

$$\dot{\underline{x}} = DF^x.(\dot{\underline{x}}^m, \dot{\underline{e}}^m), \quad \dot{\underline{x}} \in TM(X^x),$$

$$\dot{\underline{\lambda}} = DF^\lambda.(\dot{\underline{x}}^m, \dot{\underline{e}}^m), \quad \dot{\underline{\lambda}} \in TM(X^\lambda). \tag{35}$$

Let F^x be the space of externally applied node forces $\underline{f}^x$ and let F^λ be the space of moment components $\underline{f}^\lambda$, defined by eq. (25). The spaces F^x and F^λ are dual with the tangent spaces $TM(X^x)$ and $TM(X^\lambda)$, that is, both of the spaces are connected by a bilinear form < , > called the scalar product. In order to describe the loading state of the mechanism elements, for every element a vector of generalized stresses $\underline{\sigma}^k \in \Sigma^k$ is introduced, where Σ^k is the element stress space. These spaces form together the space of generalized stress vectors Σ for the mechanism. The space Σ is dual with the tangent space TM(E). Note that in the $\underline{\sigma}$ vector also the vector $\underline{\sigma}^\lambda$, associated with the λ-elements is included. Let M and $\bar{J}$ be the mechanism mass matrix and the rotational inertia matrix calculated by direct addition of the element matrices

$$M = \sum_k M_\ell^k + M_c^k, \qquad \bar{J} = \sum_k \bar{J}_\ell^k, \text{ and let } \bar{\Lambda} = \sum_k \bar{\Lambda}^k,$$

where the summation includes all elements.
According to the principle of virtual power for the external forces and torques $\underline{f}^x$, $\underline{f}^\lambda$ and the generalized stresses $\underline{\sigma}$ we then have

$$<\left(\begin{vmatrix} \underline{f}^x \\ \underline{f}^\lambda \end{vmatrix} - \begin{vmatrix} M^{xx} & M^{x\lambda} \\ M^{\lambda x} & M^{\lambda\lambda} \end{vmatrix} \begin{vmatrix} \ddot{\underline{x}} \\ \ddot{\underline{\lambda}} \end{vmatrix} - \begin{vmatrix} 0 \\ 8\dot{\bar{\Lambda}}^T \bar{J}\bar{\Lambda}\dot{\underline{\lambda}} \end{vmatrix}\right), \begin{vmatrix} \dot{\underline{x}} \\ \dot{\underline{\lambda}} \end{vmatrix} > = < \underline{\sigma}, \dot{\underline{e}} >, \tag{36}$$

for all $\dot{\underline{x}}$, $\dot{\underline{\lambda}}$ and $\dot{\underline{e}}$ defined in eq. (35) and eq. (9). From eq. (35) we can calculate the accelerations $\ddot{\underline{x}}$ and $\ddot{\underline{\lambda}}$ by

$$\begin{aligned} \ddot{\underline{x}} &= (D^2F^x.(\dot{\underline{x}}^m,\dot{\underline{e}}^m))\cdot(\dot{\underline{x}}^m,\dot{\underline{e}}^m) + DF^x.(\ddot{\underline{x}}^m,\ddot{\underline{e}}^m) \\ \ddot{\underline{\lambda}} &= (D^2F^\lambda.(\dot{\underline{x}}^m,\dot{\underline{e}}^m))\cdot(\dot{\underline{x}}^m,\dot{\underline{e}}^m) + DF^\lambda.(\ddot{\underline{x}}^m,\ddot{\underline{e}}^m). \end{aligned} \tag{37}$$

Substituting eq. (35), eq. (9) and eq. (37) into eq. (36) yields with the duals of DF^x, DF^λ and DF^e

$$\begin{aligned} &|DF^{xT}\ DF^{\lambda T}| \begin{vmatrix} M^{xx} & M^{x\lambda} \\ M^{\lambda x} & M^{\lambda\lambda} \end{vmatrix} \begin{vmatrix} DF^x \\ DF^\lambda \end{vmatrix} \cdot(\ddot{\underline{x}}^m,\ddot{\underline{e}}^m) = |DF^{xT}\ DF^{\lambda T}| \begin{vmatrix} \underline{f}^x \\ \underline{f}^\lambda - 8\dot{\bar{\Lambda}}^T \bar{J}\bar{\Lambda}\dot{\underline{\lambda}} \end{vmatrix} - DF^{eT}\underline{\sigma} + \\ &-|DF^{xT}\ DF^{\lambda T}| \begin{vmatrix} M^{xx} & M^{x\lambda} \\ M^{\lambda x} & M^{\lambda\lambda} \end{vmatrix} \begin{vmatrix} (D^2F^x.(\dot{\underline{x}}^m,\dot{\underline{e}}^m))\cdot(\dot{\underline{x}}^m,\dot{\underline{e}}^m) \\ (D^2F^\lambda.(\dot{\underline{x}}^m,\dot{\underline{e}}^m))\cdot(\dot{\underline{x}}^m,\dot{\underline{e}}^m) \end{vmatrix}. \end{aligned} \tag{38}$$

These are the generalized equations of motion for the mechanism. Their number equals the number of degrees of freedom (x^m, e^m). The degrees of freedom can chosen such that the equations of motion describe the rigid link mechanism. The force vector contains the forces $\underline{f}^x$ and $\underline{f}^\lambda$ applied to the mechanism nodes, but also the generalized stress vector $\underline{\sigma}$. Elastic deformations of flexible elements are characterized by the generalized Hook's law defined by the linear transformation [3].

$$S^k : E^k \to \Sigma^k, \text{ or } \underline{\sigma}^k = S^k \underline{\varepsilon}^k, \tag{39}$$

where S^k is a symmetric matrix containing the stiffness coefficients. In order to model an active element (e.g. actuator), the equations describing the dynamic behaviour of the actuator can be added to the system by means of special constitutive equations for the element. This is demonstrated in section 5 for a translational hydraulic servo actuator. The equations of motion are reduced to the first order form by change of variable and solved by a predictor/corrector algorithm with variable order and stepsize [7].

5. Dynamic simulation of a hydraulically actuated servo mechanism

Fig. 4 shows a finite element model of a plane positioning mechanism which is controlled by three identical hydraulic servo actuators. The positioning task implies to transfer a 300 kg mass, attached at the tip of the 5th link along prescribed trajectories denoted by the lines A-A' and B-B'.

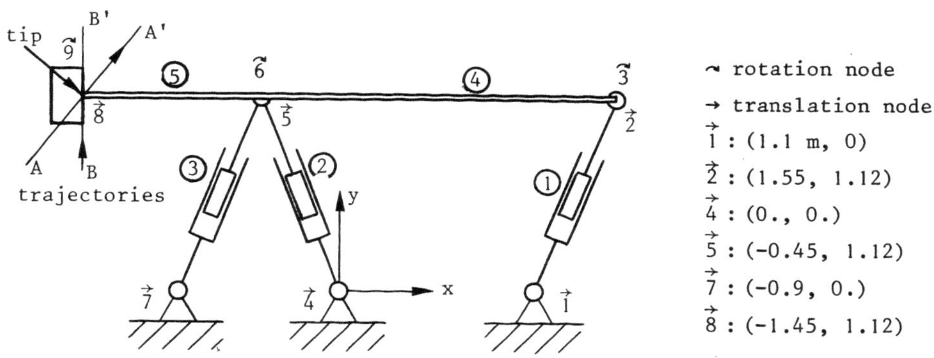

Fig. 4. Elastic mechanism model and trajectories: A : (-1.63 m, 0.88 m), A' : (-1.27, 1.36), B : (-1.45, 0.95), B' : (-1.45, 1.29).

Table 2 presents the dynamic parameters associated with the mechanism.

Lumped mass of bearing assembly and actuators	M^2 = 52 kg, M^5 = 70 kg
Lumped mass at the tip of beam segment 5	M^8 = 300 kg, J^9 = 11 kgm^2
Stiffness of beam segments	EI = 180000 Nm^2
Weight density of beam segments	m = 25 kg/m

Table 2. Dynamic parameters of the mechanism.

A consistent formulation is employed to describe the mass properties of the beam elements.

The actuators are modeled as active slider-truss elements. The actuator force σ^k of the kth actuator is related to the actuator speed $\dot{e}^k$ and the input signal u^k by the linearized equations, [5]

$$\sigma^k = - A\, P_L^k + w\, \dot{e}^k, \text{ where } \dot{P}_L^k + K_m^k \frac{c_o}{c_h} P_L^k = \frac{c_o}{A} (K_m^k K_s^k G_s^k u^k - \dot{e}^k). \quad (40)$$

The system parameters are the velocity gain K_m^k and the load pressure P_L^k of the kth actuator, the effective piston area A, the hydraulic stiffness c_h, the oilspring stiffness c_o and the viscous friction coefficient w. The effect of Coulomb friction has been neglected since the actuators are provided with hydrostatic bearings. The product $K_s^k G_s^k$ is the transfer function of the electro hydraulic value, where G_s^k represents the dynamic part. It is assumed that the servo bandwidth is enough large so that its dynamics does not influence the behaviour of the whole system, i.e. $G_s^k = 1$. The numerical values of the hydraulic parameters of the applied actuators are $c_o = 12.25.10^6$ N/m, $c_h = 12.25.10^7$ N/m and w = 830 N sec/m.
Fig. 5 shows the block diagram for the conventional servo control. The assumed control law is given by the expression

$$u^k = K_{pos}^k (e^{rk} - e^k) - K_{press}^k \, A\, P_L^k. \quad (41)$$

where e^{rk} denotes a setpoint position corresponding with the prescribed nominal trajectory. K_{pos}^k and K_{press}^k are the usual position- and pressure feedback gains.

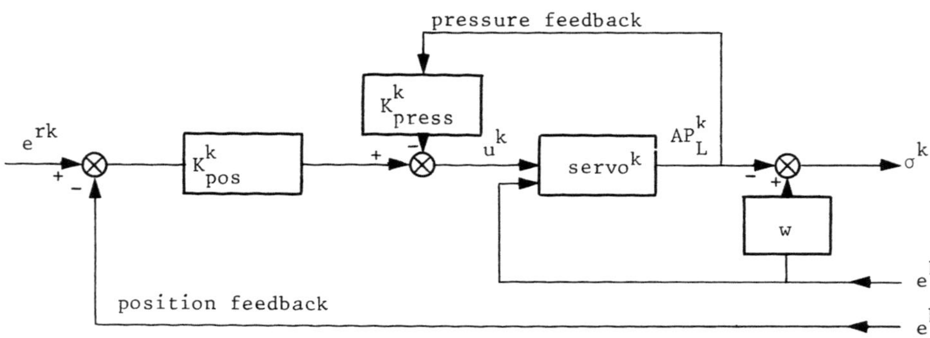

Fig. 5. Block diagram of hydraulic servo control.

Table 3 shows the numerical values of the system parameters and the feedback gain parameters used in the simulation. Specific details for choosing the values of the feedback gain parameters are presented in [6].

Actuator	K_s^k	K_m^k	K_{pos}^k	K_{press}^k [m/N]
k = 1	1	43	0.875	$1.43.10^{-7}$
k = 2	1	52	0.615	$1.0\ .10^{-7}$
k = 3	1	54	0.43	$0.7\ .10^{-7}$

Table 3. Values of system parameters and feedback gains.

The desired trajectories of the tip of the 5th link are given by a 340 mm straight vertical line AA' and a 600 mm straight sloping line BB' in the cartesian space, fig. 4. The velocity profile is trapeziodal with a maximum speed of 0.5 m/sec and an acceleration of 2 m/sec^2. Computation of the nominal (openloop) trajectories involves the solution of the inverse problem of kinematics, i.e. the determination of the setpoints (e^r) to realize the prescribed nominal trajectory. The inverse kinematic problem has been carried out by setting the setpoints of the first actuator as $e^{1r} = 0$. In fig. 6 trajectory tracking errors are shown for both the flexible- and the rigid beam model; ⊥ is the distance of the tip from the straight line on which the desired trajectory lies and // is the spatial delay along the direction of the trajectory.

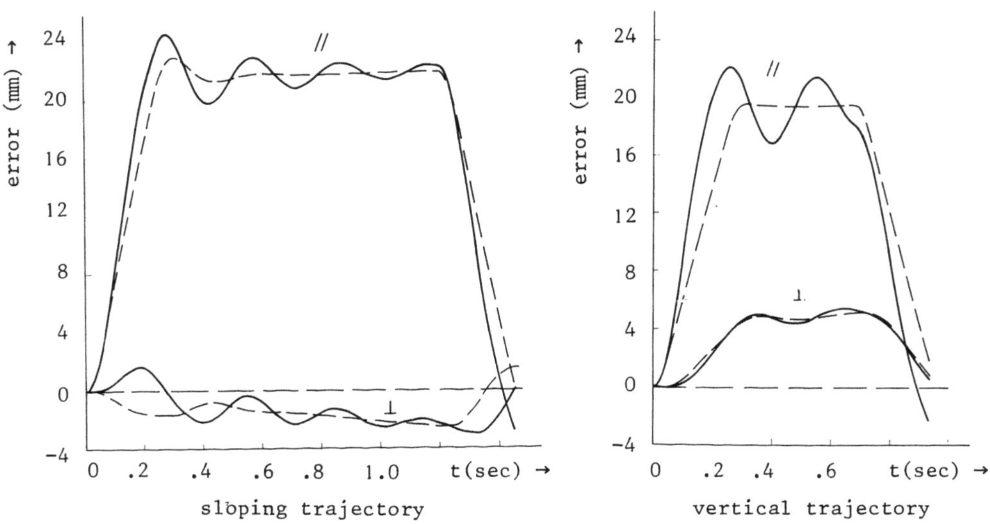

Fig. 6. Tracking errors. —— flexible beam model --- rigid beam model

It can be observed that the tracking errors along the direction of the trajectory are much larger then those measured along the direction perpendicular to the trajectory.

Acknowledgement

The author wishes to thank F.B. Sperling for critical discussions and valuable suggestions.

References

1. van der Werff, K.: Kinematic and dynamic analysis of mechanisms, a finite element approach. Doctors thesis, Delft 1977.
2. van der Werff, K. and Jonker, J.B.: Dynamics of flexible mechanisms, Proc. NATO Avanced Study Inst. on Computer Aided Analysis and Optimization of Mechanical System Dynamics, IOWA-City (1983) 381-400.
3. Jonker, J.B.: Dynamics of Spatial Mechanisms with flexible Links. Delft University of Technology, Lab. Eng. Mech., (1984), WTHD 171.
4. Schwab, A.L.: Dynamica van mechanismen met vervormbare schakels, Engineer's Thesis, Lab. Eng. Mech., (1983), rapport nr. 758.
5. Viersma, T.J.: Analysis, synthesis and design of hydraulic servo-systems and pipelines. Amsterdam, Elsevier 1980.
6. Sperling, F.B. and Jonker, J.B.: Dynamic simulation of a hydraulically actuated servomechanism. Delft University of Technology, (1985), WTHD (in preparation).
7. Shampine L.F. and Gordon, M.K.: Computer solution of ordinary differential equations. The initial value problem, W.J. Freeman, San Francisco, Calif., 1975.

Dynamics of Gyroscopic Systems for Indicating the Vertical on Ships

D.M.KLIMOV

The Institute for Problems in Mechanics
Moscow, USSR

Summary

Knowledge of vertical direction on ships is of importance for many applications. M.Schuler was the first who had proposed the use of gyroscopic systems for indicating the vertical. A.Yu.Ishlinsky obtained the conditions for the ideal performance of the raumcompass and developed its theory.
In the paper the following theorem is proved : the line connecting the suspension point and the mass center of an arbitrary gyroscopic system is always directed along the vertical if and only if the vector of the angular momentum of the system in absolute motion about the suspension point is constant in inertial space. This theorem provides the conditions for the unperturbable gyroscopic systems. The analogy between various unperturbable systems is established.
The known studies of the unperturbable systems are connected with a case when the value of system angular momentum is zero. The paper presents the investigation for the case when the system angular momentum is constant in inertial space and not zero. The equations of motions for a new class of unperturbable systems are derived and their motion is investigated. In particular it is shown that the systems studied have both Schuler's period and other periods of oscillations which depend upon the value of angular momentum.

1. Schuler's pendulum.

The problem of inertial navigation of a ship and the problem for indicating the vertical on ships are closely connected. If we knew the vertical direction on a ship we would be able to solve the navigation problem easily. To find the latitude and the longitude of a ship one should only trace the vertical motion in the inertial space using gyroscopes. For the full solution of the navigation problem one should construct by gyroscopes the plane in which the North Star and the vertical are situated.
In the paper the unperturbable systems are investigated. Here we introduce the definition of unperturbable pendulum.

Dynamics of Multibody Systems
IUTAM/IFToMM Symposium Udine 1985
Editors: G. Bianchi and W. Schiehlen

Consider the ship motion in the equator plane. The ship position in the inertial space is defined through the angle Λ, while the pendulum position with a suspension point O on the ship is defined through the angle α (fig.1).

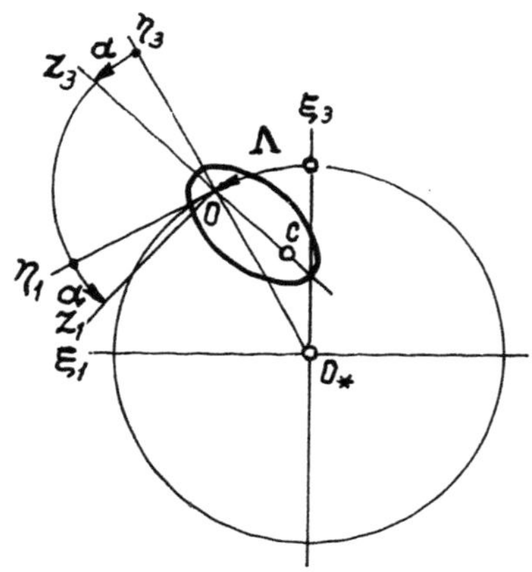

Fig.1. The pendulum on a ship.

The equation of pendulum motion is

$$I(\ddot{\Lambda}+\ddot{\alpha})=ml A, \quad A=R\ddot{\Lambda}\cos\alpha+(R\dot{\Lambda}^2-g)\sin\alpha \qquad (1.1)$$

A pendulum is unperturbable if α stay zero for any ship motions. So we call a pendulum as unperturbable (Schuler pendulum) if the line connecting the suspension point O and the pendulum mass center C is directed along the vertical despite of arbitrary ship motion. The vertical direction is changing in the inertial space due to Earth rotation and ship motion. One can find from equation (1.1) that $\alpha=0$ iff

$$I=mlR, \quad \alpha(0)=0, \quad \dot{\alpha}(0)=0 \qquad (1.2)$$

For small oscillations

$$\ddot{\alpha}+\nu^2\alpha=0 \qquad (1.3)$$

Here $\nu=\sqrt{g/R}$ is Schuler frequency.

The motion of Schuler pendulum and its stability were investigated in [1,2].

2.The general conditions of unpertubability of gyroscopic systems.

Consider the motion of a system suspended at the point 0 on ship. The system consists of several rigid bodies including gyros. The system is called unperturbable if the line connecting the suspension point 0 and the mass center C is always directed along the vertical (fig.2).

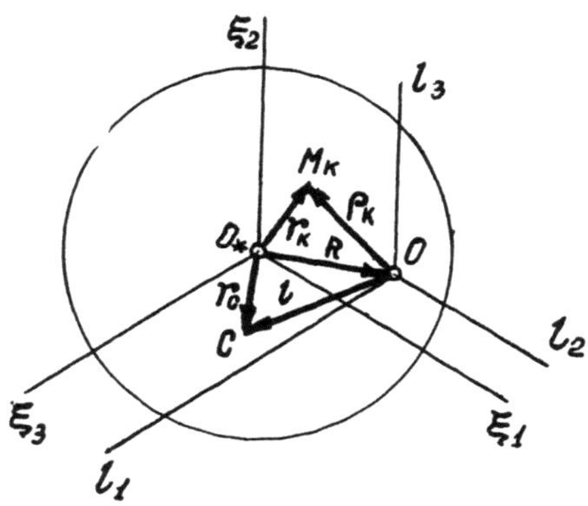

Fig. 2. The system of material points.

Using the theorem of angular moment of system we may write

$$\dot{G}_{O_*} = \sum M^e_{O_*}, \quad G_{O_*} = \sum r_K \times m_K \dot{r}_K = \qquad (2.1)$$

$$= \sum (R + \rho_K) \times m_K \dot{r}_K = R \times m \dot{r}_C + G_O$$

$$m = \sum m_K, \quad G_O = \sum \rho_K \times m_K \dot{r}_K$$

Taking into account the theorem of mass center motion

$$m \ddot{r}_C = \sum F^e_K$$

relations (2.1) are reduced to

$$\dot{R} \times m \dot{r}_C + \dot{G}_O = \sum \rho_K \times F^e_K \qquad (2.1')$$

Suppose the system is unperturbable that is the point C is situated on the line $O\,O_*$.

Then

$$R \times m \dot{r}_C = 0, \quad \sum \rho_k \times F_k^e = 0$$

and from equation (2.1') we get the general condition of unperturbability as

$$G_0 = B \qquad (2.2)$$

Here B is a constant vector in the inertial space.

The condition (2.2) is also sufficient since the vector equation (2.2) has one and only one solution for given initial conditions. The angular moment G_0 can be reduced to the form

$$G_0 = K_0 + \ell \times m V, \quad K_0 = \sum \rho_k \times m_k \dot{\rho}_k, \quad V = \dot{R} \qquad (2.3)$$

Introduce now the coordinate system $\zeta_1\ \zeta_2\ \zeta_3$ so that the axis ζ_3 is directed along the line $O\,O_*$ and the component Ω_3 of angular velocity of $\zeta_1\ \zeta_2\ \zeta_3$ along the axis ζ_3 is zero. Projecting the vector condition on the axes $\zeta_1\ \zeta_2\ \zeta_3$ yields three scalar conditions of unperturbability

$$K_1 = m\ell R\,\Omega_1 + B_1, \quad K_2 = m\ell R\,\Omega_2 + B_2, \quad K_3 = B_3 \qquad (2.4)$$

which for Schuler pendulum ($B_1 = B_2 = B_3 = 0$) have the following form

$$I_1 = I_2 = m\ell R, \quad K_3 = 0$$

<u>3. The analogy between various types of unperturbable gyroscopic systems.</u>

Consider the case $B = 0$ and introduce the coordinate system $z_1\ z_2\ z_3$ so that the axis z_3 is directed along the line CO the component ω_3 of angular velocity of $z_1\ z_2\ z_3$ is zero. In perturbed motion instead relations (2.4) we have

$$K_1 = m\ell R\,\omega_1, \quad K_2 = m\ell R\,\omega_2, \quad K_3 = 0, \quad \omega_3 = 0 \qquad (3.1)$$

Here ω_1 and ω_2 are the angular velocity components of the system $z_1\ z_2\ z_3$. Projecting the equation

$$\dot{K}_0 + \omega \times K_0 = l \times m(g - W_0)$$

on the axes $z_1\ z_2\ z_3$ and using the following relations introduced

$$A_1 = W_1 - g_1\ ,\quad A_2 = W_2 - g_2$$

$$\Omega_1' = -\frac{1}{R}\int_0^t A_2\,dt + \Omega_1(0),\quad \Omega_2' = \frac{1}{R}\int_0^t A_1\,dt + \Omega_2(0)$$

we derive the general motion equations of unperturbable gyroscopic systems for the case $B = 0$

$$(\omega_1 - \Omega_1')^{\cdot} = 0,\ (\omega_2 - \Omega_2')^{\cdot} = 0,\ \omega_3 = 0 \qquad (3.2)$$

The form of equations (3.2) does not depend upon the construction of unperturbable gyroscopic systems so this fact proves the analogy between those systems. Equations (3.2) prove the analogy between the unperturbable gyroscopic systems and the navigation systems since the navigation system equations have the following form [3]

$$\omega_1 - \Omega_1' = 0,\ \omega_2 - \Omega_2' = 0,\ \omega_3 = 0 \qquad (3.3)$$

4. A new type of unperturbable gyroscopic systems.

Consider the motion of unperturbable gyroscopic systems when the vector $B \neq 0$. We define the position of ship(point O) and the system $\zeta_1\ \zeta_2\ \zeta_3$ on the nonrotating sphere with the radius R through the angles Λ, φ, χ (fig.3,4)

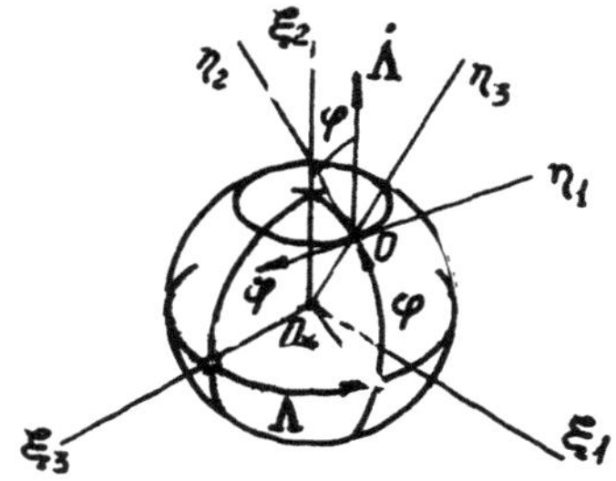

Fig.3. The geografic coordinate system.

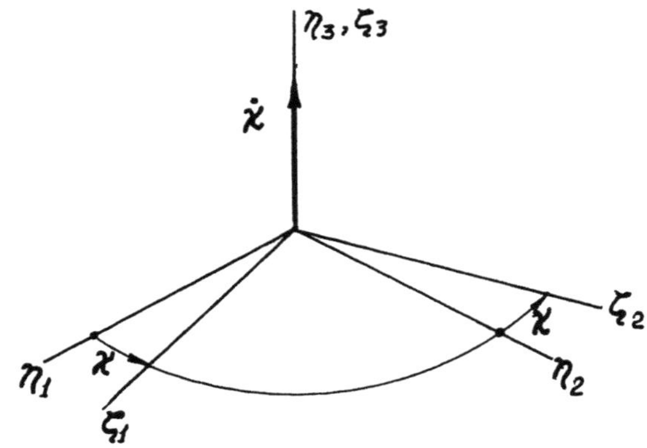

Fig. 4. The orientation of ideal axes $\zeta_1 \zeta_2 \zeta_3$

We shall consider the gyroscopic system as a body with three rotors (fig.5).

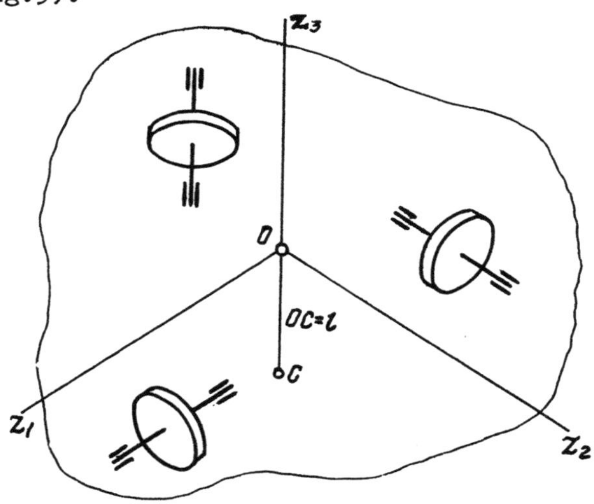

Fig. 5. The pendulum with three rotors.

Let the unperturbability conditions (2,4) and the initial conditions required are satisfied. Then the axes $z_1 z_2 z_3$ repeat the motion of axes $\zeta_1 \zeta_2 \zeta_3$ at all time and from (2.4')

$$(-ml\ell + I_1)\Omega_1 + h_1 = B\cos\varphi\sin\chi$$
$$(-mlR + I_2)\Omega_2 + h_2 = B\cos\varphi\cos\chi \qquad (4.1)$$
$$h_3 = B\sin\varphi, \quad \Omega_3 = 0$$

Here h_1, h_2, h_3 are the angular momenta of rotors. To find Ω_1 and Ω_2 we have to fix two accelerometers and using their data a_1, a_2 to solve the following equations

$$\Omega_1 = -\frac{1}{R}\int_0^t a_2\,dt + \Omega_1(0), \quad \Omega_2 = \frac{1}{R}\int_0^t a_1\,dt + \Omega_2(0) \qquad (4.2)$$
$$\Omega_3 = 0, \quad \dot{\varphi} = -\Omega_1\cos\chi + \Omega_2\sin\chi,$$
$$\dot{\chi} = -(\Omega_1\sin\chi + \Omega_2\cos\chi)\,tg\,\varphi$$
$$\dot{\Lambda}\cos\varphi = \Omega_1\sin\chi + \Omega_2\cos\chi$$

For the disturbed motion

$$h_1 \to H_1' = (mlR - I_1)\Omega_1' + B\cos\varphi'\sin\chi'$$
$$h_2 \to H_2' = (mlR - I_2)\Omega_2' + B\cos\varphi'\cos\chi'$$
$$h_3 \to H_3' = B\sin\varphi'$$
$$a_1 \to A_1, \quad a_2 \to A_2 \qquad (4.3)$$
$$\Omega_1' = -\frac{1}{R}\int_0^t A_2\,dt + \Omega_1(0)$$
$$\Omega_2' = \frac{1}{R}\int_0^t A_1\,dt + \Omega_2(0)$$
$$\Omega_3' = 0$$

$$\dot\varphi' = -\Omega_1'\cos\chi' + \Omega_2'\sin\chi', \quad \dot\chi' = -(\Omega_1'\sin\chi' + \Omega_2'\cos\chi')\operatorname{tg}\varphi', \quad \dot\Lambda'\cos\varphi' = \Omega_1'\sin\chi' + \Omega_2'\cos\chi'$$

$$\Omega_1 \to (\omega_1, \Omega_1), \quad \Omega_2 \to (\omega_2, \Omega_2')$$

and the following equations of gyroscopic system

$$\begin{aligned} &(I_1 + C_1)(\omega_1 - \Omega_1')^{\bullet} + B(\omega_2 - \Omega_2')\sin\varphi' = 0 \\ &(I_2 + C_2)(\omega_2 - \Omega_2')^{\bullet} - B(\omega_1 - \Omega_1')\sin\varphi' = 0 \\ &\omega_3 = 0 \end{aligned} \tag{4.4}$$

If the angles $\alpha_1, \alpha_2, \alpha_3$ (fig.6) are small then the equations (4.4) are simplified

$$\begin{aligned} &(I_1 + C_1)\dot\beta_1 + \beta_2 B\sin\varphi(t) = 0 \\ &(I_2 + C_2)\dot\beta_2 - \beta_1 B\sin\varphi(t) = 0 \\ &\beta_1 = \dot\alpha_1 + \int_0^t \nu^2\alpha_1\,dt, \quad \beta_2 = \dot\alpha_2 + \nu^2\int_0^t \alpha_2\,dt \end{aligned} \tag{4.5}$$

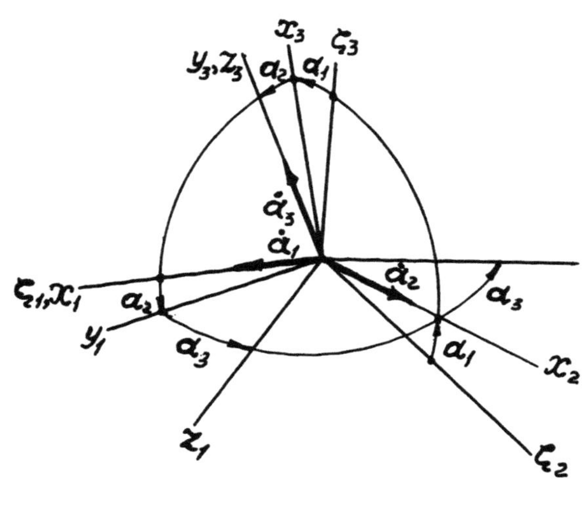

Fig.6. The orientation of axes $x_1\ x_2\ x_3$ relative to the ideal axis $\zeta_1\zeta_2\zeta_3$.

The angle φ is defined through the ship motion. For the case $\varphi = const$ we can find from equations (4.5) that the gyroscopic system described has besides Schuler frequency $\nu = \sqrt{g/R}$ the additional frequency

$$\mu = B \sin\varphi / \sqrt{(I_1 + C_1)(I_2 + C_2)}$$

This fact is of importance, since for the case $B = 0$ such a frequency does not exist.

References

I. Ишлинский А.Ю. Об относительном равновесии физического маятника с подвижной точкой опоры. ПММ, I956, т.20, вып.4

2. Климов Д.М. Об устойчивости движения невозмущаемого физического маятника. ПММ, I964, т.28, вып.2

3. Климов Д.М. Инерциальная навигация на море. Изд."Наука". Москва, I984.

Dynamics of Elastic Multibody Systems Involving Closed Loops

L. K. LILOV

Institute of Mechanics and Biomechanics
1113 Sofia, ul."Acad. G. Bonchev", bl.8, Bulgaria

Summary

A formalism for investigation of an arbitrary system of bodies, interconnected by joinjs with arbitrary (holonomic, nonholonomic, scleronomic or rheonomic) constraints is presented. This formalism extends the ideas developed in [1,2,3,4] for systems of rigid bodies. It considers systems of flexible bodies, as well. The state variables include generalized coordinates characterising the bodies' attitudes, nonholonomic variables characterising their velocities and a finite number of deformation coordinates representing the deformation fields of the bodies. Special attention is devoted to the attainment of an full description of the structure of the interconnections in the system and to the compact expression of the constraint equations due to the existence of closed loops. Thus, the formulation of the equations of motion is easily completed. The equations of motion are being derived from the variational principles of mechanics [5], while the emphasis is on the principle of Jourdain (the principle of virtual power) and the principle of Gauss.The mathematical apparatus employed is convenient for algorithmization and can be easily developed into numerical algorithms for implementation in computer programmes.

Matrix description of the interconnection structure

Let the multibody system consist of (n+1) bodies interconnected by m hinges. The bodies and the hinges are numbered from 0 to n and from 1 to m, respectively. Usually one body of the system plays a special role and its motion is prescribed or known in advance. This body always gets the number 0.The remaining numeration is arbitrary [6]. We represent the system structure by a graph, whose vertices s_i $(i=0,1,\dots,n)$ and edges u_a $(a=1,\dots,m)$ symbolize respectively the bodies and the hinges of the system.

The mechanics of a multibody system is here considered by resolving the general motion whose character usually is extremely complicated, into basic motions allowing relatively simple description and investigation. Afterwards these motions are brought together by taking into account all the interconnections in the system. The basic motions are chosen in the most natural way - these motions are the relative motions of the contiguous bodies in the hin-

Dynamics of Multibody Systems
IUTAM/IFToMM Symposium Udine 1985
Editors: G. Bianchi and W. Schiehlen

ges. In each hinge one of the two contiguous bodies (no matter which) is chosen as a reference body. After completing this choice we define two integer functions $i^+(a)$ and $i^-(a)$ the first of which means the number of the reference body and the second one is the number of the neighboring body [1,6]. With functions $i^+(a)$ and $i^-(a)$ we can transform the system's graph into an oriented graph assigning as a direction of the edge with number a the direction from the vertex $i^+(a)$ to the vertex $i^-(a)$. We later call the oriented edges arcs.

The transformation of the system's graph into a graph with a tree structure plays a major role in the investigation of the interconnection structure.The system graph is generally an arbitrary graph and its transformation can be reduced either to the removing of $\hat{n}=m-n$ appropriately selected edges from the graph or to the bifurcation of $\hat{n}$ appropriately selected vertices. The first procedure is classical and it is well familiar in the graph theory[7]. We assume that the removed edges are $u_{n+1},\dots,u_m$. By means of the second procedure if one vertex belongs simultaneously to several loops and if it is selected for a bifurcation in all of them, it is bifurcated as many times as the number of the fundamental loops containing it is[2]. Upon bifurcation all images of one vertex remain that are not connected directly and belong to different branches of the tree obtained. This tree, which we will designate a secondary graph, possesses (m+1) vertices. In the secondary graph there is a correspondence between each image and inversed image of a bifurcated vertex. All such pairs, connected with additional edges, define a new graph.We call it a secondary supplemented graph. The initially built graph when considered simultaneously with the secondary graph is called a primary graph.

One of the basic matrices describing the structure of the introduced graphs is the incidence matrix of the oriented graph $\underline{I} = (S_{ia})$, where $S_{ia} = 1$ if $i=i^+(a)$, $S_{ia} = -1$ if $i=i^-(a)$ and $S_{ia} = 0$ otherwise $(i=0,1,\dots,n;\ a=1,\dots,m)$. The matrix $\underline{I}$ can be written in the form $\underline{I} = (\underline{S}_o^T,\underline{S}^T)^T$, where $\underline{S}_o = (S_{oa})$ $(a=1,\dots,m)$, $\underline{S} = (S_{ia})$ $(i=1,\dots,n;\ a=1,\dots,m)$.

By the consideration of the relative motions between two arbitrary bodies i and j it is necessary to determinate the transition between the coordinate systems attached to the bodies. By the determination of the paths connecting two vertices we are interessted in the best path in some sence. For an optimality criterion can serve the requirment to pass the least number of arcs. Another optimality criteria are for instance, that the number of the generalised coordinates in the hinges belonging to the path be the minimal, or that the transformation matrices between the contiguous bodies be the simplest form and so on. By determining the weightes of the edges in a proper way we can formulate the considered problem as a problem of finding the minimal in the sence of weight path between two given vertices. This is a well investigated problem in the graph theory for which a number of algorithms have been suggested [7]. Of particular importance are the paths connecting the vertex s_0 with the remaining vertices. These paths are used further for the determination of the transformation matrix from the coordinate system attached to the given body to the coordinate system in the body 0. The choice of the path from s_0 to s_i which we will denote by $[s_0, s_i]$ and call an effective path is comitted entirely to the user. This path can be minimal in weight but can be chosen for reasons which are rather subjective in their nature and cannot be algorithmized. After determining the effective paths we define the matrix of the effective paths $\underline{\Psi} = (\Psi_{ai})$ $(a=1,\ldots,m;\ i=1,\ldots,n)$ with elements $\Psi_{ai}=1$ if $u_a \in [s_0, s_i]$ and is directed toward s_0, $\Psi_{ai} = -1$ if $u_a \in [s_0, s_i]$ and is directed from s_0 and $\Psi_{ai} = 0$ otherwise. The underlying exposition substentially exploits the following relationship between the incidence matrix $\underline{I}$ and the introduced matrix of the effective paths $\underline{\Psi}$

$$\underline{\Psi}^T \underline{I}^T = (-\underline{1}_n, \underline{E}_n). \tag{1}$$

Here and further $\underline{E}_n$ is the unit n by n matrix and $\underline{1}_n$ is a n-dimensional vector with all elements 1. By determining the constraint equations of the system we use the well familiar in the graph theory relation between the incidence matrix $\underline{I}$ and the fundamental loops matrix (cyclomatic matrix) $\underline{\Phi}$ [7]

$$\underline{I\Phi}^T = \underline{0}_{(n+1)\times\hat{n}} \tag{2}$$

where $\underline{0}_{(n+1)\times\hat{n}}$ is a scalar (n+1) by $\hat{n}$ matrix with all elements zero.

Kinematics of multibody system

In many problems concerning multibody systems the flexibility of the bodies plays an important role and the bodies cannot be treated as rigid. Often in the investigation of such systems one accepts that the influence of the flexibility is reduced to the appearance of small perturbations of the nominal motion experienced by the system, considered as composed of rigid bodies only and that the nominal motion itself is not under the influence of deformations.

We will derive motion equations which take into account the mutual relationship between the bodies deformations and the bodies motions. This relationship is expressed in the change of the mass center positions and inertia moments of the bodies caused by the deformation deflextions and, on the other hand, in the influence of the inertia forces on the bodies deformations. We will consider the deformable body as a hibrid system whose motion kinematically is determined both by the motion of an absolutely stiff "skeleton" and a motion relative to the skeleton of a continuously distributed mass [8,9].We assume that the motion of the continuum is determined by a finite number of variables describing the deformations. It should be noted that such an approach is being applied often in the investigation of the continuum. There exist different methods for discretization. The basic method is that of Reley-Ritz-Galerkin by which the special displacements of a point of the continuum are represented as a finite series of appropriately chosen vector functions (of the spacial point coordinates) multiplied by time dependent generalized coordinates, identical for all points of the continuum.

We will locate the attitude of an arbitrary body i relative to the inertial space $O\xi\eta\zeta$ with the radius vector $\underline{\bar{R}}_i$ of a point C_i and with an orthonormal triad $\underline{\bar{e}}^{(i)} = (\bar{e}_1^{(i)}, \bar{e}_2^{(i)}, \bar{e}_3^{(i)})^T$ (i=0,...,n). The frame $C_i\underline{\bar{e}}^{(i)}$ plays the role of a skeleton frame for the body i, relative to which the deformation of the the body will be considered. As a point C_i we choose the mass center of the

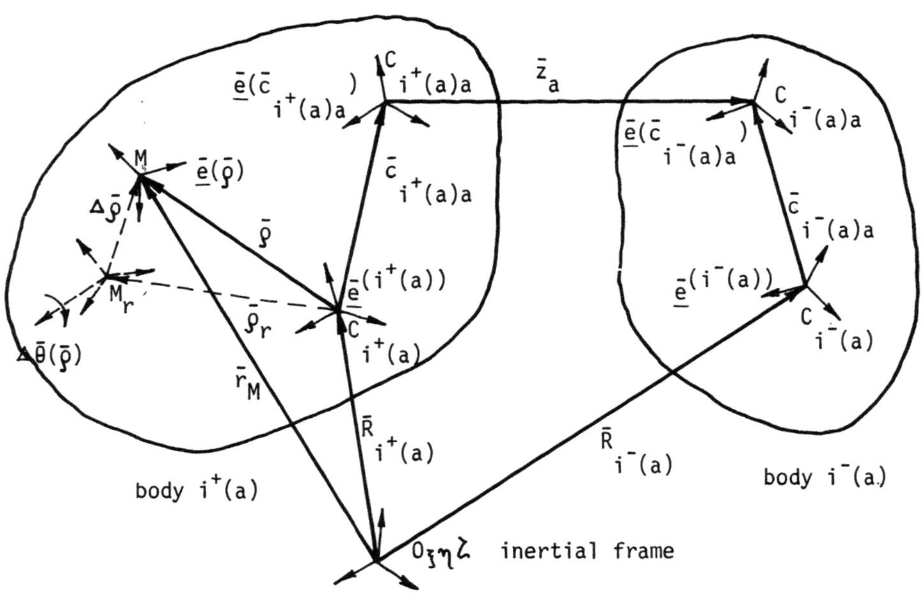

Fig.1. Vectors describing the kinematics of contiguous bodies

body in undeformed state. We represent the body deformation as a displacement field $\Delta\bar{\rho}$ which is to be considered in the floating frame $C_i\underline{\bar{e}}^{(i)}$. In the Fig.1 a point M is shown in a deformed configuration specified by the instanteneous local position vector $\bar{\rho}$. The corresponding point M_r in the undeformed state is specified by the initial position vector $\bar{\rho}_r$ fixed in the floating system $C_i\underline{\bar{e}}^{(i)}$. Thus

$$\bar{\rho} = \bar{\rho}_r + \Delta\bar{\rho} \tag{3}$$

As previously noted we consider the vector $\Delta\bar{\rho}$ as a function of the point position in the undeformed state specified by the by the radius vector $\bar{\rho}_r$ and by a set of a finite number s_i time depending variables $(u_{i1},\dots,u_{is_i})^T = \underline{u}_i$, $\Delta\bar{\rho} = \Delta\bar{\rho}(\bar{\rho}_r,\underline{u}_i)$, $\Delta\bar{\rho}(\bar{\rho}_r,\underline{0}) = \bar{0}$ $(i=0,1,\dots,n)$. For simplicity in the notation we will consider only the case of linear dependance of $\Delta\bar{\rho}$ on $\underline{u}_i$. The general case can be considered analogously. Thus

$$\Delta\bar{\rho}(\bar{\rho}_r,\underline{u}_i) = \sum_{j=1}^{s_i} u_{ij}\bar{n}_{ij}(\bar{\rho}_r) = \underline{\bar{n}}_i^T(\bar{\rho}_r)\underline{u}_i, \tag{4}$$

where $\underline{\bar{n}}_i(\bar{\rho}_r) = (\bar{n}_{i1}(\bar{\rho}_r),\dots,\bar{n}_{is_i}(\bar{\rho}_r))^T$ and the functions $\bar{n}_{ij}(\bar{\rho}_r)$ are specified in advance.

We suppose that simultaneously with the vector functions $\bar{n}_{ij}(\bar{\rho}_r)$ another vector functions $\bar{h}_{ij}(\bar{\rho}_r)$ are specified too, by which one can describe the deformation change of an orthonormal triad $\underline{\bar{e}}(\bar{\rho})$, previously fixed in the point $\bar{\rho}_r$. Since the deformation displacements are small, this deformation change can be written in the form $\underline{\bar{e}}(\bar{\rho}) = \Delta\bar{\theta}(\bar{\rho}) \times \underline{\bar{e}}(\bar{\rho})$. Now the vector $\Delta\bar{\theta}(\bar{\rho})$ is specified for each point $\bar{\rho}$ and analogously to (4) has the form

$$\Delta\bar{\theta}(\bar{\rho}) = \underline{\bar{h}}_i^T(\bar{\rho}_r)\underline{u}_i, \quad \underline{\bar{h}}_i = (\bar{h}_{i1},\dots,\bar{h}_{is_i})^T \tag{5}$$

The transformation matrix from the deformed local frame $\underline{\bar{e}}(\bar{\rho})$ to the skeleton frame $\underline{\bar{e}}^{(i)}$ is

$$\underline{\bar{e}}_i.\underline{\bar{e}}^T(\bar{\rho}) = \underline{\bar{e}}^{(i)}.\underline{\bar{e}}^T(\bar{\rho}_r)(\underline{E}_3 + \Delta\underline{\tilde{\theta}}), \quad \Delta\underline{\theta} = \Delta\bar{\theta}(\bar{\rho}).\underline{\bar{e}}(\bar{\rho}_r) \tag{6}$$

Here and further if $\underline{a} = (a_1,a_2,a_3)^T$, then $\underline{\tilde{a}}$ denotes the matrix $\underline{\tilde{a}} = (a_{ij})$ $(i,j=1,2,3)$, where $a_{ii}=0$, $a_{ij}=-a_{ji}$, $a_{12}=-a_3$, $a_{13}=a_2$, $a_{23}=-a_1$. The matrix (6) is used for the transformation of quantities such as concentrated forces when they are given in the local frame attached to a certain point of the deformable body. Let $\bar{h}_i(\bar{\rho})$ be the angular velocity of the frame $\underline{\bar{e}}(\bar{\rho})$ with respect to the frame $\underline{\bar{e}}^{(i)}$. We have analogously to (5)

$$\bar{h}_i(\bar{\rho}) = \underline{\bar{h}}_i^T(\bar{\rho}_r)\underline{\mathring{u}}_i \tag{7}$$

The basic considerations begin with the description of the relative motions in the hinges. For this purpose in each hinge a two hinge points C_{ia}, $i=i^{\pm}(a)$ in the corresponding contiguous bodies are specified and the hinge vector $\bar{z}_a = \overline{C_{i^+(a)a}C_{i^-(a)a}}$ is introduced (Fig.1). We denote the radius vectors of the hinge points C_{ia} in the corresponding skeleton frames by $\bar{c}_{ia} = \overline{C_iC_{ia}}$, $i=i^{\pm}(a)$. The relative motion in the hinge a is determined by the motion of the frame $C_{i^-(a)a}\underline{\bar{e}}^{(i^-(a)a)}$ with respect to the frame $C_{i^+(a)a}\underline{\bar{e}}^{(i^+(a)a)}, \underline{\bar{e}}^{(ia)} = \underline{\bar{e}}(\bar{c}_{ia})$. Because of the deformations the frames $C_{ia}\underline{\bar{e}}^{(ia)}$ have an own motion with respect to the corresponding skeleton frames $C_i\underline{\bar{e}}^{(i)}$, $i = i^{\pm}(a)$. Let $\underline{G}_a = \underline{\bar{e}}^{(i^-(a)a)}.\underline{\bar{e}}^{(i^+(a)a)T}$ be the transformation matrix between the frames $\underline{\bar{e}}^{(i^+(a)a)}$ and $\underline{\bar{e}}^{(i^-(a)a)}$. We will describe the motion of the frame $C_{i^-(a)a}$ $\underline{\bar{e}}^{(i^-(a)a)}$ with respect to the frame $C_{i^+(a)a}\underline{\bar{e}}^{(i^+(a)a)}$ in terms of generalized parameters $\underline{q}_a$ (which need not be independent, e.g. they might be redundant variables) and nonholonomic variables (derivatives of quasi-coordinates or quasi-velocities) $\underline{x}_a = (x_{a1},\dots,x_{an_a})^T$, where n_a is the number of the degrees of relative freedom [5,10]. In particular the formulas $\mathring{\bar{z}}_a = \underline{\bar{t}}_a^T\underline{x}_a + \bar{t}_{ao}$, $\underline{\bar{t}}_a = (\bar{t}_{a1},\dots,\bar{t}_{an_a})^T$; $\bar{\Omega}_a = \underline{\bar{p}}_a^T\underline{x}_a + \bar{p}_{ao}$, $\underline{\bar{p}}_a = (\bar{p}_{a1},\dots,\bar{p}_{an_a})^T$ and

$$\underline{\mathring{q}}_a = \underline{\mathring{q}}_a(t,\underline{q}_a,\underline{x}_a) \quad (a=1,\dots,m) \tag{8}$$

are true, where $\mathring{\bar{z}}_a$ and $\bar{\Omega}_a$ are respectively the relative translation and angular velocities of the frame $C_{i^-(a)a}\underline{\bar{e}}^{(i^-(a)a)}$ with respect to the frame $C_{i^+(a)a}\underline{\bar{e}}^{(i^+(a)a)}$. The angular velocities of the frames $C_{ia}\underline{\bar{e}}^{(ia)}$ with respect to the frames $C_i\underline{\bar{e}}_i$, $i=i^{\pm}(a)$ will be denote with $\bar{\Omega}_{ia}$. According to (7) $\bar{\Omega}_{ia} = \bar{h}_i(\bar{c}_{ia}) = \underline{\bar{h}}_i^T(\bar{c}^r_{ia})\underline{\mathring{u}}_i$, $i=i^{\pm}(a)$. The transformation from the frame $C_{ia}\underline{\bar{e}}^{(ia)}$ to the skeleton frame $C_i\underline{\bar{e}}^{(i)}$, $i=i^{\pm}(a)$ is given with the matrix $\underline{G}_{ia} = \underline{G}(\bar{c}_{ia}) = (\underline{\bar{e}}(\bar{c}_{ia})).\underline{\bar{e}}^{(i)T}$, which in accordance with (6) has the form $\underline{G}_{ia} = (\underline{E}_3 - \Delta\underline{\tilde{\theta}})$ X $\underline{\bar{e}}(\bar{c}^r_{ia}).\underline{\bar{e}}^{(i)T}$. Here $\bar{c}^r_{ia}$ is the radius vector of the point C_{ia} in an undeformed state, $\Delta\underline{\theta} = \Delta\bar{\theta}(\bar{c}_{ia}).\underline{\bar{e}}(\bar{c}^r_{ia})$ and the vector $\Delta\bar{\theta}(\bar{c}_{ia})$ is given with (5).Thus, the transformation matrix between the two contiguous skeleton systems $\underline{B}_a = \underline{\bar{e}}^{(i^-(a))} . \underline{\bar{e}}^{(i^+(a))T}$ can be written in the form $\underline{B}_a = \underline{G}^T_{i^-(a)a} \underline{G}_a \underline{G}_{i^+(a)a}$. The matrices $\underline{B}_a$ are used for calculations in the coordinate system $C_o\underline{\bar{e}}^{(o)}$. By these matrices and the matrix of the effective paths $\underline{\Psi}$ are constructed the transformation matrices $\underline{A}^{(i)} = \underline{\bar{e}}^{(o)} . \underline{\bar{e}}^{(i)T}$ $(i=1,\dots,n)$ by means of wich the orientations of the skeleton frames $C_i\underline{\bar{e}}^{(i)}$ with respect to the frame $C_o\underline{\bar{e}}^{(o)}$ are being specified. Let the effective path $[s_o, s_i]$ consist from the arcs with numbers $a_{i1},\dots,a_{im_i}$, then

$$\underline{A}^{(i)} = \prod_{k=1}^{m_i} \underline{B}_{a_{ik}}^{\Psi_{a_{ik}i}} . \tag{9}$$

The matrices (9) have the optimal for the user structure because of the extremal properties of the effective paths. For instance, they are functions of a minimal number of generalized coordinates if the effective paths are defined as the minimal (in terms of weight) paths in the weighted graph in which the weights of the edges u_a are equal to the number of the relative motion parameters in the hinges a. If all edge weights are selected to be of equal number, then $\underline{A}^{(i)}$ are products of a minimal number of matrices $\underline{B}_a$. Besides, the effective paths can be selected in such a way that $\underline{A}^{(i)}$ are products of matrices with optimal in sense of other criteria structure. It is the formula (9) that must be taken into account when choosing the effective paths.

The quantities describing the relative motions kinematics are connected with the quantities describing the kinematics of the bodies in the inertial frame by means of the structure matrices $\underline{I}$ and $\underline{\Psi}$. In particular, if $\dot{\bar{R}}_i$, $\bar{\omega}_i$ $(i=0,1,\dots,n)$ are the absolute translation and angular velocities of the skeleton frames relative to the inertial frame $O\xi\eta\zeta$, then for $\underline{\dot{\bar{R}}} = (\dot{\bar{R}}_1,\dots,\dot{\bar{R}}_n)^T$, $\underline{\bar{\omega}} = (\bar{\omega}_1,\dots,\bar{\omega}_n)^T$ the formulas

$$\underline{I}^T \begin{bmatrix} \dot{\bar{R}}_o \\ \underline{\dot{\bar{R}}} \end{bmatrix} + (\mathrm{diag}\underline{\bar{t}} - \mathrm{diag}\underline{\bar{p}}\underline{\Psi} \times (\underline{\bar{C}} + \underline{\bar{C}}^*))^T \underline{x} + (\underline{\bar{n}} + \underline{\bar{h}}^+ -$$

$$\underline{\bar{h}}\underline{\Psi} \times (\underline{\bar{C}} + \underline{\bar{C}}^*))^T \underline{\dot{u}} - (\underline{\bar{J}} + \underline{\bar{J}}^*)^T \underline{1}_{n+1} \times \bar{\omega}_o + \underline{\bar{g}} = \underline{\bar{0}}_{mx1} \tag{10}$$

$$\underline{I}^T \begin{bmatrix} \bar{\omega}_o \\ \underline{\bar{\omega}} \end{bmatrix} + \mathrm{diag}\underline{\bar{p}}^T \underline{x} + \underline{\bar{h}}^T \underline{\dot{u}} + \underline{\bar{p}}_o = \underline{\bar{0}}_{mx1}$$

take place, where $\underline{\bar{J}} = (S_{ia}\bar{c}_{ia})$, $\underline{\bar{J}}^* = (S^+_{ia}\bar{z}_a)$ $(i=0,1,\dots,n;\ a=1,\dots,m)$ $(S^+_{ia}=1$

if $i=i^+(a)$ and $S^+_{ia}=0$ otherwise); $\underline{\bar{C}} = (S_{ia}\bar{c}_{ia})$, $\underline{\bar{C}}^* = (S^+_{ia}\bar{z}_a)$ $(i=1,\ldots,n;\ a=1,\ldots,m)$; $\underline{\bar{n}} = (S_{ia}\underline{\bar{n}}_i(\bar{c}^r_{ia}))$, $\underline{\bar{h}} = (S_{ia}\underline{\bar{h}}_i(\bar{c}^r_{ia}))$, $\underline{\bar{h}}^+ = (S^+_{ia}\underline{\bar{h}}_i(\bar{c}^r_{ia}) \times \bar{z}_a)$ $(i=0,1,\ldots,n;\ a=1,\ldots,m)$; $\underline{\bar{g}}$ and $\underline{\bar{p}}_0$ are certain expressions not depending on $\underline{x}$ and $\underline{\dot{u}}$. The quasi-diagonal matrix along whose principal diagonal the elements $A_1, \ldots, A_n$ of the quantity A are located, is denote by the symbol diagA. These elements can be scalars, vectors, matrices, tensors, etc. In the formulas (10) $\underline{\bar{J}}$ and $\underline{\bar{J}}^*$ and in particular $\underline{\bar{C}}$ and $\underline{\bar{C}}^*$ are dependent not only on $\underline{q}$ but on $\underline{u}$, as well, by means of $\bar{c}_{ia}$ which with (3),(4) have the form $\bar{c}_{ia} = \bar{c}^r_{ia} + \underline{\bar{n}}^T_i(\bar{c}^r_{ia})\underline{u}_i$. The same is true for the member $\underline{\bar{g}}$. By multiplying (10) from the left with $\underline{\Psi}^T$ and taking into account (1) we obtain the following compact presentations of the translational and angular velocities of the separate skeleton frames:

$$\underline{\dot{\bar{R}}} = -\underline{\Psi}^T((\text{diag}\underline{\bar{t}} - \text{diag}\underline{\bar{p}}\underline{\Psi} \times (\underline{\bar{C}} + \underline{\bar{C}}^*))^T\underline{x} + (\underline{\bar{n}} + \underline{\bar{h}}^+ - \underline{\bar{h}}\underline{\Psi} \times (\underline{\bar{C}} + \underline{\bar{C}}^*))^T\underline{\dot{u}}) + \ldots \tag{11}$$

$$\underline{\bar{\omega}} = -\underline{\Psi}^T(\text{diag}\underline{\bar{p}}^T\underline{x} + \underline{\bar{h}}^T\underline{\dot{u}}) + \ldots$$

Only the terms depending on $\underline{x}$ and $\underline{\dot{u}}$ are given in the above relations. In a similar way the quantities $\underline{\ddot{\bar{R}}}$, $\underline{\dot{\bar{\omega}}}$ are being represented and the relative motion between arbitrarily chosen bodies of the system is being determined.

The formulas derived for an arbitrary system can be simplified in the case of the system with tree structure. However the results for a system with tree structure are applicable for an arbitrary system, as well. For this purpose we consider imaginatively as "bifurcated" those bodies of the system to which the bifurcated vertices of the primary graph correspond upon transforming it in the tree-like graph (secondary graph)[2]. The body is bifurcated with complete geometric identity conserved between the original and the image. In particular, the centers of mass and basis reference frames are identically arranged in both bodies. Here a certain part of the hinge points is conserved in any of the images, namely, the hinge points of those hinges whose corresponding arcs in the secondary graph connect the vertex corresponding to this image to the other vertices. The rest of the hinge points of the original figure is on the other image. We consequently obtain a system consisting of (m+1) bodies and having no closed loops. We call this system an extended system. It can be proved that the kinematical description of an arbitrary system using the notion of extended system is actually a particular case of the first procedure of description using the primary graph. This particular case is realized by a special choice of the matrix of the effective paths, which we will not adduce for lack of place.

One of the most essential problems in the investigation of multibody systems is the determination of constraint equations. With them are related the basic problems arising by the modelling of systems with closed loops. The suggested by many authors methods for investigation of multibody systems containing closed loops have not a universal character mainly for lack of a formal apparatus for a derivation of the constraint equations and for the differentiation of these equations necessary for the formulation of the motion equations. In the analytical mechanics there are no general methods for derivation of the constraint equations and obviously such methods cannot be created for an arbitrary mechanical system. However, the here considered type of mechanical systems, namely, multibody systems allows because of its specific character a formal description of the constraints resulting from the availability of closed loops. For this purpose the matrix $\underline{\Phi}$ of the fundamental loops and the derived relations between the structural elements of the system are substancially employed. Using (2) and multiplying (10) from the left with $\underline{\Phi}$ one gets the relations:

$$\begin{aligned}&\underline{\Phi}(\mathrm{diag}\underline{\bar{t}} - \mathrm{diag}\underline{\bar{p}\Psi} \times (\underline{\bar{C}} + \underline{\bar{C}}^*))^T\underline{x} + \underline{\Phi}(\underline{\bar{n}} + \underline{\bar{h}}^+ - \\ &\underline{\bar{h}\Psi} \times (\underline{\bar{C}} + \underline{\bar{C}}^*))^T\underline{\dot{u}} + \underline{\Phi}\underline{\bar{g}} = \underline{\bar{0}}_{\hat{n}\times 1} \\ &\underline{\Phi}\mathrm{diag}\underline{\bar{p}}^T\underline{x} + \underline{\Phi}\underline{\bar{h}}^T\underline{\dot{u}} + \underline{\Phi}\underline{\bar{p}}_o = \underline{\bar{0}}_{\hat{n}\times 1}\end{aligned} \tag{12}$$

which are the differentiated constraint equations. In a similar way twice differentiated constraint equations can be found. The constraint equations can be written in an another equivalent form using the secondary supplemented graph and quantities related to the extended system, as well.

Motion equations

The method here applied for the derivation of differential equations of motion conciders the fundamental basis - the use of the variational principles of mechanics. Let us consider now the most general case when constraints including accelerations $\overset{(s)}{\bar{r}_j} = d^s\bar{r}_j/dt^s$ of arbitrary but finite order s are being imposed on the discrete generally nonlinear mechanical system [5]

$$f_i(t,\bar{r}_j,\dot{\bar{r}}_j,\ldots \overset{(p-1)}{\bar{r}_j}) = 0 \quad (i=1,\ldots,l) \tag{13}$$

Let $\bar{r}_j = \bar{r}_j(t,\beta)$ be kinematical admissible trajectories of the points of the system, which satisfy identically for each meaning of the parameter β the constraint equations (13). We define a virtual variation of the acceleration $\overset{(s)}{\bar{r}}_j(t,\beta)$ as a quantity $\delta \overset{(s)}{\bar{r}}_j = \overset{(s)}{\bar{r}}_j(t,\beta + \delta\beta) - \overset{(s)}{\bar{r}}_j(t,\beta)$ $(s=0,1,\ldots)$. According to this definition the virtual variations of the points positions (s=0),their velocities (s=1) and accelerations of an arbitrary order are being defined for a given moment t and a given curve β. Let $\bar{r}_j = \bar{r}_j(t) = \bar{r}_j(t,\beta_o)$ be the actual motion experienced by the system subject to the given forces $\bar{F}_j(t,\bar{r}_i,\dot{\bar{r}}_i)$ and without violating the constraints imposed. Let us designate $\bar{A}_j(t) = m_j\ddot{\bar{r}}_j(t) - \bar{F}_j(t,\bar{r}_i(t),\dot{\bar{r}}_i(t))$. In accordance to the d'Alembert's principle in the form of Lagrange the real motion of the system is specified by the property $\sum_j \bar{A}_j(t).\delta\bar{r}_j(t,\beta_o) = 0$. If we apply this principle for the moment $(t + dt)$ and expand $\bar{A}_j(t + dt)$ and $\delta\bar{r}_j(t + dt,\beta_o)$ in a power series of dt we find that the main term of the expansion is: $\sum_j (m_j\ddot{\bar{r}}_j - \bar{F}_j).\delta\bar{r}_j = 0$ for $\delta\bar{r}_j \neq \bar{0}$ - d'Alembert's principle; $\sum_j (m_j\ddot{\bar{r}}_j - \bar{F}_j).\delta\dot{\bar{r}} = 0$ for $\delta\bar{r}_j = \bar{0}$, $\delta\dot{\bar{r}}_j \neq \bar{0}$ - Jourdain's principle (principle of virtual power); $\sum_j (m_j\ddot{\bar{r}}_j - \bar{F}_j).\delta\ddot{\bar{r}} = 0$ for $\delta\bar{r}_j = \delta\dot{\bar{r}}_j = \bar{0}$, $\delta\ddot{\bar{r}}_j \neq \bar{0}$ - Gauss' principle; ... ; $\sum_j (m_j\ddot{\bar{r}}_j - \bar{F}_j).\delta\overset{(k)}{\bar{r}} = 0$ for $\delta\bar{r}_j = \delta\dot{\bar{r}}_j = \ldots = \delta\overset{(k-1)}{\bar{r}}_j = \bar{0}$, $\delta\overset{(k)}{\bar{r}}_j \neq \bar{0}$ - generalized principle of d'Alembert. The generalized principle of d'Alembert can be written in the form

$$\delta Z_k = 0, \quad Z_k = \sum_j (m_j\ddot{\bar{r}}_j - \bar{F}_j).\overset{(k)}{\bar{r}}_j, \quad \delta = \delta\overset{(k)}{\bar{r}}_j \tag{14}$$

if the quantities $(m_j\ddot{\bar{r}}_j - \bar{F}_j)$ don't deppend on $\overset{(k)}{\bar{r}}_i$. The variation of Z_k is over $\overset{(k)}{\bar{r}}_j$, e.g. $\delta\bar{r}_j = \delta\dot{\bar{r}}_j = \ldots = \delta\overset{(k-1)}{\bar{r}}_j = \bar{0}$. Under the assumed in the analytical mechanics independence of the forces with respect to the accelerations $\bar{F}_j = \bar{F}_j(t,\bar{r}_i,\dot{\bar{r}}_i)$ the form (14) is always possible for $k \geqq 3$. Such a form is possible for $k = 2$, as well, if we take for Z_2 the function $Z_2=(1/2)\sum_j m_j\ddot{\bar{r}}_j^2 - \sum_j \bar{F}_j.\ddot{\bar{r}}_j$ introduced by Gauss (Zwang). In this way the generalized principle of d'Alembert can be interpreted as a condition for stationarity of the function Z_k. This function must be considered as a function of the quantities $\overset{(k)}{\bar{r}}_j$ only and the stationarity is implied without violation of the conditions (13). After once differentiating these conditions we receive constraints which are linear with respect to the last derivatives $\overset{(p)}{\bar{r}}_j$. Thus,if k is large enough ($k \geqq (p - 1)$ in the case of linesrity of the equations (13) and $k \geqq p$ in the case of nonlinearity) it is always possible to consider the stationarity of Z_k with respect to $\overset{(k)}{\bar{r}}_j$ by linear in $\overset{(k)}{\bar{r}}_j$ constraints [5].

The function Z_2 is of the special interest between all the functions Z_k. The condition $\delta Z_2 = 0$ here means minimum in $\ddot{\bar{r}}_j$. This implies that the function Z_2 can be employed conveniently in the application of numerical methods for the investigation of the motion [3].

It follows from the above statement, that by the holonomic constraints and linear nonholonomic constraints of the relative motions in the hinges of a multibody system it is preferable to employ the Jourdain's principle which for a multibody system has a form

$$\sum_{i=0}^{n} \int_{m_i} (\ddot{\bar{r}} dm - d\bar{F}) . \delta\dot{\bar{r}} = 0, \quad \delta\bar{r} = \bar{0} \tag{15}$$

where $\bar{r}$ is the radius vector of the differential element of mass in the inertial space and the variation is over the velocities while the positions remain unchanged (Fig.1). The formulas $\dot{\bar{r}} = \dot{\bar{R}}_i + \bar{\omega}_i \times \bar{\rho} + \underline{\bar{n}}_i^T \underline{\dot{u}}_i$, $\ddot{\bar{r}} = \ddot{\bar{R}}_i + \dot{\bar{\omega}}_i \times \bar{\rho} + \underline{\bar{n}}_i^T \underline{\ddot{u}}_i + \bar{\omega}_i \times (\bar{\omega}_i \times \bar{\rho}) + 2\bar{\omega}_i \times \underline{\bar{n}}_i^T \underline{\dot{u}}_i$, $\delta\dot{\bar{r}} = \delta\dot{\bar{R}}_i + \delta\bar{\omega}_i \times \bar{\rho} + \underline{\bar{n}}_i^T \delta\underline{\dot{u}}_i$ for the velocities, accelerations and virtual variations of the velocities by $\delta\bar{r} = \bar{0}$ of an arbitrary point $\bar{r}$ of body i take place, where according to (11) $\delta\dot{\underline{\bar{R}}} = -\underline{\Psi}^T((\text{diag}\underline{\bar{t}} - \text{diag}\underline{\bar{p}\Psi} \times (\underline{\bar{C}} + \underline{\bar{C}}^*))^T \delta\underline{x} + (\underline{\bar{n}} + \underline{\bar{h}}^+ - \underline{\bar{h}\Psi} \times (\underline{\bar{C}} + \underline{\bar{C}}^*))^T \delta\underline{\dot{u}})$, $\delta\underline{\bar{\omega}} = -\underline{\Psi}^T(\text{diag}\underline{\bar{p}}^T \delta\underline{x} + \underline{\bar{h}}^T \delta\underline{\dot{u}})$.

Let us consider closely the calculation of the virtual work of the external forces and forces in the hinges. If we have a concentrated external force $\bar{F}_M$ and torque $\bar{T}_M$ acting in point M of a given body, then the virtual work done by these forces (by fixed points positions) is specified by the expression $\bar{F}_M . \delta\dot{\bar{r}}_M + \bar{T}_M . \delta\bar{\omega}_M$, where $\delta\dot{\bar{r}}_M$ and $\delta\bar{\omega}_M$ are the variations of the point velocity and angular velocity of the triad fixed to the point respectively. If the point belongs to the body i, then we obtain the expression $(\delta\dot{\bar{R}}_i + \delta\bar{\omega}_i \times (\bar{\rho}_M^r + \underline{\bar{n}}_i^T(\bar{\rho}_M^r)\underline{\dot{u}}_i) + \underline{\bar{n}}_i^T(\bar{\rho}_M^r)\delta\underline{\dot{u}}_i) . \bar{F}_M + (\delta\bar{\omega}_i + \underline{\bar{h}}_i^T(\bar{\rho}_M^r)\delta\underline{\dot{u}}_i) . \bar{T}_M$ for the virtual work. We will use this form for the internal forces doing virtual work in the hinges, as well. Let $\bar{X}_a$ and $\bar{Y}_a$ be respectively the resultant force and moment applied on body $i^+(a)$ by body $i^-(a)$ and acting at the reference point $C_{i^-(a)a}$. It follows that $-\bar{X}_a$ and $-\bar{Y}_a$ are acting on body $i^-(a)$ by body $i^+(a)$. The total

virtual work of all internal forces acting in the hinges can be presented in the form $\delta W = \delta\dot{\underline{\bar{R}}}^T.\underline{S}\underline{\bar{X}} + \delta\underline{\bar{\omega}}^T.(\underline{S}\underline{\bar{Y}} + (\underline{\bar{C}} + \underline{\bar{C}}^*) \times \underline{\bar{X}}) + \delta\underline{\dot{u}}^T((\underline{\bar{n}} + \underline{\bar{h}}^+).\underline{\bar{X}} + \underline{\bar{h}}.\underline{\bar{Y}})$. All quantities in (15) are now represented as explicit functions of $t,\underline{q},\underline{u},\underline{x},\underline{\dot{u}},\underline{\dot{x}}$, $\underline{\ddot{u}}$. The principle (15) and constraints (12) will be rewritten for short in the form $\delta\underline{y}^T(\underline{A}\underline{\dot{y}} + \underline{b}) = \underline{0}$, $\underline{K}\delta\underline{y} = \underline{0}$, $\underline{K}\underline{y} = \underline{k}$, where $\underline{y} = (\underline{x}^T,\underline{\dot{u}}^T)^T$. The motion equations are found straightforward by determining that submatrix of the matrix $\underline{K}$ which is a carrier of its rank [1,2]. These equations with the equations (8) provide a complete system for the variables $\underline{q}$, $\underline{u}$, $\underline{x}$. The motion equations can be derived using the Lagrangian multipliers, as well. It seems worthwhile to mention that this method is applicable only then, when the matrix $\underline{K}$ possesses the full rank which rarely happens. That is why in the numerical integration of the motion equations it is better to use the principle of Gauss, which can be considered as a condition for minimum of the function $Z_2 = (1/2)\underline{\dot{y}}^T\underline{A}\underline{\dot{y}} + \underline{\dot{y}}^T\underline{b}$ under the constraints $\underline{K}\underline{\dot{y}} = \underline{k}_1$ [3]. The problem is reduced in this way to a minimization of a square functional under linear constraints. There exist a lot of methods for solving this problem.

Acknowledgment. The main part of this work was written during a research visit in FRG, 1984, sponsored by DAAD (The German Academic Exchange Service).

References

1. Lilov, L.; Wittenburg, J.: Equations of motion for systems of rigid bodies with arbitrary hinges (in german). Z. für ang. Math. und Mech. (ZAMM), 57 (1977), H.3.
2. Lilov, L.; Chirikov, V.: Dynamics of rigid body systems. Prikl. Mat. i Mech. (PMM), 45 (1981), v.3.
3. Lilov, L.; Lorer, M.: Dynamic analysis of multirigid-body systems based on the Gauss principle. Z. für ang. Math. und Mech. (ZAMM), 62 (1982), H.10.
4. Lilov, L.: Structure, kinematics and dynamics of systems of rigid bodies (in russian). Advances in Mechanics, 6 (1983), Nr.1/2.
5. Lilov, L.: Variational principles in the rigid body mechanics (in german). Z. für ang. Math. und Mech. (ZAMM), 64 (1984), T166, H.4/5.
6. Roberson, R.E.; Wittenburg, J.: A dynamical formalism for an arbitrary number of interconnected rigid bodies with reference to the problem of satellite attitude control. Proc. of the 3rd IFAC congress, London, 1966. London: Butterworth and Co., Ltd., 1967.
7. Christofides, N.: Graph theory. An algorithmic approach. New York, London, San Francisco: Academic Press, 1975.
8. Likins P.W.: Modal method for analysis of free rotstions of spacecraft. AIAA Journal, 5 (1967), Nr.7.
9. Lilov, L.: Motion equations of flexible body. Comptes rendus de l'Académie bulgare des Sciences, 36 (1983), Nr.10.
10. Kane, T.R.: Dynamics. New York: Holt, Reinehart and Winston, Inc., 1968.

Dynamics of Chains of Rigid Bodies and Elastic Rods with Revolute and Prismatic Joints

L. Lilov

Bulgarian Acad. Scie., Sofia and
J. Wittenburg, Inst. of Mechanics
University of Karlsruhe (FRG)

Summary

Subject of this paper is the dynamics of articulated multibody systems. The individual bodies have a rigid part and rod-shaped, deformable appendages. Joints between bodies are located on these appendages. The joints are either revolute joints or prismatic joints. Relative motions of contiguous bodies are assumed to be large whereas the deformations of appendages are treated as small so that only first-order terms need be considered.

1. Introduction

The nonlinear kinematics and dynamics of articulated-rigid-body systems has been the subject of intensive research over the last twenty years. Much less well understood is the dynamics of articulated multibody systems in which the individual bodies are elastically deformable. The need for research in this field was first recognized in the area of spacecraft simulation (ref. [1] - [5]). More recently increased accuracy requirements in the performance of industrial robots have renewed interest in these problems (ref. [6] - [9]). It is obvious that combined effects of large displacements in joints and of small deformations of bodies occur in other branches of mechanical engineering as well.

In the existing literature on the dynamics of articulated systems of deformable bodies joints between bodies are always revolute joints. The present paper takes into consideration prismatic joints as well. A minimal set of differential equations is developed for a finite set of hybrid coordinates. One subset $\underline{q}$ of these coordinates describes large relative displacements in joints and the remaining coordinates $\underline{u}$ describe small body deformations. The differential equations are nonlinear with

Dynamics of Multibody Systems
IUTAM/IFToMM Symposium Udine 1985
Editors: G. Bianchi and W. Schiehlen

respect to $\underline{q}$ and linearized with respect to $\underline{u}$. In developing equations of motion the same notations are used that have been used in treating articulated-rigid-body systems (cf. Wittenburg [10] and Wittenburg/Wolz [11]). It is therefore possible to point out where and in which way the additional deformation variables $\underline{u}$ come into play. Although the present investigation is restricted to rod-shaped elastic parts and to two types of joints the chosen method should be applicable to more general systems as well.

2. Dynamics

The dynamics part in the development of equations of motion is based on the principle of virtual power. For an assembly of n elastically deformable bodies this principle can be written

$$\sum_{i=1}^{n}\left[\int_{m_i} \delta\dot{\vec{z}}\cdot(\ddot{\vec{z}}dm - d\vec{F}) + \delta P_i\right] + \delta P = 0. \tag{1}$$

In this expression $\vec{z}$ is the radius vector of a mass particle dm belonging to the mass m_i of body i, $d\vec{F}$ is the externally applied force on dm, δP_i is the virtual power of internal elastice forces in body i and δP represents the total virtual power of (spring and damper) forces acting in joints between bodies. Virtual power is the result of virtual velocity changes $\delta\dot{\vec{z}}$.

Let $\underline{\vec{e}}^{(i)} = \left[\vec{e}_1^{(i)}\ \vec{e}_2^{(i)}\ \vec{e}_3^{(i)}\right]^T$ (i=1...n) be a reference frame of mutually orthogonal unit base vectors which is fixed on the rigid part of body i. In the undeformed state the origin of $\underline{\vec{e}}^{(i)}$ is located at the body i center of mass. In what follows $\underline{\vec{e}}^{(i)}$ will be referred to as floating reference frame because it is floating relative to the non-rigid parts of body i.

For the radius vector $\vec{z}$ of a mass element on body i we write

$$\vec{z} = \vec{r}_i + \vec{\rho} + \underline{\vec{\phi}}_i^T\underline{u}_i. \tag{2}$$

The vector $\vec{r}_i$ locates the origin of the floating reference frame $\underline{\vec{e}}^{(i)}$ and $\vec{\rho}$ points from there to the mass element in the undeformed state. The last term describes the displacement field relative to the floating reference frame as product of two column matrices $\vec{\underline{\phi}}_i$ and $\underline{u}_i$, the latter containing a set of suitably selected scalar deformation variables and the former an associated set of pre-selected shape functions.

The term δP_i in Eq.(1) is known from structural mechanics to be

$$\delta P_i = \delta \dot{\underline{u}}_i^T \underline{K}_i \underline{u}_i \tag{3}$$

where $\underline{K}_i$ is the stiffness matrix of body i. Based on the shape functions $\vec{\underline{\phi}}_i$ it is determined in the usual manner known from finite element methods.

From Eq.(2) follows

$$\left.\begin{aligned} \delta\dot{\vec{z}} &= \delta\dot{\vec{r}}_i + \delta\vec{\omega}_i \times (\vec{\rho} + \vec{\underline{\phi}}_i^T \underline{u}_i) + \delta\dot{\underline{u}}_i^T \vec{\underline{\phi}}_i \\ \ddot{\vec{z}} &= \ddot{\vec{r}}_i + \dot{\vec{\omega}}_i \times (\vec{\rho} + \vec{\underline{\phi}}_i^T \underline{u}_i) + \vec{\underline{\phi}}_i^T \ddot{\underline{u}}_i + \\ &\quad + \vec{\omega}_i \times \left[\vec{\omega}_i \times (\vec{\rho} + \vec{\underline{\phi}}_i^T \underline{u}_i)\right] + 2\vec{\omega}_i \times \dot{\underline{u}}_i^T \vec{\underline{\phi}}_i . \end{aligned}\right\} \tag{4}$$

Eqs.(3) and (4) are substituted into Eq.(1). The integral is evaluated term by term. In this process all second-order terms in $\underline{u}_i$ and $\dot{\underline{u}}_i$ are neglected. This results in the equation

$$\begin{aligned} \sum_{i=1}^{n} \Big\{ &\delta\dot{\vec{r}}_i \cdot (m_i \ddot{\vec{r}}_i - \vec{a}_i \times \dot{\vec{\omega}}_i + \vec{\underline{A}}_i^T \ddot{\underline{u}}_i + \vec{f}_{i1} - \vec{F}_i) + \\ &+ \delta\vec{\omega}_i \cdot (\vec{a}_i \times \ddot{\vec{r}}_i + (\mathbb{J}_i + \mathbb{J}_i^*) \cdot \dot{\vec{\omega}}_i + \vec{\underline{N}}_i^T \ddot{\underline{u}}_i + \vec{f}_{i2} - \vec{M}_i) + \\ &+ \delta\dot{\underline{u}}_i^T (\vec{\underline{A}}_i \cdot \ddot{\vec{r}}_i + - \vec{\underline{N}}_i \cdot \dot{\vec{\omega}}_i + \underline{M}_i \ddot{\underline{u}}_i + \vec{f}_{i3} + \underline{K}_i \underline{u}_i) - \delta P = 0 \end{aligned} \tag{5}$$

with the following abbreviations

$$\left.\begin{aligned} &\vec{\underline{A}}_i = \int_{m_i} \vec{\underline{\phi}}_i \, dm, && \vec{a}_i = \vec{\underline{A}}_i^T \underline{u}_i, && \mathbb{J}_i^* = \underline{u}_i^T \underline{\mathbb{B}}_i + \hat{\underline{\mathbb{B}}}_i^T \underline{u}_i \\ &\underline{\mathbb{B}}_i = \int_{m_i} (\vec{\underline{\phi}}_i \cdot \vec{\rho}\,\mathbb{E} - \vec{\underline{\phi}}_i \vec{\rho}) \, dm, && \hat{\underline{\mathbb{B}}}_i = \underline{\mathbb{B}}_i \text{ conjugate}, && \mathbb{E} = \text{unit dyadic} \\ &\vec{\underline{N}}_i^T = \vec{\underline{N}}_{i1}^T + \underline{u}_i^T \vec{\underline{N}}_{i2}^T && \vec{\underline{N}}_{i1}^T = \int_{m_i} \vec{\rho} \times \vec{\underline{\phi}}_i^T \, dm, && \vec{\underline{N}}_{i2}^T = \int_{m_i} \vec{\underline{\phi}}_i \times \vec{\underline{\phi}}_i^T \, dm \\ &\vec{f}_{i1} = \vec{\omega}_i \times (\vec{\omega}_i \times \vec{a}_i) + 2\vec{\omega}_i \times \dot{\vec{a}}_i && && \underline{M}_i = \int_{m_i} \vec{\underline{\phi}}_i \cdot \vec{\underline{\phi}}_i^T \, dm \end{aligned}\right\} \tag{6}$$

$$\vec{f}_{i2} = \vec{\omega}_i \times (\mathbf{J}_i + \mathbf{J}_i^*) \cdot \vec{\omega}_i + 2\, \hat{\underline{B}}_i^T \dot{\underline{u}}_i \cdot \vec{\omega}_i - \underline{u}_i^T \int \vec{\underline{\Phi}}_i \times d\vec{F}$$

$$\vec{f}_{i3} = \vec{\omega}_i \cdot \left[\underline{\mathbf{B}}_i + \int_{m_i} (\vec{\underline{\Phi}}_i \vec{\underline{\Phi}}_i^T - \vec{\underline{\Phi}}_i \cdot \vec{\underline{\Phi}}_i^T \mathbf{E})\, dm\, \underline{u}_i \right] \cdot \vec{\omega}_i - $$

$$- 2\vec{\omega}_i \cdot \vec{\underline{N}}_{i2}^T \dot{\underline{u}}_i - \underline{u}_i^T \int \vec{\underline{\Phi}}_i \cdot d\vec{F}.$$

$\vec{F}_i$ and $\vec{M}_i$ are the resultant external force and torque, respectively, on body i. Eq.(5) was first formulated for a single deformable body by Lilov [8]. Eq.(5) is re-written in the matrix form

$$\left[\delta\dot{\vec{\underline{r}}}^T \cdot \;\; \delta\vec{\underline{\omega}}^T \cdot \;\; \delta\dot{\underline{u}}^T\right] \left\{ \underbrace{\begin{bmatrix} \underline{m} & -\vec{\underline{a}}\times & \vec{\underline{A}}^T \\ \vec{\underline{a}}\times & (\mathbf{J} + \mathbf{J}^*) \cdot & \vec{\underline{N}}^T \\ \vec{\underline{A}} & -\vec{\underline{N}} \cdot & \underline{M} \end{bmatrix} \begin{bmatrix} \ddot{\vec{\underline{r}}} \\ \dot{\vec{\underline{\omega}}} \\ \ddot{\underline{u}} \end{bmatrix}}_{\underline{G}} + \underbrace{\begin{bmatrix} \vec{\underline{f}}_1 - \vec{\underline{F}} \\ \vec{\underline{f}}_2 - \vec{\underline{M}} \\ \vec{\underline{f}}_3 + \underline{K}\underline{u} \end{bmatrix}}_{\underline{H}} \right\} - \delta P = 0. \quad (7)$$

Here, $\delta\dot{\vec{\underline{r}}}$, $\delta\vec{\underline{\omega}}$, $\delta\dot{\underline{u}}$, $\ddot{\vec{\underline{r}}}$, $\dot{\vec{\underline{\omega}}}$, $\ddot{\underline{u}}$, $\vec{\underline{f}}_1$, $\vec{\underline{f}}_2$, $\vec{\underline{f}}_3$, $\vec{\underline{F}}$, $\vec{\underline{M}}$ and $\underline{u}$ are column matrices of n elements each, e.g.

$$\ddot{\vec{\underline{r}}} = \left[\ddot{\vec{r}}_1 \; \ddot{\vec{r}}_2 \; \ldots \; \ddot{\vec{r}}_n\right] \quad , \quad \underline{u} = \left[\underline{u}_1^T \; \underline{u}_2^T \; \ldots \; \underline{u}_n^T\right]^T.$$

Each element in the (3×3) matrix as well as $\underline{K}$ is itself a matrix in which the respective elements with indices i=1...n are arranged along the diagonal. Multiplication symbols · and × are placed immediately behind the leading factor. Eq.(7) contains two well-known special cases. One case is structural mechanics. In this case the virtual power of the external forces and torques, i.e.

$$-\delta\dot{\vec{\underline{r}}}^T \cdot \vec{\underline{F}} - \delta\vec{\underline{\omega}}^T \cdot \vec{\underline{M}},$$

is expressed in the form $-\delta\dot{\underline{u}}^T\underline{Q}$ with a column matrix $\underline{Q}$ of generalized forces. Eq.(7) then reduces to

$$\delta\dot{\underline{u}}^T (\underline{M}\,\ddot{\underline{u}} + \underline{K}\,\underline{u} - \underline{Q}) = 0$$

whence follow the equations of motion

$$\underline{M}\,\ddot{\underline{u}} + \underline{K}\,\underline{u} = \underline{Q}.$$

The second special case is dynamics of articulated-rigid-body systems. Here Eq.(7) reduces to

$$\left[\delta\dot{\vec{\underline{r}}}^T \;\; \delta\vec{\underline{\omega}}^T\right] \cdot \left\{ \begin{bmatrix} \underline{m} & \underline{0} \\ \underline{0} & \mathbf{J} \cdot \end{bmatrix} \begin{bmatrix} \ddot{\vec{\underline{r}}} \\ \dot{\vec{\underline{\omega}}} \end{bmatrix} + \begin{bmatrix} -\vec{\underline{F}} \\ \vec{\underline{V}} - \vec{\underline{M}} \end{bmatrix} \right\} - \delta P = 0 \qquad (\vec{V}_i = \vec{\omega}_i \times \mathbf{J}_i \cdot \vec{\omega}_i).$$

3. Kinematics

3.1. Outline of the General Strategy

Kinematical constraints in joints between bodies have the effect that the radius vectors in the matrix $\vec{\underline{r}}$ as well as the angular velocities in the matrix $\vec{\underline{\omega}}$ of Eq.(7) are not independent. If, in addition to the deformation variables $\underline{u}$, a minimal set of joint variables $\underline{q}$ is chosen then $\vec{\underline{r}}$ can be expressed as a function of $\underline{q}$ and of $\underline{u}$, and $\vec{\underline{\omega}}$ can be expressed as a function of $\underline{q}$, $\dot{\underline{q}}$, $\underline{u}$ and $\dot{\underline{u}}$. More specifically it is possible to write equations of the form

$$\begin{aligned} &\vec{\underline{r}} = \vec{\underline{r}}(\underline{q},\underline{u}) \\ &\delta\dot{\vec{\underline{r}}} = \vec{\underline{\alpha}}_1\delta\dot{\underline{q}} + \vec{\underline{\alpha}}_2\delta\dot{\underline{u}}\,, \qquad && \delta\vec{\underline{\omega}} = \vec{\underline{\beta}}_1\delta\dot{\underline{q}} + \vec{\underline{\beta}}_2\delta\dot{\underline{u}} \\ &\ddot{\vec{\underline{r}}} = \vec{\underline{\alpha}}_1\ddot{\underline{q}} + \vec{\underline{\alpha}}_2\ddot{\underline{u}} + \vec{\underline{s}}(\underline{q},\dot{\underline{q}},\underline{u},\dot{\underline{u}})\,, \qquad && \dot{\vec{\underline{\omega}}} = \vec{\underline{\beta}}_1\ddot{\underline{q}} + \vec{\underline{\beta}}_2\ddot{\underline{u}} + \vec{\underline{w}}(\underline{q},\dot{\underline{q}},\underline{u},\dot{\underline{u}}). \end{aligned} \tag{8}$$

How the variables $\underline{q}$ are defined will be shown further below. It will also be shown how the matrices $\vec{\underline{\alpha}}_1$, $\vec{\underline{\alpha}}_2$, $\vec{\underline{\beta}}_1$, $\vec{\underline{\beta}}_2$, $\vec{\underline{s}}$ and $\vec{\underline{w}}$ are determined. From Eq.(8) follow the matrix equations

$$\begin{bmatrix} \delta\dot{\vec{\underline{r}}}\cdot \\ \delta\vec{\underline{\omega}}\cdot \\ \delta\dot{\underline{u}} \end{bmatrix} = \underbrace{\begin{bmatrix} \vec{\underline{\alpha}}_1\cdot & \vec{\underline{\alpha}}_2\cdot \\ \vec{\underline{\beta}}_1\cdot & \vec{\underline{\beta}}_2\cdot \\ \underline{0} & \underline{E} \end{bmatrix}}_{\underline{J}} \begin{bmatrix} \delta\dot{\underline{q}} \\ \delta\dot{\underline{u}} \end{bmatrix}, \qquad \begin{bmatrix} \ddot{\vec{\underline{r}}} \\ \dot{\vec{\underline{\omega}}} \\ \ddot{\underline{u}} \end{bmatrix} = \begin{bmatrix} \vec{\underline{\alpha}}_1\cdot & \vec{\underline{\alpha}}_2\cdot \\ \vec{\underline{\beta}}_1\cdot & \vec{\underline{\beta}}_2\cdot \\ \underline{0} & \underline{E} \end{bmatrix} \begin{bmatrix} \ddot{\underline{q}} \\ \ddot{\underline{u}} \end{bmatrix} + \underbrace{\begin{bmatrix} \vec{\underline{s}} \\ \vec{\underline{w}} \\ \underline{0} \end{bmatrix}}_{\underline{D}}. \tag{9}$$

It should be noted that the matrix $\underline{J}$ is composed of submatrices which are in part vector and in part scalar submatrices and that each vector submatrix carries its own multiplication symbol. The same holds for the matrix $\underline{G}$ in Eq.(7)

The virtual power δP in Eq.(7) can be expressed in the form

$$\delta P = \delta\dot{\underline{q}}^T\underline{Q}_1 + \delta\dot{\underline{u}}^T\underline{Q}_2 \tag{10}$$

where $\underline{Q}_1$ and $\underline{Q}_2$ are column matrices of generalized forces. When Eqs.(9) and (10) are introduced into Eq.(7) we get

$$\begin{bmatrix} \delta\dot{\underline{q}}^T & \delta\dot{\underline{u}}^T \end{bmatrix} \left\{ \underline{J}^T \left\{ \underline{G}\left(\underline{J}\begin{bmatrix} \ddot{\underline{q}} \\ \ddot{\underline{u}} \end{bmatrix} + \underline{D} \right) + \underline{H} \right\} - \begin{bmatrix} \underline{Q}_1 \\ \underline{Q}_2 \end{bmatrix} \right\} = 0. \tag{11}$$

By assumption the elements of $\delta\dot{\underline{q}}$ and of $\delta\dot{\underline{u}}$ are independent whence follows that the expression in curled brackets is identically zero. This results in the minimal set of differential equations

$$\underline{J}^T\underline{G}\,\underline{J}\begin{bmatrix}\ddot{\underline{q}}\\ \ddot{\underline{u}}\end{bmatrix} = -\underline{J}^T(\underline{G}\,\underline{D}+\underline{H}) + \begin{bmatrix}\underline{Q}_1\\ \underline{Q}_2\end{bmatrix}. \tag{12}$$

This concludes the outline of the general strategy. It is seen that the equations of motion are known as soon as the Eqs.(8) are explicitly available. In the following section the problem of formulating these equations will be discussed in general terms.

3.2. The Kinematics of a Single Pair of Contiguous Bodies

Fig.1 depicts two bodies with one deformable rod each. The end points of the rods are connected by a revolute joint. The joint axis is perpendicular to the terminal tangents of both rod axes. The common end point will be referred to as the articulation point of the joint. The joint and the two bodies are labeled a, i and j, respectively. Altogether four reference frames are shown. Two of them are the floating reference frames $\underline{\vec{e}}^{(i)}$ and $\underline{\vec{e}}^{(j)}$ already defined on the rigid parts of the bodies. The other two are called $\underline{\vec{e}}^{(ia)}$ and $\underline{\vec{e}}^{(ja)}$, respectively. They have their origin at the articulation point. Each of these two reference

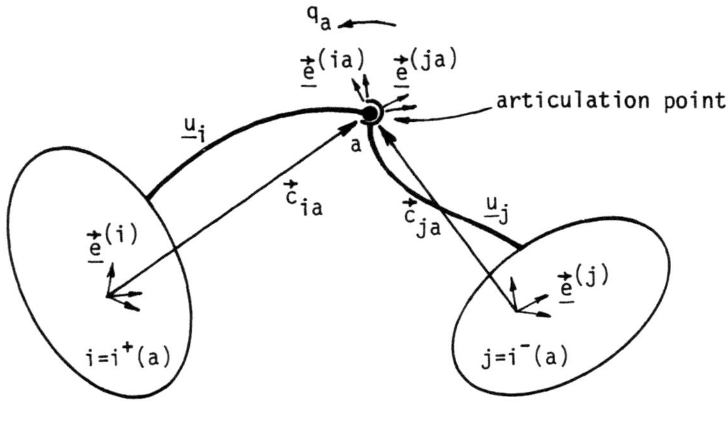

Fig.1: A revolute joint connecting two deformable rods

frames is rigidly attached to the terminal cross section of one rod with one unit base vector being tangent to the rod and with another base vector being parallel to the joint axis.

For an unambiguous description of relative motion in joint a it is necessary to specify which of the two bodies serves as reference body. Let $i^+(a)$ be the label of the chosen reference body and let $i^-(a)$ be the label of the other body. Thus, i^+ and i^- are the names of two integer functions with integer argument. In the following discussion of Fig.1 we will assume that $i^+(a)=i$ and $i^-(a)=j$.

Fig.2 depicts a prismatic joint or rather a particular mathematical model of a prismatic joint. In reality a prismatic joint connecting two rods is built such that the rod axes are made to coincide at two or more points by bearings which allow only relative translational motion. In the mathematical model shown in Fig.2 a single bearing is located at a point fixed on the rod of body $j=i^-(a)$. This point will be referred to as the articulation point of the joint. The rod of body i is free to move through the articulation point but the bearing keeps the two rods tangent to one another at this point. The articulation point is the origin of two reference frames $\underline{\vec{e}}^{(ia)}$ and $\underline{\vec{e}}^{(ja)}$ which always coincide and which have one unit base vector tangent to the rods.

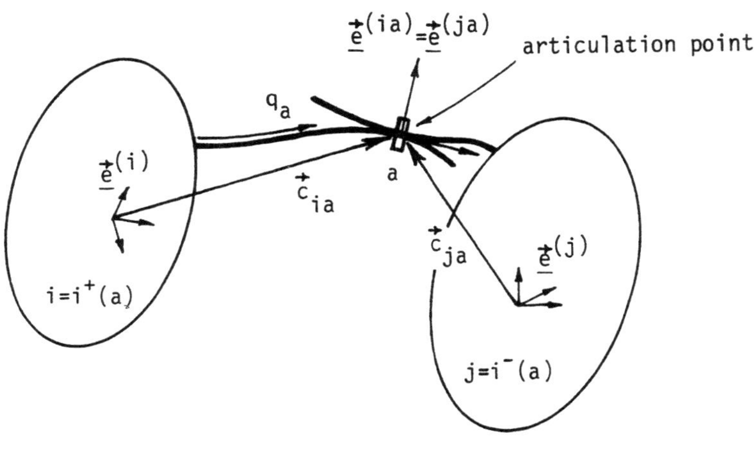

Fig.2: A prismatic joint connecting two deformable rods

It is now possible to define the joint variables denoted $\underline{q}$ in Eqs.(8) through (12). For the revolute joint of Fig.1 a single variable q_a exists. It is the rotation angle of $\vec{\underline{e}}^{(ja)}$ relative to $\vec{\underline{e}}^{(ia)}$. Also for the prismatic joint in Fig.2 a single joint variable q_a exists, namely the arc length along the rod on body i which defines the position of the articulation point.

The description of the motion of body j relative to body i requires the following quantities:

- the location $\vec{c}_{ia}$, the velocity $\vec{v}_{ia}$ and the acceleration $\vec{a}_{ia}$ of the articulation point relative to $\vec{\underline{e}}^{(i)}$,
- the location $\vec{c}_{ja}$, the velocity $\vec{v}_{ja}$ and the acceleration $\vec{a}_{ja}$ of the articulation point relative to $\vec{\underline{e}}^{(j)}$,
- the angular orientation $\underline{G}_{ia}$, the angular velocity $\vec{\Omega}_{ia}$ and the angular acceleration $\vec{\varepsilon}_{ia}$ of $\vec{\underline{e}}^{(ja)}$ relative to $\vec{\underline{e}}^{(i)}$,
- the angular orientation $\underline{G}_{ja}$, the angular velocity $\vec{\Omega}_{ja}$ and the angular acceleration $\vec{\varepsilon}_{ja}$ of $\vec{\underline{e}}^{(ja)}$ relative to $\vec{\underline{e}}^{(j)}$.

For Fig.1 $\vec{c}_{ia}$ is a known function of the deformation variables $\underline{u}_i$ while $\vec{c}_{ja}$ is a known function of $\underline{u}_j$. To be precise these functions are known in the respective reference frames $\vec{\underline{e}}^{(i)}$ and $\vec{\underline{e}}^{(j)}$. Differentiation yields the velocities and accelerations:

$$\begin{aligned} \vec{c}_{ia} &= \vec{c}_{ia}(\underline{u}_i) & \vec{c}_{ja} &= \vec{c}_{ja}(\underline{u}_j) \\ \vec{v}_{ia} &= \vec{\underline{k}}^T_{ia}\dot{\underline{u}}_i & \vec{v}_{ja} &= \vec{\underline{k}}^T_{ja}\dot{\underline{u}}_j \\ \vec{a}_{ia} &= \vec{\underline{k}}^T_{ia}\ddot{\underline{u}}_i+\vec{s}_{ia} & \vec{a}_{ja} &= \vec{\underline{k}}^T_{ja}\ddot{\underline{u}}_j+\vec{s}_{ja}. \end{aligned} \tag{13}$$

In these expressions $\vec{\underline{k}}_{ia}$ and $\vec{\underline{k}}_{ja}$ are column matrices of known vectors.

For the revolute joint on Fig.2 the relationships are more complicated. For one thing, $\vec{c}_{ia}$ depends explicitly on q_a. Secondly, and this is a more serious problem, $\vec{c}_{ia}$ depends on a subset of the variables $\underline{u}_i$ which is monitored by the value of q_a. Eqs.(13) will therefore take the form

$$\left.\vec{c}_{ia} = \vec{c}_{ia}(q_a,\underline{u}_i) \qquad \vec{c}_{ja} = \vec{c}_{ja}(\underline{u}_j)\right\} \tag{14}$$

$$\vec{v}_{ia} = \vec{k}_a\dot{q}_a + \underline{\vec{k}}_{ia}^T\underline{\dot{u}}_i \qquad \vec{v}_{ja} = \underline{\vec{k}}_{ja}^T\underline{\dot{u}}_j$$

$$\vec{a}_{ia} = \vec{k}_a\ddot{q}_a + \underline{\vec{k}}_{ia}^T\underline{\ddot{u}}_i + \vec{s}_{ia} \qquad \vec{a}_{ja} = \underline{\vec{k}}_{ja}^T\underline{\ddot{u}}_j + \vec{s}_{ja}.$$

Similar observations are true for the remaining kinematical quantities:

$$\underline{G}_{ia} = \underline{G}_{ia}(q_a,\underline{u}_i) \qquad \underline{G}_{ja} = \underline{G}_{ja}(\underline{u}_j)$$

$$\vec{\Omega}_{ia} = \vec{p}_a\dot{q}_a + \underline{\vec{p}}_{ia}^T\underline{\dot{u}}_i \qquad \vec{\Omega}_{ja} = \underline{\vec{p}}_{ja}^T\underline{\dot{u}}_j \tag{15}$$

$$\vec{\varepsilon}_{ia} = \vec{p}_a\ddot{q}_a + \underline{\vec{p}}_{ia}^T\underline{\ddot{u}}_i + \vec{w}_{ia} \qquad \vec{\varepsilon}_{ja} = \underline{\vec{p}}_{ja}^T\underline{\ddot{u}}_j + \vec{w}_{ja}$$

For the prismatic joint q_a monitors the subset of $\underline{u}_i$ on which the direction cosine matrix $\underline{G}_{ia}$ depends from which $\vec{\Omega}_{ia}$ and $\vec{\varepsilon}_{ia}$ are determined.

The previous equations are greatly simplified in the special case of rigid bodies. All quantities $\underline{u}_i$, $\underline{\dot{u}}_i$, $\underline{u}_j$ and $\underline{\dot{u}}_j$ are then identically zero. The remaining formulas which involve only q_a are those given in refs. [10] and [11].

3.3. Motions Relative to Inertial Space

Fig.3 depicts a system of bodies labeled i=0...n (n=5) and with joints labeled a=1...n. Revolute and prismatic joints as well as the pertinent elastic rods are indicated by dashed lines.

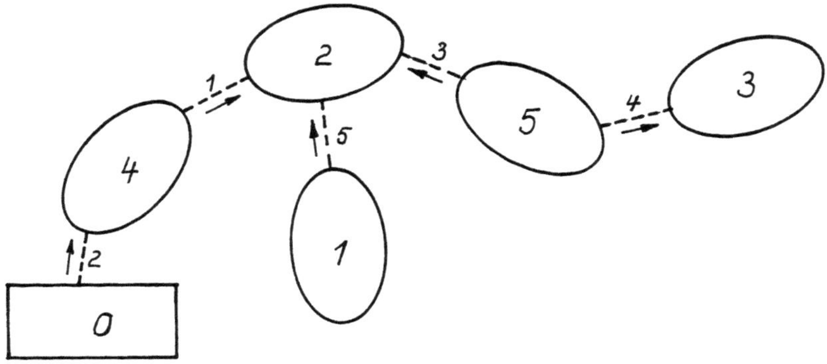

Fig.3: A system of n=5 interconnected deformable bodies

Arrows associated with the joints a=1...n indicate which of the two contiguous bodies is chosen as reference body $i^+(a)$ and which one as moving body $i^-(a)$. Each arrow points from body $i^+(a)$ toward body $i^-(a)$. The labeling as well as the choice of arrow directions is completely arbitrary except for the following. The motion of body 0 in inertial space is prescribed as function of time. This body is a moving platform or carrier vehicle on which the system of bodies 1...n is mounted. The system is supposed to have tree structure, i.e. no closed kinematic chains. We define the (n×n) incidence matrix $\underline{S}$ and the (n×n) path matrix $\underline{T}$ by their elements

$$S_{ia} = \begin{cases} +1 & \text{if } i=i^+(a) \\ -1 & \text{if } i=i^-(a) \\ 0 & \text{else} \end{cases} \tag{16}$$

$$T_{ai} = \begin{cases} +1 \ (-1) & \text{if joint a is located along the direct path from body 0 to body i and if the arrow of joint a is pointing toward body i (toward body 0)} \\ 0 & \text{else.} \end{cases} \tag{17}$$

In terms of $\underline{T}$ the absolute angular velocity of body k (k=1...n) is

$$\vec{\omega}_k = \vec{\omega}_o(t) - \sum_{a=1}^{n} T_{ak}(\vec{\Omega}_{i^+(a),a} - \vec{\Omega}_{i^-(a),a}). \tag{18}$$

It follows that the column matrix $\underline{\vec{\omega}}$ of all n angular velocities is

$$\underline{\vec{\omega}} = \vec{\omega}_o \underline{1}_n - \underline{T}^T (\underline{\vec{\Omega}}^+ - \underline{\vec{\Omega}}^-) \tag{19}$$

with $\underline{1}_n = \begin{bmatrix} 1 & 1 & \dots & 1 \end{bmatrix}^T$ and with obvious definitions of $\underline{\vec{\Omega}}^+$ and $\underline{\vec{\Omega}}^-$. According to Eq.(15) we can write

$$\underline{\vec{\Omega}}^+ = \underline{\vec{p}}^T \underline{\dot{q}} + \underline{\vec{p}}^{+T} \underline{\dot{u}}, \qquad \underline{\vec{\Omega}}^- = \underline{\vec{p}}^{-T} \underline{\dot{u}}. \tag{20}$$

In these expressions $\underline{\dot{q}}$ is the column matrix of all joint variable derivatives, while $\underline{\dot{u}} = \begin{bmatrix} \underline{\dot{u}}_1^T & \dots & \underline{\dot{u}}_n^T \end{bmatrix}^T$ combines the deformation velocity variables of all bodies. The matrices $\underline{\vec{p}}$, $\underline{\vec{p}}^+$ and $\underline{\vec{p}}^-$ contain the associated vectors shown in Eq.(15). Eqs.(19) and (20) yield

$$\underline{\vec{\omega}} = -\underline{T}^T \left[\underline{\vec{p}}^T \underline{\dot{q}} + (\underline{\vec{p}}^+ - \underline{\vec{p}}^-)^T \underline{\dot{u}} \right] + \vec{\omega}_o \underline{1}_n. \tag{21}$$

Comparison with Eq.(8) shows that

$$\underline{\vec{\beta}}_1 = -(\underline{\vec{p}}\,\underline{T})^T , \qquad \underline{\vec{\beta}}_2 = -\left[(\underline{\vec{p}}^+ - \underline{\vec{p}}^-)\underline{T}\right]^T . \tag{22}$$

These are the first two of the matrices which were anticipated in Eq.(8) and which immediately lead to the equations of motion (12). The matrix $\underline{\vec{w}}$ defined by Eq.(8) is found in a straightforward way by differentiating Eq.(20).

The equivalent to Eq.(18) for the radius vector $\vec{r}_k$ reads

$$\vec{r}_k = \vec{r}_o(t) - \sum_{a=1}^{n} T_{ak}(\vec{c}_{i^+(a),a} - \vec{c}_{i^-(a),a}) . \tag{23}$$

This yields

$$\dot{\vec{r}}_k = \dot{\vec{r}}_o(t) - \sum_{a=1}^{n} T_{ak}\Big[(\vec{v}_{i^+(a),a} - \vec{c}_{i^+(a),a}\times\vec{\omega}_{i^+(a)} - \\ -\vec{v}_{i^-(a),a} + \vec{c}_{i^-(a),a}\times\vec{\omega}_{i^-(a)}\Big] . \tag{24}$$

The equivalent to Eq.(19) for $\dot{\underline{\vec{r}}}$ reads

$$\dot{\underline{\vec{r}}} = \dot{\vec{r}}_o \underline{1}_n - \underline{T}^T(\underline{\vec{v}}^+ - \underline{\vec{v}}^- - \underline{\vec{C}}^T\times\underline{\vec{\omega}}) \tag{25}$$

where $\underline{\vec{C}}$ is an (n×n) matrix with the elements

$$\vec{C}_{ia} = S_{ia}\vec{c}_{ia} \qquad i,a=1\ldots n. \tag{26}$$

According to Eq.(14) we have the equivalent to Eq.(20)

$$\underline{\vec{v}}^+ = \underline{\vec{k}}^T\dot{\underline{q}} + \underline{\vec{k}}^{+T}\dot{\underline{u}}, \qquad \underline{\vec{v}}^- = \underline{\vec{k}}^{-T}\dot{\underline{u}}. \tag{27}$$

This together with Eqs.(21) and (25) yields

$$\dot{\underline{\vec{r}}} = (\underline{\vec{p}}\times\underline{T}\,\underline{\vec{C}}\,\underline{T} - \underline{\vec{k}}\,\underline{T})^T\dot{\underline{q}} + \left[(\underline{\vec{p}}^+ - \underline{\vec{p}}^-)\times\underline{T}\,\underline{\vec{C}}\,\underline{T} - (\underline{\vec{k}}^+ - \underline{\vec{k}}^-)\underline{T}\right]^T\dot{\underline{u}} + \\ + \left[\vec{\omega}_o\times(\underline{\vec{C}}\,\underline{T})^T + \dot{\vec{r}}_o\right]\underline{1}_n . \tag{28}$$

Comparison with Eq.(8) shows that

$$\begin{aligned} \underline{\vec{\alpha}}_1 &= (\underline{\vec{p}}\times\underline{T}\,\underline{\vec{C}}\,\underline{T} - \underline{\vec{k}}\,\underline{T})^T \\ \underline{\vec{\alpha}}_2 &= \left[(\underline{\vec{p}}^+ - \underline{\vec{p}}^-)\times\underline{T}\,\underline{\vec{C}}\,\underline{T} - (\underline{\vec{k}}^+ - \underline{\vec{k}}^-)\underline{T}\right]^T . \end{aligned} \tag{29}$$

These are two more matrices for the construction of the equations of motion (12). The matrix $\underline{\vec{s}}$ defined by Eq.(8) is found by differentiating Eq.(28) once more. As was to be expected it is seen that the matrices $\underline{\vec{\alpha}}_1$ and $\underline{\vec{\beta}}_1$ are identical with those for systems of rigid bodies.

Conclusion

A formalism has been outlined which generalizes an existing theory for articulated-rigid-body systems to include effects of small elastic deformations. The formalism results in a minimal set of differential equations which are nonlinear with respect to joint variables and linear with respect to deformation variables. Particular problems are pointed out which occur when elastic rods are coupled by prismatic joints.

References

1. Likins, P.W.; Wirsching, P.H.: The Use of Synthetic Modes in Hybrid Coordinate Dynamic Analysis.AIAA J.6(1968)1867-73
2. Likins, P.W.: Finite Element Appendage Equations for Hybrid Coordinate Dynamic Analysis. Int.J.Solids Structures 8(1972)
3. Likins, P.W.: Analytical Dynamics and Nonrigid Spacecraft Simulation. JPL Techn.Rep.32-1593,1974
4. Hooker, W.W.: Equations of Motion for Interconnected Rigid and Elastic Bodies: A Derivation Independent of Angular Momentum. Celestial Mechanics 1974
5. Frisch, H.P.: A Vector-Dyadic Development of the Equations of Motion for N-Coupled Flexible Bodies and Point Masses. NASA TN D-8047,1975
6. Truckenbrodt, A.:Motion Behavior and Control of Hybrid Multibody Systems With Applications to Industrial Robots (in German). Fortschrittber. VDI Zeitschr.,s.8,33(1980)
7. Barraco, A.: Dynamique des systèmes mécaniques déformables. Diss.Univ.Paris VI,1983
8. Lilov, L.: Equations of Motion for an Elastic Body (in Russ.) Dokl.Bulgarian Aca.Scie.v.36,10(1983)
9. Wielenga, T.J.: Simplifications in the Simulation of Mechanisms Containing Flexible Members.Phd Thes.Univ.Michigan 1984
10. Wittenburg, J.: Analytical Methods in Mechanical System Dynamics. appeared in: Computer Aided Analysis and Optimization of Mechanical Systems (ed. Haug, E.J.). Springer 1984
11. Wittenburg, J.; Wolz, U.: MESA VERDE: A Symbolic Program for Nonlinear Articulated-Rigid-Body Dynamics. Proc.ASME Conf. Mechanical Vibration and Noise, Cincinnati 1985

Symbolic Generation of Efficient Simulation/ Control Routines for Multibody Systems

PAUL E. NIELAN

Stanford University
and
Sandia National Laboratories
Livermore, California 94550
USA

THOMAS R. KANE

Stanford University
Stanford, California 94305
USA

Summary

Techniques for improving the efficiency of simulation subroutines generated with the aid of computer algebra are presented in the form of MACSYMA utility functions.

Introduction

Dynamical equations governing the behavior of devices which can be modeled as interconnected systems of rigid bodies play an important role in the design and operation of such devices. In design, dynamical equations are employed to carry out simulations for the purpose of predicting the performance of a device. For example, the response of a device to various loading conditions, or the relative merits of competing control schemes, can be studied by means of simulations. As regards robot operations, dynamical equations are frequently used to compute the forces and torques needed to drive members of a manipulator in such a way as to achieve desired end-effector motions, a task that must be performed repeatedly and, in most cases, rapidly. Consequently, it is important to construct the most efficient possible computational algorithms.

During the past 15 to 20 years, it has become a widespread practice to employ general-purpose computer programs to perform dynamical simulations, thus freeing the user from the burden of dynamical analysis. Some of these programs, hereafter called numerical multibody programs, are constructed by deriving once and for all, the equations of motion (dynamical and kinematical) of a very general system of interconnected bodies. In such derivations, the number of bodies in the system, as well as

Dynamics of Multibody Systems
IUTAM/IFToMM Symposium Udine 1985
Editors: G. Bianchi and W. Schiehlen

the type of connection between adjacent bodies are treated as parameters; this leads to iteration loops and branching statements in the resulting program. While use of such programs leads to a savings in analysts' time, it tends to entail computational inefficiencies that can become so severe as to preclude satisfactory operation of the device being modeled. This difficulty can be alleviated by formulating explicit equations of motion manually. Kane and Levinson [6], illustrating a procedure for doing this, obtain explicit equations of motion for the six degree of freedom manipulator known as the Stanford Arm. In the course of forming the equations, they introduce new symbols to replace recurring combinations of old ones. This procedure eliminates redundant computations and keeps the size of intermediate expressions from becoming unmanageable, at the cost of maintaining a list of symbol definitions. The result is a computational algorithm that requires fewer operations than the best available general-purpose Lagrangian and Newton-Euler approaches, a property that is clearly desirable. The drawback of this approach is the amount of analyst's effort required to derive equations of motion and produce a correct computer program. Hence, the question that arises is this: Is it possible to employ a computer to produce explicit equations of motion, or more usefully, such equations in the form of a FORTRAN/PASCAL subroutine, that exhibit at least the same level of computational efficiency as the subroutine generated manually?

In recent years, a number of symbolic multibody programs that perform the task in question, have been created. [7, 9, 11]. However, in much of the literature describing these programs, the computational efficiency of the subroutines produced is not discussed and, in indeed, does not appear to be a consideration. This is surprising in view of the fact that computational efficiency is often advanced as the primary reason for preferring symbolic multibody programs to their numerical counterparts. For example, Levinson [8] used FORMAC to produce a simulation subroutine containing the following code fragment:

```
X(4,6) = MR*(-CQ*R(3)-SQ*R(2))
X(4,7) = MR*(CQ*R(2)-SQ*R(3))
X(5,1) = 0
X(5,2) = MR*(-B(3)*CQ*R(2)+SQ*R(2))
```

which is clearly less efficient than it might be; and the pioneering program NEWEUL of Kreuzer and Schiehlen [7] produced the following code:

```
RM(3,3)=-C**2*M2**2*RMGI+2.*C*M2**2*R*RMGI+2.*B*C
       *M2**2*RMGI*COS(TET)-M2**2*R**2*RMGI-2.*B
       *M2**2*R*RMGI*COS(TET)-B**2*M2**2*RMGI+
       C**2*M2-2.*C*M2*R-2.*B*C*M2*COS(TET)+M2
       *R**2+2.*B*M2*R*COS(TET)+B**2*M2+IZ
RM(3,4)=B*M2**2*RMGI*SIN(TET)-B*M2*SIN(TET)
```

which contains redundant computations a human programmer would undoubtedly eliminate.

Occasionally, an apparently innocent symbolic algorithm can result in dynamic subroutines that are more expensive to use than a well-coded numerical multibody program. For example, Hussain and Noble [5] incorrectly report that

> " MACSYMA can simplify the Kane-Levinson end result, reducing the number of arithmetic operations required to compute numerical results",

and also say,

> " because Kane and Levinson are carrying out the algebra by hand, it is convenient for them to introduce intermediate symbols z_1, z_2, ...z_{196}, and similarly 36 X's and 31 W's. MACSYMA has no difficulty in generating the end result in explicit form. These end results are no more complex than the complexity of the equations in Kane-Levinson."

However, when one runs the actual MACSYMA program given in Appendix VIII of [5], one finds that using the equations produced to calculate the terms in the dynamical equations would require 5320 multiplications and 1837 additions. In contrast, these numbers are reduced to 1541 and 1190 when the general purpose Newton-Euler subroutine reported by Walker and Orin [13] is used, and to 646 and 394 by Kane and Levinson (all operation counts are summarized in the appendix). In drawing their conclusions, Hussain and Noble appear to have considered only the operations required to compute the mass matrix, for which their program requires 502 multiplications and 272 additions, but seem to have ignored the number of operations required to construct the inertial forces due to gyroscopic effects. This large number of arithmetic operations arises in part because Hussain and Noble deem the introduction of intermediate symbols unnecessary and so produce code such as:

```
X(1,1) = -(((mf+me+md+mc)*cos(q2)**2-mf-me-md-mc)*q6**2
 +(((2*13*mf+2*16*me)*cos(q2)**2-2*13*mf-2*16*me)
 + ...
```

A description [11] of the more recent symbolic multibody program of Rosenthal and Sherman indicates a definite concern with computational efficiency, but only compares computation times using a subroutine produced by their symbolic multibody program to computation times using an existing numerical multibody program [3]. Clearly, the relative merits of different symbolic multibody programs could be more accurately judged if more detailed information regarding the code they produce were available. We have augmented our symbolic program with the capability to automatically keep track of and report the number of operations required by each section of subroutine code.

<u>Concepts leading to generation of efficient simulation subroutines</u>

In this section, a method for using the computer algebra system MACSYMA to generate efficient simulation subroutines is explained. While, at present, MACSYMA requires too much computational overhead to be used as the implementation language for a "production" symbolic multibody program, it is a useful workbench for the prototyping of algorithms and procedures that can be used in a production code. Additionally, MACSYMA is useful for dealing with features of dynamics problems not easily handled by production codes. The method introduce techniques which reduce the occurrence of intermediate expression swell*, a problem which has severely limited the size of the systems that could be analyzed with the aid of general-purpose computer algebra programs like MACSYMA.

Our objective was to augment MACSYMA with a series of utilities that would permit the analysis of, at least, moderately sized (5-15 bodies) dynamic systems via the general method of attack illustrated in the Kane-Levinson paper, but would relieve the analyst of algebraic drudgery. Hussain and Noble attempted the same task, but produced unwieldy and inefficient equations, primarily because, as mentioned earlier, they did not deem it necessary to implement one of the central features of the Kane-Levinson approach: the systematic introduction of intermediate variables. The MACSYMA utilities to be described automatically introduce intermediate variables in a fashion that mimics the behavior of a human analyst.

In order to facilitate the presentation, a short summary of relevant portions of MACSYMA will now be given. A complete description can be found in [10].

The MACSYMA assignment operator is denoted by ":". For example, the statement a : x + y assigns the value x + y to the MACSYMA symbol a. The additional assignment b : z, followed by the statement c : a + b results in the value x + y + z being assigned to the symbol c. MACSYMA provides many functions which can be used to manipulate expressions of the type above. RATCOEFF(E1,V1,1) returns the coefficient of symbol V1 in the expression E1. For example, w : RATCOEFF(a*b*u1 + c*d, u1,1) results in w being assigned the value a*b. DIFF(E1, V1) returns the derivative of E1 with respect to V1, thus w : DIFF(a*b*u1 + c*d,d) assigns the value c to w. GRADEF(S1,S2,E1) adds the fact that expression E1 is the derivative of symbol S1 with respect to S2 to the internal database of derivatives that DIFF accesses. Since we will only be interested in time

*Intermediate expression swell refers to the phenomena of intermediate computations producing expressions so large that their storage requirements exceed the capacities of the computer, even though the system being analyzed may be composed of only a modest number of elements. This problem is exacerbated by the fact that both the memory and processor time used by many important algebraic functions tend to grow exponentially while the size of the expressions operated on grows only linearly [2].

derivatives, we will consider this database to be a list containing entries of the form S1' = E1, hereafter called the derivative list. MACSYMA's powerful substitution capability can be accessed by using the command RATSUBST. RATSUBST(E1, E2, E3) substitutes E1 for E2 in E3. It knows something about the meaning of expressions, so that, for example, RATSUBST (xx1,a+b+c+d,a+b+c) returns xx1-d. The user can also define his own functions in terms of those predefined in MACSYMA. A vector can be represented in MACSYMA by a list of three measure numbers; thus the vector v = (q1+q2)$\underline{b}_1$ + h$\underline{b}_2$ + q3$\underline{b}_3$ could be represented as v : [q1+q2, h, q3]. These representations can be manipulated as entities, so that w : v + v results in w being assigned the value [2*q1 + 2*q2, 2*h, 2*q3] and the dot product u : v.v results in u being assigned the value (q1+q2)**2 + h**2 + q3**2.

Our MACSYMA function LAYER(E1,S) reduces the size of an algebraic expression by introducing new symbols to replace combinations of existing symbols occurring within it and simultaneously recording the definitions of these new symbols in a list (hereafter called the symbol list) for future reference. E1 is an expression to be layered and S is a set of "sacred" symbols, which are not allowed to be present in any combination of symbols that is to be replaced. For example, if U denotes the set of symbols representing generalized speeds, ([u1, u2, u3, u4, u5, u6] for the Stanford Arm), and w3 (which may be, for example, an angular velocity measure number) has the value

u3**c2*xx19 + u1*s2*xx18 (1)

then LAYER(w3,U) returns

u3*xx20 + u1*xx21 (2)

and adds the definitions xx20=c2*xx19 and xx21=s2*xx18 to the symbol list.

LAYER automatically creates new symbols as needed by appending numbers to the two-letter prefixes xx and yy. The yy preface is used when the symbol is to replace an expression containing generalized speeds and/or other yy symbols; the xx preface is used otherwise.

If E is the empty set, then LAYER(w3,E) returns yy8 and the definition yy8=u3*c2*xx19 + u1*s2*xx18 is added to the symbol list. The assignment w3 : LAYER(w3,S) caused the layered expression to replace the original expression as the value of w3 in memory; we refer to this as replacement layering.

Layering allows MACSYMA to easily form two different expressions for central and angular velocities of bodies, one involving the generalized speeds implicitly, the other explicitly, a facility that is important in automating the Kane/Levinson approach.

Frequently, quantities formed at an early stage of an analysis must later be differentiated with respect to time (e.g., angular velocities are differentiated to form angular accelerations.) When these quantities are layered, information must be recorded about the derivatives of the new symbols introduced, so that derivatives of the expressions containing them can be computed correctly. We proceed by computing the derivative of a new symbol at the time it is defined (assume that the derivatives of all existing symbols are known and stored in the derivative list), layering the derivative itself, and adding the result to the derivative list. Our MACSYMA function DLAYER accomplishes this. For example, if w3 has the value of the expression in (1), and the derivative list already contains the information {...,c2' = -s2*u2, s2' = c2*u2, ..., xx18' = yy12, xx19' = yy13, ...}, DLAYER(w3,U) proceeds by introducing the symbol xx20 as before, computing its derivative as d/dt(c2*xx19) = -s2*u2*xx19 + c2*yy13, and adding the definition xx20' = LAYER(-s2*u2*xx19 + c2*yy13,E) to the derivative list. Then, the symbol xx21 is introduced as before, the derivative of xx21 is computed as d/dt(x2*xx18) = x2*u2*xx18+x2*yy12, and the definition xx21' = LAYER(c2*u2*xx18+s2*yy12,E) is added to the derivative list. Finally, the expression u3*xx20 + u1*xx21 is returned. The symbol list and derivative list are augmented by the following additional entries: {xx20 = c2*xx19, xx21 = s2*xx18, yy25 = -s2*u2*xx19 + c2*yy13, yy26 = x2*u2*xx18 + s2*yy13} and {xx20' = yy25, xx21' = yy26}, respectively.

If w3 is assigned the value of u3*xx20 + u1*xx21, computed as just described, the derivative of w3 with respect to time can be obtained as DIFF(w3, t), which returns u3*yy25 + u3'*xx20 + u1*yy26 + u1'*xx21, whereas if w3 has not been layered and had the value of the expression in (1), then evaluating DIFF(w3, t) would have returned u3'*c2*xx19 - u3*s2*u2*xx19 + u3*c2*yy13 + u1'*s2*xx18 + u1*c2*u2*xx18 + u1*s2*yy12, which is significantly more complicated.

Oftentimes, an expression to be layered contains subexpressions that have already been assigned symbols during the layering of previous expressions. In this event, it is advantageous to use the existing symbols to replace the subexpressions, rather than to create new ones. This will allow MACSYMA to recognize possible cancellations in later combinations of layered expressions and to avoid repetitious calculations in the simulation code being generated. For example, if w3 has the value of the expression given in (1), and LAYER(w3,U) is evaluated twice in succession, we would like it to return u3*xx20 + u1*xx21 both times, rather than u3*xx20 + u1*xx21, followed by u3*xx22 + u1*xx23. To accomplish this, LAYER and DLAYER are designed so that when they encounter a subexpression to be replaced by a symbol, they check to see if it is equal to the right hand side of any of the definitions in the symbol list. If so, they use the symbol on the left hand side of the appropriate definition to replace the subexpression, rather than generating a new symbol. An analyst does the same thing when following the procedure of Kane and Levinson by hand and, in fact, sometimes goes one step further, when realizing that a subexpression to be replaced, while not being exactly equal to an existing right hand side in the symbol list, may be related to it. For example, suppose w3, as given in (1), has already been layered and we now wish to layer a quantity w4, which has the value u3*c2*xx19*xx3 + u1*s3*xx18. We would like LAYER(w4,U) to return u3*xx22 +

ul*xx21, with the definition xx22 = c2*xx19*xx3. We call this type of layering "condensation", and implement it by adding LAYER (and DLAYER) an enhanced version of the following algorithm:

Define a function EXPENSE(E) which returns a measure of the computation time required to evaluate E numerically. This measure takes in to account the number of multiplications, additions, etc., required to compute a desired quantity, as well as the relative costs of these operations. The user can modify the definition to reflect the operational characteristics of the computer to be used. Now let el designate a subexpression to be replaced, let $S_i = D_i$ denote a generic entry in the symbol list and search through the symbol list as follows.

```
lc : EXPENSE(el);
best : el;
FOR i : 1 thru  number of entries in symbol list  DO
    ( new : RATSUBST(Si, Di, el);
      cost : EXPENSE(new);
      IF cost   lc + {cost of a FORTRAN assignment} THEN
        ( lc : cost;
          best : new;
        )
    )
```

The built-in MACSYMA function RATSUBST is used to determine whether or not the expressions D_i and el are related. After following this procedure, LAYER then uses the value of best as the right hand side of the new symbol definition.

While the heuristic algorithm* above will catch many repeated subexpressions, it does not fully mimic the behavior of the human analyst because it only determines whether or not the subexpression to be replaced is related to any ONE of the right hand sides of the definitions in the symbol list. It does not try to determine whether or not the subexpression is related to a combination of right hand sides, because searching through all the possible combinations would be prohibitively expensive.

*More involved heuristic algorithms for condensation have been put forth by Breuer [1] and van Hulzen [12], and have been applied to automatic code generation of finite element programs by Wang [15]. This problem is of special interest to computer scientists constructing optimizing compilers, where it is known as common subexpression elimination.

In fact, even the simple search used in the program fragment above can be extremely time consuming when the length of the symbol list is large. Although better techniques based on hash coding could be used to reduce the search time, it turns out that most expressions to be layered are related to the most recent additions to the symbol list. Because of this, the search cost can be greatly reduced by "emptying" the symbol list at strategic intervals, after first printing out each definition in the form of a FORTRAN assignment statement. Together, these assignment statements form the body of the simulation subroutine.

A MACSYMA program using the above utilities has been developed and used to generate a simulation subroutine for the Stanford Arm; unfortunately, both the program and the subroutine are too long to be listed in this paper. Some parts of the utility functions are implemented in LISP, the base language underlying MACSYMA. Running the MACSYMA program requires about 40 minutes of VAX 11/780 CPU time. The simulation subroutine produced requires 772 multiplications and 456 additions to calculate the terms in the equations of motion. While the subroutine requires more operations than does the code of Kane and Levinson, it is much more efficient than that of Hussain and Noble. SD/EXACT [11] generated a simulation subroutine of equivalent generality requiring only 632 multiplications and 400 additions. It require approximately 2 minutes of VAX 11/780 CPU time to produce the subroutine. SD/EXACT requires less operations than our program because the algorithm employed takes better advantage of the chain topology of the Stanford Arm. The CPU costs are less because it is written in Pascal and avoids the computational overhead associated with MACSYMA.

The authors would be interested in seeing simulation subroutines for this problem, as described in [6], produced by other symbolic multibody programs.

Closure

Techniques for improving the efficiency of simulation subroutines generated by symbolic multibody programs have been presented in the form of MACSYMA functions. While application of these techniques can result in substantial reductions in computational expense, recent papers by Horak [4] and Wampler [14] indicate there is still substantial room for improvement.

For example, following a recursive/iterative algorithm applicable only to chain systems and based on Kane's equations, Wampler manually produced a subroutine that requires only 468 multiplications and 332 additions to calculate the terms in the equations of motion for the Stanford Arm. A symbolic multibody program that automatically generates subroutines of comparable efficiency will need to employ condensation techniques that are more effective than the one shown here, as well as dynamical algorithms that take maximum advantage of the structure of the system at hand. Realization of such a program is a goal of our current research.

Acknowledgements

This work was partially supported by the U.S. Department of Energy under contract DE-AC04-76DP00789.

Appendix

This appendix summarizes the operation counts given in the body of the paper.

Program Authors	Multiplications	Additions
Hussain and Noble [5]	5320	1837
Walker and Orin [13]	1541	1190
Kane and Nielan	772	456
Kane and Levinsón [6]	646	394
Rosenthal and Sherman [11]	632	400
Wampler [14]	468	332

References

1. Breuer, M. A., Generation of Optimal Code for Expressions via Factorization. Communications of the ACM, Vol. 12, No. 6, June 1969.

2. Cook, G. O., Jr., Development of a Magnetohydrodynamic Code Code for Axisymmetric, High-B Plasmas with Complex Magnetic Fields. Ph.D. Thesis, Brigham Young University, December 1982.

3. Fleischer, G. E., and Likins, P. W., Attitude Dynamics Simulation Subroutines for Systems of Hinge-Connected Rigid Bodies. Technical Report 32-1592, Jet Propulsion Laboratory, Pasadena, CA, 1974.

4. Horak, D. T., A fast computational scheme for dynamic control of manipulators. Proc. of the ACC, San Diego, July 1984.

5. Hussain, M. A. and Noble, B., Application of MACSYMA to Kinematics and Mechanical Systems, in Proc. of the 1984 MACSYMA Users' Conference, Schenectady, New York, July 1984.

6. Kane, T. R. and Levinson, D. A., The Use of Kane's Dynamical Equations in Robotics, International Journal of Robotics Research, Vol. 2, 1983, pp. 3-21.

7. Kreuzer, E. J., and Schiehlen, W. O., Generation of Symbolic Equations of Motion for Complex Spacecraft using Formalism NEWEUL. Advances in the Astronautical Sciences, V. 54, 1983.

8. Levinson, D. A., Equation of Motion for Multiple-Rigid-Body Systems via Symbolic Manipulation. Journal of Spacecraft and Rockets, Vol. 14, No. 8, August 1977.

9. Macala, G. A., SYMBOD: A Computer Program for the Automatic Generation of Symbolic Equations of Motion for Systems of Hinge-Connected Rigid Bodies. AIAA Paper No. 83-0013, American Inst. Aeron. Astron. (AIAA), New York, 1983.

10. MACSYMA Reference Manual. M.I.T. Laboratory for Computer Science, Version 10, December 1983.

11. Rosenthal, D. E. and Sherman, M. A., Symbolic Multibody Equations via Kane's Method. AAS/AIAA Astrodynamics Specialist Conference Paper 83-803, Lake Placid, NY 1983.

12. van Hulzen, J. A., Code Optimization of Multivariate Polynomial Schemes: A Pragmatic Approach. Proc. EUROCAL '83, Springer-Verlag LNCS No. 162, pp. 286-300, 1983.

13. Walker, M. W., and Orin D. E., Efficient Dynamic Computer Simulations of Robotic Mechanisms. Journal of Dynamics Systems, Measurement and Control, Vol. 104, pp. 205-211. Sept. 1982.

14. Wampler, C. W., Computer Methods in Manipulator Kinematics, Dynamics and Control: A Comparative Study. Doctoral Thesis, Stanford University, December, 1984.

15. Wang, P. S., Chang, T.Y.P., and van Hulzen, J. A., Code Generation and Optimization for Finite Element Analysis. EUROSAM 1984, London, England, July 9-11, 1984.

Identification of Models for Dynamic Systems

PAULI PEDERSEN

Department of Solid Mechanics
The Technical University of Denmark
Lyngby, Denmark

Abstract

To perform a dynamic analysis a model has to be chosen. The model is described by some model parameters and the selection of model and model parameters are by no means unique. The question then arise: which is model "best"? A mathematical formulation of this problem is given, and the similarities with problems of optimal design are pointed out.

From this it follows that the problem can be solved by a unified approach to optimal design. This approach solves the non-linear problem by a sequence of linear problems, each containing a dynamic analysis, a sensitivity analysis, and a linear programming solution.

Special attention is paid to the sensitivity analysis, and the general results for non-conservative (damped) systems are presented. Interpretations in terms of "mutual energies" are put forward, and hopefully this gives the presentation an engineering slant.

The application of the procedure for identification of a stiffness model for a shaft will be discussed.

1. Introduction

Engineering analyses are inherently connected with the model concept. The physical realities are so complex that we are forced to simplify, i.e. choose models, and this selection is by no means unique. The question then arise: what is a "good" model? or more extreme: which is the "best" model? The answer to this is the goal of the identification discussed in this paper.

Lecture for IUTAM/IFToMM Symposium on Dynamics of Multibody Systems, CISM, Udine, Sept. 1985.

Dynamics of Multibody Systems
IUTAM/IFToMM Symposium Udine 1985
Editors: G. Bianchi and W. Schiehlen

To evaluate a model we need a reference and although other possibilities are present (like alternative models), a proper justification must be based on laboratory experiments. Again we are confronted with the non-uniqueness. Experiments and experimental outputs have to be chosen and bounded.

Next, with chosen model parameters and experimental references we have to arrive at an order of priority. The model parameters are probably not of equal importance, and the experimental results do not have the same reliability and importance.

This all indicates that the mathematical identification should be carried through with a proper engineering background. Thus we assume an adaptive proces with feedback to experiments from the identification procedure and vice versa.

In the case of dynamic systems, the models may finally be described by a stiffness matrix and a mass matrix (possibly also a damping matrix). Identification directly from experiments to these matrices is possible, see BARUCH [1], but here we shall advocate more physically related model parameters prior to the discretization. The example in section 4 illustrates such an approach.

The reason for choosing this apparently roundabout way is a belief that the model will behave more resonably in relation to other situations than the experimental ones on which the identification is based.

Fig. 1.1 gives an overview of the suggested approach, consisting of the three levels: I , II and III. According to the arguments just presented, level I contains the adaptive selection of model parameters, of experimental measures, and of identification criterion. We may term this level a kind of "try and learn".

Levels II and III are identical to those of the unified approach to optimal design discussed in [2]. This means that the iden-

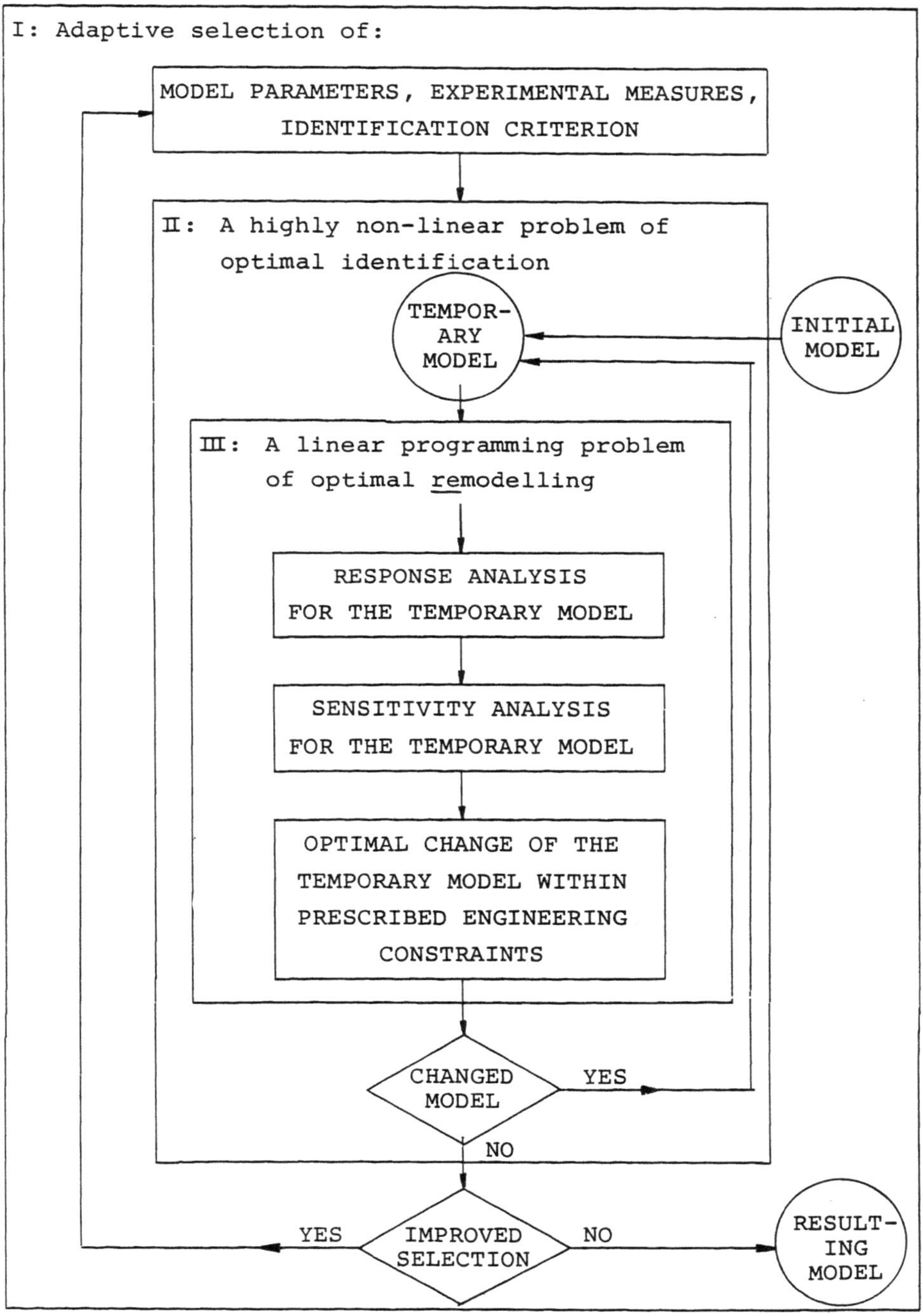

Fig. 1.1: Overview of the three-level identification approach.

tification problem is formulated as a problem of optimal modelling. Then this highly non-linear problem is solved by a sequence of remodelling problems, which, in terms, are formulated as linear programming problems.

In a recent M.Sc. thesis by Ingason [3] the procedure in fig. 1.1 is applied to the identification of a stiffness model for a shaft, and we shall finally discuss some of these results.

2. Optimal Identification by a Sequence of Optimal Remodelling

The problem of optimal remodelling is formulated as a linear programming problem with the goal that it can be solved directly by an available, well-documented and stable numerical method, i.e. the Simplex method.

As this formulation is based on linear expansions of criterion and constraints for a temporary model we must restrict the model changes so that the linear expansions are acceptable. This is done by so-called move-limits, which are of major importance because they are the key to controlling the entire optimal identification, i.e. the non-linear problem.

The usefulness of a model is measured by a scalar Φ , which is a functional of the n model parameters h_i , gathered in the vector $\{H\}$. We may then interpret the functional ϕ as an "error-functional" that quantifies the difference between the experimental response and the calculated response of the actual model. Thus our criterion is to

$$\text{Minimize} \quad \Phi = \Phi(\{H\}) \quad . \tag{2.1}$$

Now in the linearized problems of optimal remodelling from model p (i.e. $\{H\}_p$) to model $p+1$ (i.e. $\{H\}_{p+1}$) , the optimization variables are the changes $\{\Delta H\}_p$, defined by

$$\{H\}_{p+1} = \{H\}_p + \{\Delta H\}_p \quad , \tag{2.2}$$

and the criterion to be minimized is Φ_{p+1} . As Φ_p is given, our linearized criterion will be

$$\text{Minimize} \quad \Delta\Phi = \{\nabla\Phi\}^T\{\Delta H\} \ , \qquad (2.3)$$

omitting from now on the index of sequence p . The elements of the row vector $\{\nabla\Phi\}^T$ are $\partial\phi/\partial h_i$ for $i = 1,2,...n$.

The constraints of move-limits are expressed directly in relation to the change vector $\{\Delta H\}$ by the limitations

$$\{\Delta H\}_{min} \leq \{\Delta H\} \leq \{\Delta H\}_{max} \ , \qquad (2.4)$$

and by adjusting $\{\Delta H\}_{min}$, $\{\Delta H\}_{max}$ to the individual step p in the sequence of linearizations, we control the entire identification.

Additional constraints may be formulated as the functional inequalities

$$f_j = f_j(\{H\}) \leq 0 \quad , \quad j = 1,2,... \qquad (2.5)$$

and by the linearization we get

$$f_j + \{\nabla f_j\}^T\{\Delta H\} \leq 0 \quad , \quad j = 1,2,... \qquad (2.6)$$

or, for all the constraints gathered in $\{F\} \leq \{0\}$,

$$[\nabla F]\{\Delta H\} \leq \{F\} \ . \qquad (2.7)$$

In the gradient matrix $[\nabla F]$ the element $\partial f_j/\partial h_i$ expresses the partial derivative of the functional j with respect to model parameter i .

The problem defined by (2.3), (2.4) and (2.7) constitutes a linear programming problem, which can be solved by the Simplex method.

3. Analysis and Sensitivity Analysis

The term response analysis used in fig. 1.1 is a very wide one, which may include displacement, stress, vibration and stability analysis. In the present formulation we shall restrict ourselves to eigenvalue analysis.

The discretized formulation (say by the finite element method FEM for continuous problems) for the free, undamped structural eigenfrequencies is

$$\left([S] - \omega_r^2[M]\right)\{D_r\} = \{0\} \ , \tag{3.1}$$

where ω_r^2 and $\{D_r\}$ are, respectively, the squared eigenfrequency and corresponding eigenmode. The total stiffness matrix $[S]$ and the total mass matrix $[M]$ are both obtained by summation of the individual element matrices

$$[S] = \sum_e [S_e] \quad , \quad [M] = \sum_e [M_e] \ . \tag{3.2}$$

Today we usually solve the problem (3.1) by subspace iteration with shift and Sturm sequence tests, see BATHE [4]. Then let us assume that the eigenfrequency(ies) and corresponding eigenmode(s) are available. Now for the linear programming problem of optimal remodelling we need the gradients of squared eigenfrequency ω_r^2 with respect to the model parameter h_i . The notation of "," for partial differentiation is applied, i.e.

$$(\omega_r^2)_{,h_i} := \frac{\partial \omega_r^2}{\partial h_i} \ , \tag{3.3}$$

and we shall furthermore omit the numbering index r .

As pointed out by WITTRICK [5], the resulting expression for $(\omega^2)_{,h_i}$ is simple and does not involve the gradients of the eigenmodes. We find

$$(\omega^2)_{,h_i} = \{D\}^T[S]_{,h_i}\{D\} - \omega^2\{D\}^T[M]_{,h_i}\{D\} \ , \tag{3.4}$$

and for details in relation to specifically Timoshenko-beam models, see [6].

Extending our models to include damping and/or non-conservative loads, the eigenvalue λ will be complex

$$\lambda := \alpha + i\omega \quad , \quad i := \sqrt{-1} \tag{3.5}$$

and both the analysis and the sensitivity analysis will be more complicated, but we still get the result that the gradients of the eigenmodes are not involved. This is the subject of the lecture [7], from which we give solutions for the eigenvalue problem:

$$[L]\{D\} = \Big([S] + p[K] + \lambda[C] + \lambda^2[M]\Big)\{D\} = \{0\} \quad , \tag{3.6}$$

with the system matrix $[L]$ consisting of: symmetric stiffness matrix $[S]$, non-symmetric load matrix $[K]$ and load level p , damping matrix $[C]$, mass matrix $[M]$ and eigenvalue λ by (3.5).

The adjoint eigenvector $\{\tilde{D}\}$ and the total mutual potential Π are defined by

$$\Pi := \{\tilde{D}\}^T[L]\{D\} \quad ; \quad [L]^T\{\tilde{D}\} = \{0\} \quad , \tag{3.7}$$

and for self-adjoint problems we have $\{\tilde{D}\} = \{D\}$. Taking variations of Π , we find that the variations of the eigenvectors $\{D\}$, $\{\tilde{D}\}$ disappear, which is naturally the reason for introducing the adjoint eigenvector and for taking variations of the mutual energy $\delta\Pi$ instead of variations relative to equilibrium (3.6).

Treating the eigenvalue λ , the load level p (real) and the model h (real) as involved parameters, we then get:

$$\begin{aligned} &\{\tilde{D}\}^T[L]_{,\lambda}\{D\}\delta\lambda + \{\tilde{D}\}^T[L]_{,p}\{D\}\delta p + \{\tilde{D}\}^T[L]_{,h}\{D\}\delta h \\ &\qquad =: \mathcal{A}\,\delta\lambda + \mathcal{B}\,\delta p + \mathcal{C}\,\delta h = 0 \quad , \end{aligned} \tag{3.8}$$

by which the complex functionals $\mathcal{A}$, $\mathcal{B}$ and $\mathcal{C}$ are defined.

It is convenient to define "specific" mutual elastic $\mathcal{U}$, kinetic $\mathcal{T}$, dissipative $\mathcal{D}$, and external $\mathcal{W}$ energies, by

$$\mathcal{U} := \{\tilde{D}\}^T[S]\{D\} \quad ; \quad \mathcal{T} := \{\tilde{D}\}^T[M]\{D\} \quad ;$$
$$\mathcal{D} := \{\tilde{D}\}^T[C]\{D\} \quad ; \quad \mathcal{W} := -\{\tilde{D}\}^T[K]\{D\} \ , \qquad (3.9)$$

and then the $\mathcal{A}$ and $\mathcal{B}$ functionals will be

$$\mathcal{A} = \mathcal{D} + 2\lambda\,\mathcal{T} \quad ; \quad \mathcal{B} = -\,\mathcal{W} = -\,(\mathcal{U} + \lambda\,\mathcal{D} + \lambda^2\,\mathcal{T})/p \ . \qquad (3.10)$$

To determine the functional $\mathcal{C} = \{\tilde{D}\}^T[L]_{,h}\{D\}$ we have to know how the matrices [S] , [K] , [C] and [M] depend on h . A special but important case relates to finite element modelling and gives the most simple result:

$$\mathcal{C} = \left(k\,\mathcal{U}_e + \ell\lambda\,\mathcal{D}_e + m\lambda^2\,\mathcal{T}_e\right)/h_e \ , \qquad (3.11)$$

where k , ℓ , m are integers and $\mathcal{U}_e$, $\mathcal{D}_e$, $\mathcal{T}_e$ are element mutual energies defined analogously to the system mutual energies (3.9). To obtain (3.11) it is assumed that the model parameter h_e only influences the actual element e and that this dependence is to the power k , ℓ , m , respectively.

With these few comments to the practical determination of the functionals $\mathcal{A}$, $\mathcal{B}$, $\mathcal{C}$ we return to the variational equation (3.8), from which the following important results can be derived:

a) Sensitivities with respect to change in load level for unchanged model h :

$$(\partial\lambda/\partial p)_{\delta h = 0} = -\,\mathcal{B}/\mathcal{A} \ . \qquad (3.12)$$

b) Sensitivities with respect to change in model for unchanged load:

$$(\partial\lambda/\partial h)_{\delta p = 0} = - C/A \ . \qquad (3.13)$$

c) Sensitivities with respect to the flutter load level:

$$(\partial p_F/\partial h) = - \mathrm{Re}(C/A)/\mathrm{Re}(B/A) \ , \qquad (3.14)$$

and flutter frequency:

$$(\partial \omega_F/\partial h) = - \mathrm{Im}(C/B)/\mathrm{Re}(A/B) \ . \qquad (3.15)$$

The conclusion of the present section is that it is not necessary to perform repeated analysis to obtain gradient information, because this information can be subtracted directly from the eigenvalue and eigenmode(s) of the primary analysis.

4. Example

For a shaft with multiple, discontinuous change of diameter we shall discuss the identification of a stiffness model. This problem is treated in the thesis (M.Sc.) of Ingason [3], and the difference between the physical geometry and the stiffness model of a sector of diameter change is illustrated in fig. 4.1.

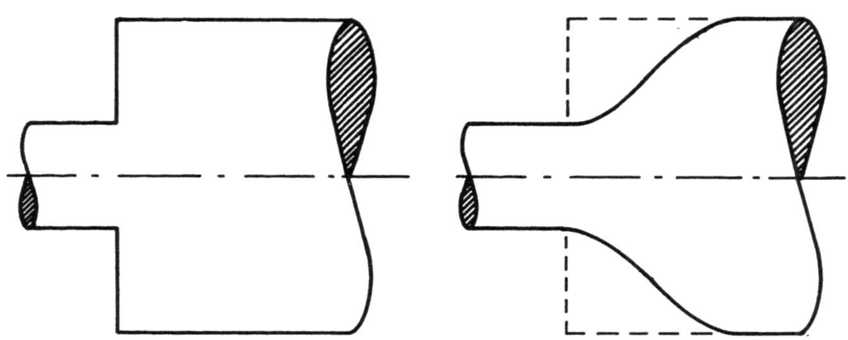

Physical model = mass model   Stiffness model

Fig. 4.1: Stiffness modelling by abrupt change of diameter for a shaft.

The preassumption was to use a one-dimensional Timoshenko-beam model, although two-dimensional modelling is certainly possible, as by SATO [8] or SANDERSON and KITCHING [9]. Also, the consistent mass model with rotational inertia was assumed to be satisfactory and was therefore not subjected to identification.

The problem was thus to identify the stiffness model that yields the best agreement with a number (three) of measured eigenfrequencies and some static displacement curves. The model was subjected to the constraint that the identified "boundary shape" should not exceed the physical geometry. We may therefore say that we identify a stiffness model and not only a stiffness matrix.

The first attempts to solve the problem were not very successful, and we recognized that the assumed experimental boundary conditions (simply supported) were not the actual ones. The failure of identification thus identified a non-accurate experiment.

However, with free-free boundary conditions of the experiments, the identification as described in the present paper worked very stably and excellent results were obtained.

5. Conclusion

The intention of this paper is to show that in solving identification problems we may effectively use the tools of automatic/optimal design, i.e. the many available results from sensitivity analysis and the methods of mathematical programming. Specifically, we advocate the use of sequential linear programming with move-limits.

The feed-back from the computational identification to the experimental basic and vice versa is of major importance.

References

[1] BARUCH, M.: Optimization Procedure to Correct Stiffness and Flexibility Matrices Using Vibration Tests, AIAA, Vol. 16, No. 11, pp. 1208-1210, 1978.

[2] PEDERSEN, P.: A Unified Approach to Optimal Design, Proc. of "Optimization Methods in Structural Design", (Eds.: Eschenauer & Olhoff), pp. 182-187, B.I. Wissenschafts-verlag, 1983.

[3] INGASON, K.: Identifikation via eksperiment af stivheds-model for aksel med multiple aftrapninger, M.Sc. thesis (in Danish), 1984.

[4] BATHE, K.J.: Finite Element Procedures in Engineering Analysis, Prentice-Hall, 1982.

[5] WITTRICK, W.H.: Rates of Change of Eigenvalues, with Reference to Buckling and Vibration Problems, J.R. Aeronaut. Soc., Vol. 66, pp. 590-591, 1962.

[6] PEDERSEN, P.: Design with Several Eigenvalue Constraints by Finite Elements and Linear Programming, J. Struct. Mech., Vol. 10, No. 3, pp. 243-271, 1982-83.

[7] PEDERSEN, P.: Sensitivity Analysis for Non-Selfadjoint Problems, Proc. of A.M.S. New York Meeting, May 1983, (Ed.: V. Komkov), pp. 119-130, Springer-Verlag, 1984.

[8] SATO, H.: Free Vibration of Beams with Abrupt Changes of Cross-section, J. of Sound and Vibration, Vol. 89, pp. 59-64, 1983.

[9] SANDERSON, N. and KITCHING, R.: Flexibility of Shafts with Abrupt Changes of Section, Int. J. of Mech. Sciences, Vol. 20, pp. 189-199, 1978.

Modelling Optimal Prescribed Manipulator Trajectories

F. PFEIFFER

Faculty of Mechanical Engineering
and Institute for Mechanics B,
Technical University, München

Summary

Following a prescribed trajectory in an ideal way results in a one degree of freedom motion whatever the robotic joint degrees of freedom might be. This concept allows to transform the set of nonlinear equations of motion to a system of equations in terms of path acceleration $\ddot{s}$, path velocity $\dot{s}$ and path coordinate s. Though still nonlinear the transformed set can be reduced to formal quadrature yielding path velocity dependent on some integrals containing the path coordinate and the joint torques. These may be optimized within given bounds applying a dynamic programming approach in connection with a certain control joint strategy. The decision which of the joints controls the trajectory and which produce constraint torques to keep the manipulator on the path can be derived from a set of switching polyhedrons to be constructed from the equations of motion. The procedure results in path velocity and joint torques along the given trajectory, optimized according to a performance criterion containing time, energy and change of acceleration/deceleration.

Moreover, arranging above polyhedrons along the given path yields a subspace in the phase space ($\ddot{s}$-$\dot{s}^2$-s) bounded by warped ruled surfaces and constraining any motion to the interior of that subspace. The construction of time-optimal extremals within this subspace gives possible motion in the time-minimum case and indicates maximum velocity-potential of the combined system path-manipulator under consideration. Theory fits well with other methods.

Introduction

Trajectory planning includes several aspects and problems: the type of paths considered, the influences of constraints and the type of manipulator. In the following we shall consider manipulators with rotational and/or translational degrees of freedom in connection with prescribed trajectories (Fig. 1). Most publications on optimal manipulator trajectories deal with point-to-point-paths, where the space between these two points may be

Dynamics of Multibody Systems
IUTAM/IFToMM Symposium Udine 1985
Editors: G. Bianchi and W. Schiehlen

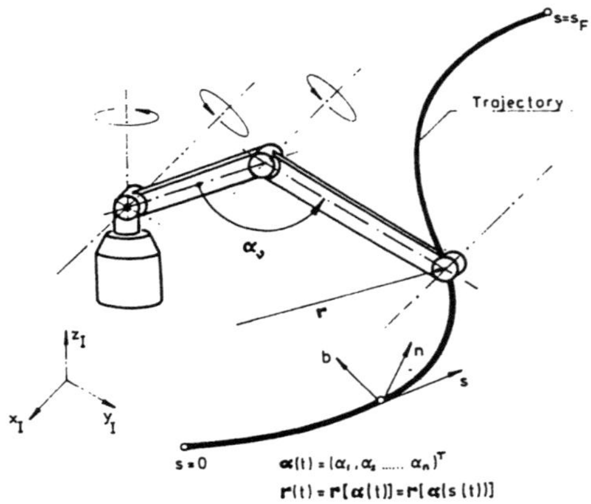

Fig. 1. Manipulator and Prescribed Trajectory

constraint in some way or not [1, 4, 5, 19]. Some authors consider prescribed trajectories starting with a state space representation of manipulator dynamics and applying some optimization algorithms [1, 18, 19]. The publications [8] and [15] contain ideas relatively near to the method presented here, where in [8] certain transformations are used to obtain an estimate for trajectory planning, whereas [15] evaluates the influence of torque constraints on trajectory velocities.

The main idea of this paper is the following: Looking at a prescribed trajectory to be followed in an ideal manner by the manipulator tip, i.e. neglecting in a first step any disturbances, corresponds to a one degree-of-freedom-motion of the manipulator. This one DOF may be represented by the path coordinate s in tangential direction of the path (Fig. 1). A transformation of the equations of motion expressed in joint coordinates to path coordinates results in a set of differential equations, which can be solved by formal quadrature. Moreover, above set allows a simple and direct representation of the torque constraints in phase-space as a closed sub-space formed by 2f ruled surfaces where f is the number of joints. The field of time-optimal extremals in the interior of that subspace permits a consistent choice of allowable trajectories, which may be additionally optimized with regard to certain performance criteria.

Mathematical Model

The equations of motion for a manipulator with a treelike structure may be established by several methods well known in analytical dynamics. The choice of the method is a question of economics and convenience and will not be discussed here. We start with the kinetical energy $T = \frac{1}{2} \mathbf{v}^T \mathbf{M}^* \mathbf{v}$, where the f-vector $\mathbf{v}$ designates the absolute velocity of the manipulator components and $\mathbf{M}^*$ is an (fxf) mass matrix. Friction phenomena will be neglected. The absolute component velocities $\mathbf{v}$ can be expressed by the time-derivatives of the relative joint coordinates $\boldsymbol{\alpha}$, and, as is well-known, this relationship is a linear one for the manipulator type considered. Thus, for applying Lagrange's formalism we use the kinetical energy in the following form:

$$T = \frac{1}{2}\mathbf{v}^T\mathbf{M}^*\mathbf{v} \quad ; \qquad \mathbf{v} = \left(\frac{\partial \mathbf{v}}{\partial \dot{\boldsymbol{\alpha}}}\right)\dot{\boldsymbol{\alpha}}$$
$$T = \frac{1}{2}\dot{\boldsymbol{\alpha}}^T\mathbf{M}\dot{\boldsymbol{\alpha}} \quad ; \qquad \mathbf{M} = \left(\frac{\partial \mathbf{v}}{\partial \dot{\boldsymbol{\alpha}}}\right)^T\mathbf{M}\left(\frac{\partial \mathbf{v}}{\partial \dot{\boldsymbol{\alpha}}}\right) \tag{1}$$

For the evaluation of the Jacobian $(\partial\mathbf{v}/\partial\dot{\boldsymbol{\alpha}})$ we have to regard $\mathbf{v} = \mathbf{v}(\dot{\boldsymbol{\alpha}})$, which usually can be derived rather easily. Applying to (1) together with the gravitation potential V Langrange's equation yields the equations of motion:

$$\mathbf{M}_\nu\ddot{\boldsymbol{\alpha}} + \dot{\boldsymbol{\alpha}}^T\left[\left(\frac{\partial \mathbf{M}_\nu^T}{\partial \boldsymbol{\alpha}}\right) - \frac{1}{2}\left(\frac{\partial \mathbf{M}}{\partial \alpha_\nu}\right)\right]\dot{\boldsymbol{\alpha}} + \left(\frac{\partial V}{\partial \alpha_\nu}\right) = T_\nu^* \tag{2}$$
$$(\nu = 1, 2, \cdots\cdots f)$$

$\mathbf{M}_\nu$ is the ν-th row of $\mathbf{M}$ and α_ν the ν-th component of the vector $\boldsymbol{\alpha}$. T_ν^* is the torque (force) acting in joint ν. The bracket term represents Christoffel's symbol [6, 7].

Applying the idea of an ideal motion along the prescribed path and transforming to path coordinate s with the help of $\dot{\boldsymbol{\alpha}} = \boldsymbol{\alpha}'\dot{s}$, $\ddot{\boldsymbol{\alpha}} = \boldsymbol{\alpha}'s + \boldsymbol{\alpha}''\dot{s}^2$ (' = d/ds) we get:

$$A_i(s)\ddot{s} + B_i(s)\dot{s}^2 + C_i(s) = T_i \operatorname{sgn}(\alpha_i' - \alpha_{i-1}') \tag{3}$$
$$(i = 1, 2, \cdots\cdots f)$$

With the abbreviations:

$$A_i(s) = \mathbf{M}_i\,\boldsymbol{\alpha}'$$

$$B_i(s) = \frac{dA_i(s)}{ds} - Z_i(s)$$

$$Z_i(s) = \frac{1}{2}\,\boldsymbol{\alpha}'^T\left(\frac{\partial \mathbf{M}}{\partial \alpha_i}\right)\boldsymbol{\alpha}' \tag{4}$$

$$G_i(s) = \left(\frac{\partial V}{\partial \alpha_i}\right)$$

A formal quadrature of eq. (3) regarding (4) results in [1o, 13, 14]:

$$(A_i\dot{s})_s^2 = \left\{ (A_i\dot{s})_{s_0}^2 + \right. \tag{5}$$

$$\left. + 2\int_{s_0}^{s} A_i\left[T_i\,\mathrm{sgn}(\alpha_i' - \alpha_{i-1}') - G_i\right] e^{-2\int_{s_0}^{\sigma}\left(\frac{Z_i}{A_i}\right)d\zeta} \cdot d\sigma \right\} e^{+2\int_{s_0}^{s}\left(\frac{Z_i}{A_i}\right)d\sigma}$$

The quantity $(A_i\dot{s})$ represents the local moment of momentum with regard to joint i at path point s. From this and eq. (4) $A_i(s)$ may be interpreted as gradient of the moment of inertia refered to joint i along the path tangent s, $(A_i\ddot{s})$ thus being the change of angular momentum around joint i. $(B_i\ \dot{s}^2)$ are gyroscopical and C_i gravitational torques (eq. 3). The dependence of motor drive on the sign-function of relative joint angle-change $(\alpha_i' - \alpha_{i-1}')$ is obvious.

There are f equations of motion (3) and f solutions (5). The special form of the equations of motion (3) permits a direct evaluation in a modified phase space containing path acceleration $\ddot{s}$, squared path velocity $\dot{s}^2$ and path coordinate s. Moreover, the solutions (5) will facilitate considerably the application of optimizing strategies.

Constraints and Extremals

The torques $T_i(s)$ driving the joints are constraint to certain limit values given by the motor design. In the following it will be assumed for simplicity but without loss of generality that upper and lower bounds are the same. Then our constraints can be written as

$$|T_i(s)| \leqslant T_{io} \tag{6}$$

For these maximum torques eqs. (3) define for each path point s a polyhedron in the $\ddot{s}$ - $\dot{s}^2$ - plane [14]:

$$A_i(s)\,\ddot{s} + B_i(s)\,\dot{s}^2 = \pm T_{io} - C_i(s) \tag{7}$$

The polyhedron is thus bounded by pairs of straight lines for each joint. Motion can only take place within this area whatever initial conditions may be (Fig. 2). Assembling these polyhedrons along the path axis s results in a bounded phase space ($\ddot{s}$, $\dot{s}^2$, s), the surfaces of which are 2f warped ruled surfaces with some interesting properties [9, 16]. First, all phase-trajectories of the manipulator tip will be contained in the interior of this bounded subspace. Even time-optimal phase-trajectories cannot be part of the ruled surfaces themselves with the exception of critical points with $A_j(s) = 0$ for any j. Second, any contact of a phase-trajectory with any ruled surface indicates the end of that trajectory, because the manipulator system has then reached its capability of acceleration or deceleration without leaving the prescribed path. Third, critical points with $A_i(s) = 0$ can be passed with one velocity only to be determined from eq. (3) with $A_i = 0$. In any case, this will be true for time-optimal motion. Using some formulas from differential geometry [9, 16] we come for a path point s to the following ruled surface properties:

Every point of the warped ruled surfaces is a hyperbolic point

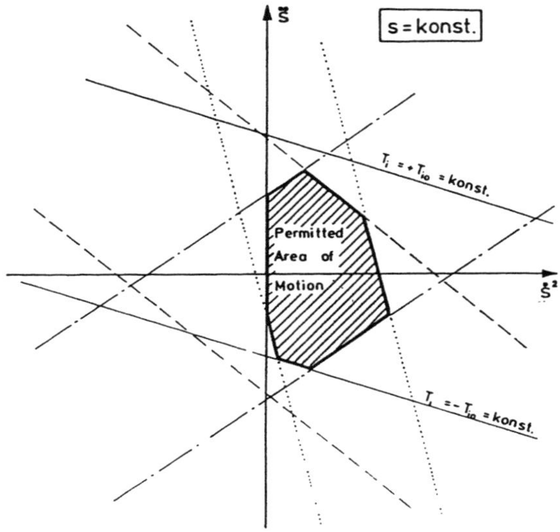

Fig. 2. Polyhedron of allowed motion for path point s

with two curvatures different in sign. Thus, in the direct neighbourhood of a surface point the surface possesses properties similar to a hyperboloid of one sheet. This follows from the Gauß curvature, which may be evaluated with standard formulas as given in [16]. A good presentation of main curvatures of a surface point delivers the Dupin indicatrix [16]. An example is shown in Fig. 3 giving the constraint phase-space of a circular path for a two DOF manipulator.

The constrained phase space developed so far already represents a very helpful tool for trajectory planning by indicating maximum possible path velocities and path accelerations or decelerations. More detailed informations can be achieved by considering extremal curves within the above constraint sub-space. Such extremals start at the beginning or any other point of the prescribed path with zero velocity and proceed with maximum joint torques in equs. (3) and (5), i.e. maximum acceleration, until they meet a constraining ruled surface. Or they start at the path end with zero velocity decelerating with maximum torques until they meet constraints. The field of these extremals within the constraining phase-space permits a direct evaluation

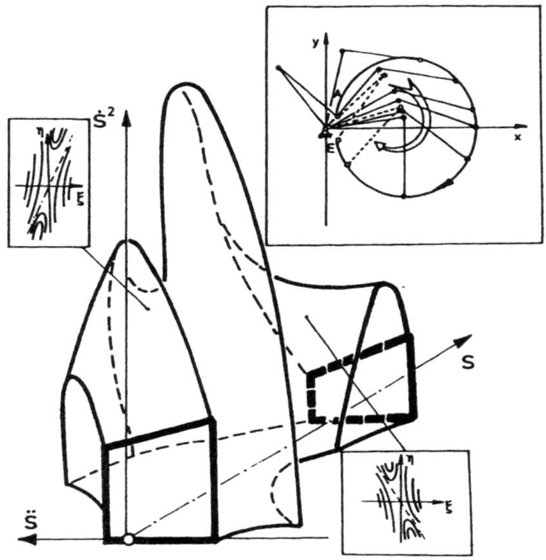

Fig. 3. Constraint phase-space for a two DOF manipulator

of time-optimal trajectories, which represent more or less the velocity or energy potential of the combined system manipulator-prescribed path [11]. Of course, such time optimal trajectories may as well be determined with the help of a bang-bang-strategy or by Bellman's dynamic programming approach [13, 14]. In this case "forbidden" areas within the phase-space have to be regarded by methods as discussed above. An example of such a field of extremals gives Fig. 4, where the constraining space is projected on the ($\dot{s}$ - s)-plane [11].

Optimization

To include an optimization process a formal trick has to be introduced to combine the assumption of a one DOF motion and the f equations of motion (3) with the f solutions (5). According to one DOF path-control can only be performed by one joint, the torque of which might be optimized. All other joints produce constraint torques keeping the manipulator tip on the prescribed

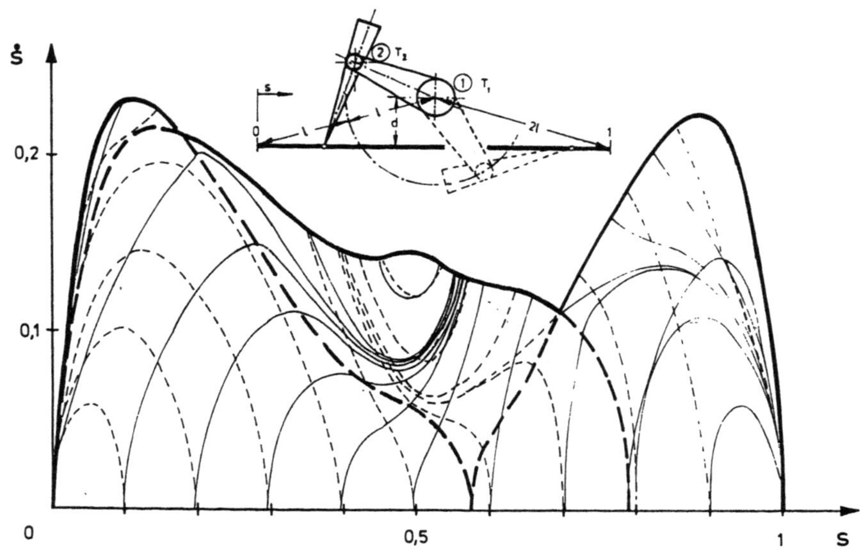

Fig. 4. Field of extremals for a two DOF-manipulator

path [13]. At the starting point $s = s_o$ control should be transferred to that joint j with a maximum rate of change of relative angular velocity $|\dot{\alpha}_j - \dot{\alpha}_{j-1}|_o$ or

$$|\alpha'_j - \alpha'_{j-1}|_o = \max_{i=1,2,\cdots f} |\alpha'_i - \alpha'_{i-1}| \tag{8}$$

It means, that this joint j with maximum change of its relative joint angle $(\alpha_j - \alpha_{j-1})$ with path progress $\dot{s}(t)$ should achieve best controllability of the fixed trajectory. Of course, this situation might change during the course of the manipulator path. But such a change will always be accompanied by a quick increase of one of the constraint torques exceeding given joint torque limits. That joint k, which first reaches its torque limit, will then take over control thus being the new heading or control joint. This leads to the decision sequence:

1. if $|T_i| < (T_i)_{Limit}$ for all $i \neq j$,
   then joint j=control joint

2. if $|T_k| > (T_k)_{Limit}$ for any $k \neq j$, then joint k=control joint (9)

3. (j=k) go to 1

The consistency of this procedure follows immediately from physical reasoning, especially the fact, that a joint with a small gradient $|\alpha_k' - \alpha_{k-1}'|$ would need too big a torque to control the trajectory and will thus better be used to produce a constraint torque for keeping the manipulator tip on the trajectory. Changes as defined by eq. (9) depend on the path type. Complicated trajectories may afford several changes between the joints [14].

On the basis of such an algorithm we are able to optimize the torque of the control joint using three weighted performance criteria [13]:

$$G = \left(\frac{1}{s_F}\right)\int_0^{s_F}\left[\frac{G_1}{\dot{s}} + G_2\dot{s}^2 + G_3\left|\frac{d\ddot{s}}{ds}\right|\right]ds \qquad (1o)$$

The first term weighted with G_1 corresponds to minimum time, the second term to the square of velocity. It is proportional to the energy. The third term regards changes in acceleration and deceleration. It produces a smoothing effect on the path-velocity-curve thus avoiding excitation of elastic vibrations [13, 14].

Regarding eqs. (3), (5), (8) to (1o) an optimization will be achieved by applying the dynamic programming algorithm [2, 3, 17]. This process suits especially well to our method, because the two ideas of first reducing the equations of motion to the path coordinate s and of introducing the concept of a control joint in a second step results without any doubt in a highly nonlinear problem but of one-dimensional character only. Therefore, it can be treated with very moderate expense by Bellman's principle. Some results are shown in Fig. 5, which indicates the consistency of different methods. Figure 6 represents an example as calculated by [19] for a given velocity distribution without any optimization. For comparison submitted theory has been applied to this case. Results are identical with [19].

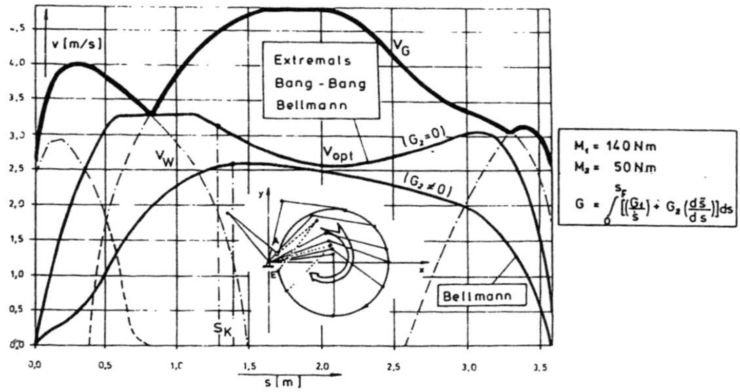

Fig. 5. Optimization Results and Comparisons

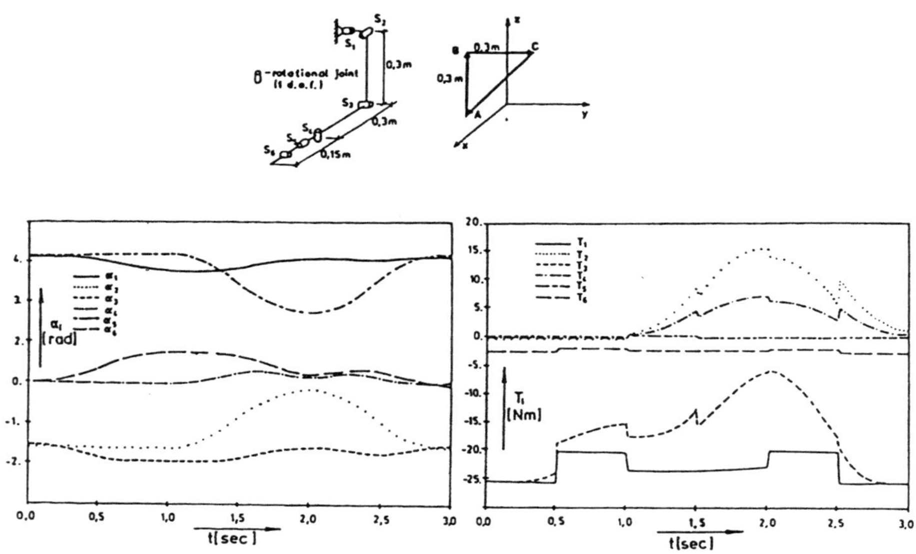

References

1. Balestrino, A.; De Maria, G.; Sciavicco, L.: An Adaptive Model Following Control for Robotic Manipulators. Journal of Dynamic Systems, Measurement and Control, Sept. 1983, Vol. 1o5, 143-151.

2. Bellman, R.E.; Dreyfus, S.E.: Applied Dynamic Programming. Princeton University Press, 1962.

3. Beveridge, G.S.G.; Schechter, R.S.: Optimization: Theory and Practice. McGraw Hill Book Company, New York 197o.

4. Bremer, H.: Parameterunempfindliche Schwenkregelung eines Industrieroboters. Regelungstechnik, 32. Jahrgang, (1984), 74-81.

5. Cvetković, V.; Vukobratović, M.: Contribution to Controlling Non-Redundant Manipulators. Mechanism and Machine Theory Vol. 16, Pergamon Press, (1981), 81-91.

6. Duschek, A.; Hochrainer, A.: Tensorrechnung in analytischer Darstellung. Springer-Verlag, Wien, 196o-1965, 3 Teile.

7. Hamel, G.: Theoretische Mechanik, Springer Verlag 1949 (1967).

8. Hollerbach, J.M.: Dynamic Scaling of Manipulator Trajectories. J. of Dynamic Systems, Measurement and Control, Vol. 1o6, March 1984, 1o2-1o6.

9. Hoschek, J.: Liniengeometrie. Hochschulskripten, BI 733/a/b, Zürich 1971.

1o. Kamke, E.: Differentialgleichungen, Lösungsmethoden und Lösungen. Akademische Verlagsgesellschaft, Leipzig 1967.

11. Müller, J.: Überlegungen zu zeitoptimalen Roboterbahnen. TU München, Lehrstuhl B für Mechanik, Diplomarbeit 1983.

12. Pfeiffer, F.: Beitrag zu optimalen Roboterbahnen. GAMM-Tagung 1984, ZAMM, Band 65, Heft 4/5, (1985), T321.

13. Pfeiffer, F.: On Optimal Control of Manipulator Trajectories. To appear in ASME, J. of Dynamic Systems, Meas. and Control.

14. Pfeiffer, F.: Schaltfunktionen für optimale Roboter-Sollbahnen. To appear in Automatisierungstechnik.

15. Shin, K.G.; McKay, N.D.: Minimum-Time Control of Robotic Manipulators with Geometric Path Constraints. IEEE Trans. on Automatic Control, Vol. AC-3o, No. 6, June 1985, 531-541.

16. Smirnow, W.I.: Lehrgang der höheren Mathematik, Teil II, VEB Deutscher Verlag der Wissenschaften, Berlin 1981.

17. Takahashi, Y., et al.: Control and Dynamic Systems. Addison-Wesley Publishing Company, Massachusetts 197o.

18. Vukobratović, M.; Kirćanski, M.: A Method for Optimal Synthesis of Manipulation Robot Trajectories. Transactions of the ASME, Vol. 1o4, June 1982, 188-193.

19. Vukobratović, M.; Potkonjak, V.: Dynamics of Manipulation Robots. Springer-Verlag, Berlin-Heidelberg-New York, 1982.

2o. Vukobratović, M.; Stokić, D.: Control of Manipulation Robots. Springer-Verlag, Berlin-Heidelberg-New York, 1982.

Analysis of Nonlinear Stochastic Multibody Systems

K. Popp*), P. C. Müller**), H. Windrich*)

*) Institute of Mechanics, University of Hanover, FRG

**) Institute of Control, University-GHS Wuppertal, FRG

1. Introduction

Improved modeling of today's technical problems often leads to nonlinear multibody systems. Additionally, in many cases external random disturbances are present and have to be considered in the analysis. Examples of this type of problem are vehicle-guideway systems, cf. [1], [2], [3], particularly wheel-rail systems for high-speed transportation. Commonly, the analysis of such systems is carried out using either Monte- Carlo simulation or equivalent statistical linearization methods, [4]. Both methods show severe drawbacks. The Monte-Carlo simulation is generally applicable. However, the procedure is very time consuming and the results have to be postprocessed in order to receive the quantities wanted. In the equivalent statistical linearization method, assumptions have to be made about the probability distribution of the system response, and furthermore, possible limit cycles of the system are not regarded. Thus, attempts have been made to circumvent these drawbacks by modifying the statistical linearization method, cf. [5,p. 183ff].

The aim of the present paper is to give modified methods for the investigation of nonlinear multibody systems with external stochastic disturbances. Such methods are:

- Combination of equivalent harmonical and statistical linearization.
- Equivalent harmonical linearization in combination with a subsequent covariance analysis.

In the first part of the paper, these methods are coherently formulated for multibody systems. In the second part, the different methods are applied to a simple system which exhibits a strong limit cycle behavior, and the results are critically compared.

Dynamics of Multibody Systems
IUTAM/IFToMM Symposium Udine 1985
Editors: G. Bianchi and W. Schiehlen

2. Nonlinear Stochastic Multibody Systems

The modeling of mechanical problems by nonlinear multibody systems is generally discussed in the case of vehicles in [2,3]. An ordinary multibody system is described by a set of f ordinary differential equations with respect to f generalized coordinates y_i, i = 1,.., f. Using matrix notation the result is a f x 1 - vector differential equation of second order,

$$\underline{M}(\underline{y})\ddot{\underline{y}}(t) + \underline{g}(\underline{y},\dot{\underline{y}}) = \underline{G}\ \underline{h}(t) \tag{2.1}$$

The generalized coordinates are summarized in the vector $\underline{y}$. The fxf-inertia matrix $\underline{M}$ is symmetric and usually positive definite, the fx1-vector $\underline{g}$ describes the generalized gyro and centrifugal forces as well as the generalized applied forces. The fxr-matrix $\underline{G}$ characterizes the input matrix of the disturbances. The disturbance vector is assumed to consist of deterministic and stochastic parts:

$$\underline{h}(t) = \underline{h}_d(t) + \underline{h}_s(t). \tag{2.2}$$

The deterministic disturbances are taken as harmonic time functions:

$$\underline{h}_d(t) = \frac{1}{\sqrt{2}}\bar{\underline{h}}_o + \bar{\underline{h}}_c \cos\omega t + \bar{\underline{h}}_s \sin\omega t. \tag{2.3}$$

The stochastic disturbances may be considered as stationary white or colored noise processes:

$$\underline{h}_s(t) = \underline{C}\ \underline{v}\ (t) + \underline{D}\ \underline{w}(t), \tag{2.4}$$

$$\dot{\underline{v}}(t) = \underline{V}\ \underline{v}(t) + \underline{W}\ \underline{w}(t), \tag{2.5}$$

where $\underline{w}(t)$ is a stationary white noise process with the intensity matrix $\underline{Q}$. If $\underline{C} = \underline{0}$, $\underline{D} \neq \underline{0}$ the disturbance $\underline{h}_s(t)$ presents also a white noise process, but if $\underline{C} \neq 0$, $\underline{D} = \underline{0}$ then $\underline{h}_s(t)$ is a colored noise process modelled by the shape filter (2.5).

Besides the description (2.1) we also use the common state space reprensentation

$$\dot{\underline{x}}(t) = \underline{a}(\underline{x}(t)) + \underline{B}\ \underline{h}(t), \tag{2.6}$$

cf. [3]. This notation also includes more general dynamical systems, e. g. general multibody systems.

Since exact solutions of nonlinear differential equations are quite rare, usually approximation techniques are applied. If the nonlinearities are weak, Taylor series expansion is applicable resulting in a linearized system, where the wellknown solution methods of the linear system theory can be applied. If the nonlinearities are strong, quasilinearization techniques are used, harmonical linearization in case of periodic solutions, and statistical linearization in case of stochastic excitations. These two methods are fully discussed in [6]. If periodic solutions and stochastic responses appear simultaneously, e. g. in the case of limit cycles and stochastic excitations, the quasilinearization techniques have to be modified.

3. Simultaneous Equivalent Harmonical and Statistical Linearization Technique

3.1 State Space Formulation

The simultaneous equivalent harmonical and statistical linearization technique was suggested by Kazakov, cf. [5, pp. 183ff]. Here, the same fundamental ideas can be applied to the deterministic and to the stochastic part of the solution. Assuming the solution of (2.6) in the form

$$\underline{x}(t) = \underline{m}(t) + \underline{\xi}(t), \tag{3.1}$$

where $\underline{m}(t) = E\{\underline{x}(t)\}$ is the mean vector and $\underline{\xi}(t)$ the stochastic deviation, then (2.6) can be separated in

$$\dot{\underline{m}}(t) = E\{\underline{a}(\underline{m}(t) + \underline{\xi}(t))\} + \underline{B}\,\underline{h}_d(t), \tag{3.2}$$

$$\dot{\underline{\xi}}(t) = \underline{a}(\underline{m}(t) + \underline{\xi}(t)) - E\{\underline{a}(\underline{m}(t) + \underline{\xi}(t))\} + \underline{B}\,\underline{h}_s(t). \tag{3.3}$$

The unknown stochstically disturbed periodic solution $\underline{x}(t)$ (3.1) is approximated by

$$\underline{m}(t) = \frac{1}{\sqrt{2}}\underline{m}_o + \underline{m}_c \cos\omega t + \underline{m}_s \sin\omega t \tag{3.4}$$

$$= [\underline{m}_o\ \underline{m}_c\ \underline{m}_s]\begin{bmatrix}\frac{1}{\sqrt{2}}\\ \cos\omega t\\ \sin\omega t\end{bmatrix} = \underline{M}_d \cdot \underline{c}(t) \tag{3.5}$$

and the deviation by a stationary process $\underline{\xi}(t)$. Then the equivalent harmonic linearization technique is applied to (3.2), cf. [6]. The nonlinearity is approximated by a linear vector function with respect to (3.5):

$$\underline{e}_d(t) = E\{\underline{a}(\underline{m}(t) + \underline{\xi}(t))\} - \underline{A}_d\,\underline{m}(t), \tag{3.6}$$

where the equivalent gain matrix $\underline{A}_d$ is determined such that the mean square approximation error is minimized:

$$\int_o^T \underline{e}_d^T(t)\,\underline{e}_d(t)\,dt \to \underset{\underline{A}_d}{\text{minimum}}, \tag{3.7}$$

where $T = \frac{2\pi}{\omega}$ is the period of (3.5). From (3.7) the gain matrix $\underline{A}_d$ is derived depending on the unknown vectors $\underline{m}_o$, $\underline{m}_c$, $\underline{m}_s$ and in the case of limit cycles on the frequency ω:

$$\underline{A}_d\,\underline{M}_d = \frac{\omega}{\pi} \int_o^T E\{\underline{a}(\underline{m}(t) + \underline{\xi}(t))\}\,\underline{c}^T(t)\,dt. \tag{3.8}$$

In a similar manner the nonlinearity in (3.3) is linearized:

$$\underline{e}_s(t) = \underline{a}(\underline{m}(t) + \underline{\xi}(t)) - E\{\underline{a}(\underline{m}(t) + \underline{\xi}(t))\} - \underline{A}_s\,\underline{\xi}(t), \tag{3.9}$$

$$\int_o^T E\{\underline{e}_s^T(t)\,\underline{e}_s(t)\}\,dt \to \underset{\underline{A}_s}{\text{minimum}}. \tag{3.10}$$

This leads to

$$\underline{A}_s = \frac{\omega}{2\pi} \int_o^T E \{\underline{a}(\underline{m}(t) + \underline{\xi}(t)) \, \underline{\xi}^T(t)\} \, dt. \, \underline{P}^{-1} , \tag{3.11}$$

where

$$\underline{P} = E \{\underline{\xi}(t) \, \underline{\xi}^T(t)\} \tag{3.12}$$

is the covariance matrix of the solution process. It has to be noted that the equivalent gain matrices depend on the deterministic values $\underline{m}_o$, $\underline{m}_c$, $\underline{m}_s$, ω as well as on the stochastic characteristics of $\underline{P}$ and on the probability density function $p(\underline{x})$ which is not yet fixed.

By the equivalent gain technique the nonlinear equations (3.2 - 3) are replaced by linearized equations

$$\dot{\underline{m}}(t) = \underline{A}_d \, \underline{m}(t) + \underline{B} \, \underline{h}_d(t), \tag{3.13}$$

$$\dot{\underline{\xi}}(t) = \underline{A}_s \, \underline{\xi}(t) + \underline{B} \, \underline{h}_s(t). \tag{3.14}$$

These pseudo-linear equations are analysed in the usual manner, cf. [6].

Introducing the notation

$$\underline{h}_d(t) = [\bar{\underline{h}}_o \;\; \bar{\underline{h}}_c \;\; \bar{\underline{h}}_s] \, \underline{c}(t) = \underline{H} \, \underline{c}(t) \tag{3.15}$$

and considering

$$\dot{\underline{c}}(t) = \begin{bmatrix} o & o & o \\ o & o & -\omega \\ o & \omega & o \end{bmatrix} \underline{c}(t) = \underline{S} \, \underline{c}(t) \tag{3.16}$$

the periodic part (3.5) of the solution is determined by the nonlinear algebraic matrix equation

$$\underline{M}_d \, \underline{S} = \underline{A}_d(\underline{M}_d,\underline{P}) \, \underline{M}_d + \underline{B} \, \underline{H} \; ; \tag{3.17}$$

while the stochastic part $\underline{\xi}(t)$ is characterized by the covariance matrix $\underline{P}$ which satisfies a nonlinear Lyapunov matrix equation

$$\underline{A}_s(\underline{M}_d,\underline{P}) \, \underline{P} + \underline{P} \, \underline{A}_s^T(\underline{M}_d,\underline{P}) + \underline{R} = \underline{O} \, . \tag{3.18}$$

The matrix $\underline{R}$ represents the correlation of the input $\underline{h}_s(t)$ and the output $\underline{\xi}(t)$. If $\underline{h}_s(t)$ is a white noise vector process, $\underline{h}_s(t) = \underline{D} \, \underline{w}(t)$, then

$$\underline{R} = \underline{B} \, \underline{D} \, \underline{Q} \, \underline{D}^T \, \underline{B}^T. \tag{3.19a}$$

On the contrary, if $\underline{h}_s(t)$ is a colored noise process (2.4 - 5), then we have

$$\underline{R} = \underline{B} \, \underline{C} \, \underline{P}_{xv}^T + \underline{P}_{xv} \, \underline{C}^T \, \underline{B}^T, \tag{3.19b}$$

where

$$\underline{A} \, \underline{P}_{xv} + \underline{P}_{xv} \, \underline{V}^T + \underline{B} \, \underline{C} \, \underline{P}_v = \underline{O}, \tag{3.20}$$

$$\underline{V} \, \underline{P}_v + \underline{P}_v \, \underline{V}^T + \underline{W} \, \underline{Q} \, \underline{W}^T = \underline{O} \, .$$

The matrix equations (3.17 - 18) are nonlinear and coupled because of the determination of $\underline{A}_d$ and $\underline{A}_s$ according to (3.8) and (3.11). Due to the expectation operator the formulas (3.8) and (3.11) cannot be evaluated without the knowledge of the probability density function $p(\underline{\xi})$. Because $p(\underline{\xi})$ is not known a priori, we have to assume the type of the probability distribution of $\underline{\xi}(t)$. Therefore, after a first step of approximation according to the quasilinearization technique a second step of approximation has to be accomplished. Usually the process $\underline{\xi}(t)$ is assumed to be Gaussian; then

$$p(\underline{\xi}) = \frac{1}{\sqrt{(2\pi)^n \det \underline{P}}} \exp(-\tfrac{1}{2}\, \underline{\xi}^T\, \underline{P}^{-1}\, \underline{\xi}) \tag{3.21}$$

is used. But since the stochastic response is really not a Gaussian process because of the nonlinearity, also other types may be suitable, e. g. a uniform distribution; then

$$p(\underline{\xi}) = p_o \text{ for all } \underline{\xi} \in \mathcal{B} \tag{3.22}$$

where

$$p_o = \int_{\mathcal{B}} d\, \underline{\xi}. \tag{3.23}$$

Summarizing the discussion so far, we have to distinguish between several cases.

1. Limit cycle problem with stochastic excitation.

In this case the deterministic excitation vanishes, $\underline{h}_d(t) \equiv \underline{0}$. Nevertheless, periodic solutions, i. e. limit cycles, may appear because of the nonlinearity, and therefore the approximation (3.5) still holds. The only modification arises in (3.17) which becomes homogeneous in $\underline{M}_d$:

$$\underline{M}_d\, \underline{S} = \underline{A}_d\, \underline{M}_d. \tag{3.24}$$

Determining $\underline{A}_d$ according to (3.8) by the assumption of a normal distribution (3.21) the equations (3.18) and (3.24) allow to calculate the unknowns $\underline{m}_o$, $\underline{m}_c$, $\underline{m}_s$, ω and $\underline{P}$.

If $\underline{A}_d$ is fixed by the assumption of a uniform distribution (3.22), the covariance matrix $\underline{P}$ is uniquely determined by the domain $\mathcal{B}$. Thereforce the unknowns to be calculated by (3.18) and (3.24) are $\underline{m}_o$, $\underline{m}_c$, $\underline{m}_s$, ω and $\mathcal{B}$. In this case, $\mathcal{B}$ has to be characterized by a parallelepiped, hypersphere, hyperellipsoid etc. with unknown dimensions.

2. Harmonic and stochastic excitation.

Here, the frequency ω is known by the harmonic excitation. If ω does not accidentally coincide with a limit cycle frequency, then the inhomogeneous equation (3.17) has to be used to calculate the amplitudes and phases of the approximate harmonic response (3.5). Depending on the choice of the probability density function the unknowns are $\underline{m}_o$, $\underline{m}_c$, $\underline{m}_s$, and $\underline{P}$ or $\underline{m}_o$, $\underline{m}_c$, $\underline{m}_s$ and $\mathcal{B}$ which are determined by (3.17 - 18).

3.2 Multibody Formulation

The general results of section 3.1 can be specialized to ordinary multibody systems (2.1) to avoid an unnecessary amount of computation. Using

$$\begin{aligned} \underline{y}(t) &= \underline{n}(t) + \underline{\eta}(t) \\ &\approx [\underline{n}_o \ \underline{n}_c \ \underline{n}_s] \ \underline{c}(t) + \underline{\eta}(t) \\ &= N_d \ \underline{c}(t) + \underline{\eta}(t) \end{aligned} \tag{3.25}$$

the equivalent linearized equations read

$$\underline{M}\ \ddot{\underline{n}}(t) + \underline{D}_d\ \dot{\underline{n}}(t) + \underline{K}_d\ \underline{n}(t) = \underline{G}\ \underline{h}_d(t), \tag{3.26}$$

$$\underline{M}\ \ddot{\underline{\eta}}(t) + \underline{D}_s\ \dot{\underline{\eta}}(t) + \underline{K}_s\ \underline{\eta}(t) = \underline{G}\ \underline{h}_s(t) \tag{3.27}$$

replacing the equations (3.13 - 14). The equivalent gain matrices have to satisfy

$$[\underline{K}_d\ \underline{D}_d] \begin{bmatrix} \underline{N}_d \\ \underline{N}_d\ \underline{S} \end{bmatrix} = \frac{\omega}{\pi} \int_o^T E\{\underline{g}(\underline{n}(t) + \underline{\eta}(t),\ \dot{\underline{n}}(t)+\dot{\underline{\eta}}(t))\}\ \underline{c}^T(t)dt, \tag{3.28}$$

$$\left[\underline{K}_s\ \underline{D}_s\right] = \frac{\omega}{2\pi} \left[\begin{array}{c|c} \int_o^T E\ \{\underline{g}(\underline{n}(t)+\underline{\eta}(t),\ \dot{\underline{n}}(t)+\dot{\underline{\eta}}(t)).\underline{\eta}^T(t)\}dt & \int_o^T E\ \{\underline{g}(\underline{n}(t)+\underline{\eta}(t),\ \dot{\underline{n}}(t)+\dot{\underline{\eta}}(t)).\dot{\underline{\eta}}^T(t)\}dt \end{array} \right] \underline{P}^{-1}, \tag{3.29}$$

where the covariance matrix (3.12) is partitioned as

$$\underline{P} = \begin{bmatrix} \underline{P}_{\eta\eta} & \underline{P}_{\eta\dot{\eta}} \\ \underline{P}^T_{\eta\dot{\eta}} & \underline{P}_{\dot{\eta}\dot{\eta}} \end{bmatrix} . \tag{3.30}$$

The linearized equations (3.26 - 27) are analysed with respect to harmonic solutions and covariance properties leading to the following nonlinear matrix equations instead of (3.17 - 18):

$$\underline{M}\ \underline{N}_d\ \underline{S}^2 + \underline{D}_d\ \underline{N}_d\ \underline{S} + \underline{K}_d\ \underline{N}_d = \underline{G}\ \underline{H}\ , \tag{3.31}$$

$$\underline{P}^T_{\eta\dot{\eta}} + \underline{P}_{\eta\dot{\eta}} = \underline{0}\ , \tag{3.32a}$$

$$\underline{P}_{\dot{\eta}\dot{\eta}}\ \underline{M} - \underline{P}_{\eta\eta}\ \underline{K}_s^T - \underline{P}_{\eta\dot{\eta}}\ \underline{D}_s^T = \underline{0}\ , \tag{3.32b}$$

$$\underline{D}_s\ \underline{P}_{\dot{\eta}\dot{\eta}}\ \underline{M} + \underline{M}\ \underline{P}_{\dot{\eta}\dot{\eta}}\ \underline{D}_s^T + \underline{K}_s\ \underline{P}_{\eta\dot{\eta}}\ \underline{M} + \underline{M}\ \underline{P}^T_{\eta\dot{\eta}}\ \underline{K}_s^T - \underline{G}\ \underline{D}\ \underline{Q}_1\ \underline{D}^T\ \underline{G}^T = \underline{0}. \tag{3.32c}$$

In (3.32b - c) a stochastic excitation by white noise is assumed in the case of colored noise excitation, (3.32b - c) have to be modified similar to (3.19b regarding the shape filter (3.20).

The solution of (3.31 - 32) requires the evaluation of (3.28 - 29). Again assumptions on the probability density function are necessary. Usually the Gaussian distribution (3.21) is applied but also the use of the uniform distribution (3.22) may be possible. In both cases the joint probability density function with respect to

$$\underline{\xi} = \begin{bmatrix} \underline{\eta} \\ \\ \dot{\underline{\eta}} \end{bmatrix} \tag{3.33}$$

has to be considered.

4. Equivalent Harmonical Linearization in Combination with a Subsequent Covariance Analysis

Usually the analysis of a nonlinear multibody system will be cumbersome because of the simultaneous averaging with respect to time and to randomness as shown in (3.8) and (3.11). Therefore, we are looking for suitable simplifications of the equivalent linearization technique of chapter 3.

Assuming that the amplitudes of the stochastic response $\underline{\xi}(t)$ will be small compared to the limit cycle amplitudes $\underline{m}(t)$, then we use a Taylor series expansion (if possible) to evaluate equation (3.8) approximately:

$$\underline{a}(\underline{m}(t) + \underline{\xi}(t)) \approx \underline{a}(\underline{m}(t)) + \left.\frac{\partial \underline{a}(\underline{x})}{\partial \underline{x}^T}\right|_{\underline{x}=\underline{m}(t)} \cdot \underline{\xi}(t) + \ldots \tag{4.1}$$

Substituting these two terms of the series expansion into (3.8), the first term leads to the common expression of the harmonical linearization technique and the second term vanishes. This yields

$$\underline{A}_d \approx \underline{A}_h \tag{4.2}$$

where $\underline{A}_h$ is the harmonical equivalent gain matrix satisfying

$$\underline{A}_h\, \underline{M}_d = \frac{\omega}{\pi} \int_0^T \underline{a}(\underline{m}(t))\, \underline{c}^T(t)\, dt. \tag{4.3}$$

To simplify matters also the statistical equivalent gain matrix is approximated by

$$\underline{A}_s \approx \underline{A}_h \,. \tag{4.4}$$

Thus, the computational amount is essentially reduced. The coupled equations (3.17 - 18) for limit cycle parameters and covariance matrix are approximately decoupled firstly in an equation

$$\underline{M}_d\, \underline{S} = \underline{A}_h\, \underline{M}_d \quad (+\, \underline{B}\, \underline{H}) \tag{4.5}$$

determining $\underline{M}_d$ and ω of a limit cycle or $\underline{M}_d$ of a harmonic response independent of the covariance matrix, and secondly in a linear Lyapunov matrix equation

$$\underline{A}_h\,\underline{P} + \underline{P}\,\underline{A}_h^T + \underline{R} = \underline{0} \qquad (4.6)$$

which yields the covariance matrix $\underline{P}$.

Of course, we have to pay for this simplification. The disadvantage of this combination of equivalent harmonical linearization and subsequent covariance analysis is that the method fails to calculate a realistic covariance matrix in the case of limit cycle. In this case the matrix $\underline{A}_h$ has purely imaginary eigenvalues $\lambda = \pm j\omega$ leading to a singularity in the Lyapunov matrix equation, cf. the stability discussion in [7]. Nevertheless, this simplified method can be applied characterizing the system response due to harmonic and stochastic excitations.

An alternative which circumvents the problems in case of limit cycles is given by

$$\underline{A}_d \approx \underline{A}_h\,(\underline{M}_d),\ \underline{A}_s = \underline{A}_s\,(\underline{P}, \underline{M}_d), \qquad (4.7)$$

where $\underline{M}_d$ follows from (4.5). The assumption (4.7) provides a decoupling of (3.13), (3.14) and, thus, simplifies the analysis.

5. Application of the Methods to a Self-Sustained Oscillator

As a simple example we investigate the system

$$y''(\tau) + [\alpha + \beta y^2(\tau) + \gamma\, y^4(\tau)]\, y'(\tau) + y(\tau) = \kappa\, w(\tau) \qquad (5.1)$$

Here, τ denotes the dimensionless time and the symbol $'$ means the corresponding derivative. Eq. (5.1) describes qualitively the lateral oscillations of railway vehicles. The right-hand side is a stationary Gaussian white noise process $w(\tau)$ with intensity q. By means of analog computer simulations the probability density function $p(y)$ and the mean square $\overline{y}^2$ of the process $y(\tau)$ have been determined. The results are shown in Fig. 1.

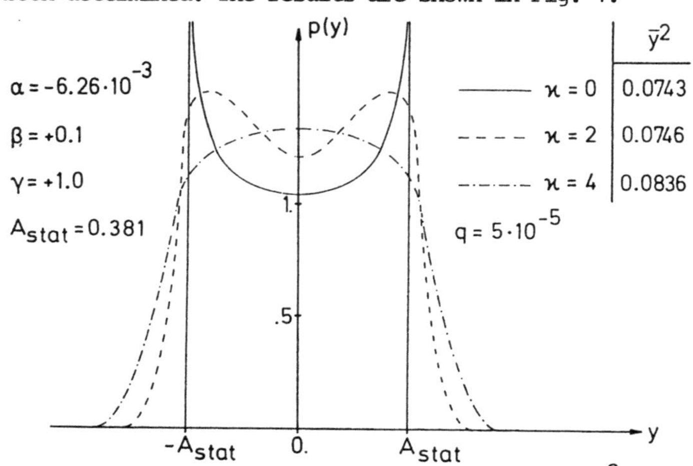

Fig. 1. Probability density function $p(y)$ and mean square $\overline{y}^2$ of stochastically disturbed self-sustained oscillations

Besides an analog computer simulation the system (5.1) has been treated with linearization methods.

5.1 Equivalent Statistical Linearization

If a Gaussian response process is assumed, then the variance $\sigma_y^2 = E\{y^2\}$ satisfies the equation

$$\sigma_y^6 + \frac{\beta}{3\gamma}\,\sigma_y^4 + \frac{\alpha}{3\gamma}\,\sigma_y^2 - \frac{\kappa^2 q}{6\,\gamma} = 0. \tag{5.2}$$

This equation yields only one positive solution $\sigma_y^2 = 0.0399$ for $\kappa = 2$ and $\sigma_y^2 = 0.0534$ for $\kappa = 4$. Starting with the assumption of an uniform distribution one obtains the equation

$$\sigma_y^6 + \frac{5}{9}\,\frac{\beta}{\gamma}\,\sigma_y^4 + \frac{5}{9}\,\frac{\alpha}{\gamma}\,\sigma_y^2 - \frac{5\,\kappa^2 q}{18\gamma} = 0 \tag{5.3}$$

which has the only positive solution

$\overline{y}^2 \equiv \sigma_y^2 = 0.046$ for $\kappa = 2$ and $\overline{y}^2 \equiv \sigma_y^2 = 0.061$ for $\kappa = 4$.

5.2 Combination of Equivalent Harmonical and Statistical Linearization

From equation (5.1) two linear equations can be derived according to equation (3.13) and (3.14). The determination of the amplitude A and the variance σ_y^2 requires the solution of two coupled nonlinear algebraic equations:

$$\alpha + \beta\left(\frac{1}{4}A^2 + \sigma_y^2\right) + \gamma\left(\frac{1}{8}A^4 + \frac{3}{2}A^2\sigma_y^2 + E\{y^4\}\right) = 0, \tag{5.4}$$

$$\alpha\,\sigma_y^2 + \beta\left(\frac{1}{2}A^2 + \sigma_y^2\right)\sigma_y^2 + \gamma\left(\frac{3}{8}A^4 + 3A^2\sigma_y^2 + E\{y^4\}\right)\sigma_y^2 - \frac{1}{2}\kappa^2 q = 0. \tag{5.5}$$

Here, the expected value reads $E\{y^4\} = 3\,\sigma_y^4$ in the case of a Gaussian distribution and $E\{y^4\} = \frac{9}{5}\,\sigma_y^4$ in the case of a uniform distribution. For the chosen parameters, cf. Fig. 1, solutions of (5.4) and (5.5) do not exist. Ambiguos solutions of these equations can be determined for $\kappa < \kappa_{crit}$, where $\kappa_{crit} \approx 1.695$ (Gaussian distribution) and $\kappa_{crit} \approx 1.785$ (uniform distribution).

5.3 Equivalent Harmonical Linearization in Combination with a Subsequent Covariance Analysis

The result of a harmonical linearization and a subsequent covariance analysis is an expression for the variance σ_y^2 depending on the amplitude A of a harmonic oscillation:

$$\sigma_y^2 = \frac{\kappa^2 q}{2\,\alpha + \frac{1}{2}\,\beta\,A^2 + \frac{1}{4}\,\gamma\,A^4} \tag{5.6}$$

The amplitude A can be chosen arbitrarily. But if A is the limit cycle amplitude A_{stat} then the denominator in (5.4) is zero. In the case $A<A_{stat}$ the variance σ_y^2 does not exist. The alternative formulation mentioned at the end of chapter 4 uses

$$\alpha + \frac{1}{4}\beta A^2 + \frac{1}{8}\gamma A^4 = 0 \qquad (5.7)$$

instead of (5.4). The resulting limit cycle amplitude $A=A_{stat} = 0.3809$ coincides with the measured one. This amplitude is substituted into (5.5) which, then, leads to a cubic equation for σ_y^2. The only nonnegative solution $\overset{*}{\sigma}_y^2$ results in mean square values $\overline{y}^2 = \frac{1}{2} A_{stat}^2 + \overset{*}{\sigma}_y^2 = 0.0801$ for $\kappa = 2$ and $\overline{y}^2 = 0.092$ for $\kappa = 4$ in case of a Gaussian and an uniform distribution as well.

5.4 Results

The results of the different linearization methods for the simple example (5.1) can be summarized as follows:

i) Equivalent statistical linearization:
The obtained variances reveal discrepancies of about 50%, compared with the results of the measurements.

ii) Combination of equivalent harmonical and statistical linearization:
It occurs that for given parameters the derived equations for A and σ_y^2 have no solution.

iii) Equivalent harmonical linearization in combination with a subsequent covariance analysis:
Only the results for the alternative formulation are meaningful. They show errors of about 7,4% ($\kappa = 2$) and 10% ($\kappa = 4$), respectively, compared with the results of the measurements.

The large errors of the results i) gained by equivalent statistical linearization are not surprising. They are found also in the literature, e.g. [4], and may be reduced by assuming appropriate probability distribution functions as shown in [8]. The failure of the combined harmonical and statistical linearization method, cf. ii), is unexpected, since this method gives satisfactory results compared to field measurements in case of a more complex system, cf. [9]. Thus, our example can be considered as a counterexample for the application of this combined method. Best results are obtained by an alternative formulation of the equivalent harmonical linearization with a subsequent covariance analysis, cf. iii), however more experience is needed with this method.

References

1. Schiehlen, W.O. (ed.): Dynamics of High-Speed Vehicles. CISM No. 274, Springer, Wien-New York 1982.

2. Popp, K.: Stochastic and Elastic Guideway Models. In [1], pp. 13 - 38.

3. Schiehlen, W.O.: Modeling by Multibody Systems. In [1], pp. 39 - 49.

4. Hedrick, J.K.; Castelazo, I.A.: Statistical Linearization of the Nonlinear Rail Vehicle Wheelset. Proc. 6th IAVSD-Symp. Swets & Zeitlinger, Lisse, 1980, pp. 164 - 178.

5. Csáki, F.: Modern Control Theories. Akadémiai Kiadó, Budapest 1972.

6. Müller, P.C.: Mathematical Methods in Vehicle Dynamics. In [1], pp. 51 - 81.

7. Müller, P.C.: Zur Stabilität von Grenzzykeln. Z. Angew. Math. Mech. 61 (1981), T 49 - T 51.

8. Arslan, A.V.; Hedrick, J.K.: Nonlinear Dynamic Response of a Locomotive Via a Nongaussian Statistical Linearization Procedure. Proc. 7th IAVSD-Symp. Swets & Zeitlinger, Lisse 1982, pp. 521 - 534.

9. Hauschild, W.; Mensen, R.; Schütz, K.; Führer, C.: On the Dynamic Behavior of a Powered Bogie at High Speed. Vehicle System Dynamics 14 (1985), pp. 160 - 165.

A near Minimum Iterative Analytical Procedure for Obtaining a Robot-Manipulator Dynamic Model

Marc RENAUD

Laboratoire d'Automatique et d'Analyse des Systèmes
du Centre National de la Recherche Scientifique
7, avenue du Colonel Roche - 31077 TOULOUSE CEDEX - FRANCE

Summary

The dynamic control synthesis of robot manipulators requires a great number of arithmetic operations, and it cannot be effected in real time unless this number is reduced. This paper presents a systematic analytical procedure for obtaining the dynamic model necessary for the dynamic control synthesis. This procedure which uses a Lagrangian formulation is applicable to all manipulators having a simple kinematic chain structure with revolute and/or prismatic joints.
An example shows that the calculation of the dynamic model requires 368 multiplications and 271 additions for a particular 6 revolute joint manipulator using the systematic procedure. The examination of the results shows that only a few simpliciations are a fortiori possible and proves that the procedure is near-minimum.

1. Introduction

The dynamic characteristics of robot manipulators are highly nonlinear. To control these systems, it is necessary to compute the forces and torques needed to drive all their joints accurately, and frequently at adequate sampling frequency. This necessitates the establishment of the so-called robot manipulator dynamic model, in other words the relationship between the generalized forces which have to be applied at each joint and the generalized coordinates, velocities, and accelerations q, $\dot{q}$ and $\ddot{q}$, such that :

$$\Gamma = \Gamma(q,\dot{q},\ddot{q})$$

The numerical calculations necessary for the dynamic control synthesis cannot be carried out in real-time unless the number of arithmetic operations is reduced.

The aim of this paper is to present a method for computing the dynamic model which minimizes this number. To achieve this, it is demonstrated that analytical methods are more effective than numerical ones.

Dynamics of Multibody Systems
IUTAM/IFToMM Symposium Udine 1985
Editors: G. Bianchi and W. Schiehlen

In order to compare analytical and numerical methods and to arrive at identical results, the same data must be used, i.e.

i) the constant parameters of the manipulator such as angles, lengths, masses, inertias, etc., and every constant quantity which can be previously calculated,

ii) the generalized coordinates, velocities and accelerations.

The unknown quantities of the problem are the generalized forces which have to be applied at each joint.

To simplify the presentation, the worst case of robot manipulators having a simple kinematic chain structure with n joints of the revolute type is studied, as it is well known that the dynamic model is simpler when the robot manipulator possesses some joints of the prismatic type.

2. Formulations for obtaining the dynamic model of robot manipulators

The most important formulations for obtaining the dynamic model of robot manipulators are :

a) the Newton-Euler formulation, and

b) the Lagrange formulation.

The generalized forces numerically obtained using the first or the second formulation are, obviously, the same : thus it is possible to establish relationships between the two sets of equations obtained [12]. The real problem is not, in the author's opinion, to choose between an analytical or numerical method.

The Newton-Euler formulation is oriented towards the numerical method and the Lagrange formulation towards the analytical method. Indeed :

a) an analytical method of computation can be obtained using the Newton-Euler formulation, but there are great difficulties in the elimination of constraint forces and torques at the different joints of the manipulator. Nevertheless, this solution is used by several authors [7-6-16],

b) a numerical method of computation using the Lagrange formulation is difficult to imagine as the calculation of the partial derivatives poses problems.

This is the reason why the most frequently used methods are :

- the numerical Newton-Euler method,
- the analytical Lagrange method.

2.1. The numerical Newton-Euler method

This method is based on the use of both the mass centre theorem (Newton's theorem) and the kinetic momentum theorem (Euler's theorem) applied to each link of the robot manipulator. The generalized forces are numerically calculated in an iterative way. This method is used by several authors [8-15-14-4]. To solve the problem presented in the introduction, this method necessitates 798 multiplications and 615 additions for a general 6 revolute joint robot manipulator.

2.2. The analytical Lagrange method [10]

This method is based on the calculation of the Lagrangian function L of the robot manipulator :

$L = T - U$

where T is the kinetic energy and U refers to the potential energy of the robot manipulator :

$$T = 1/2\, \dot{q}^t A(q) \dot{q} \quad \text{and} \quad U = U(q)$$

(q and $\dot{q}$ are the two column matrices of dimension n x 1 whose elements are the generalized coordinates and velocities of the robot manipulator, respectively) . Matrix A(q) (or more simply A) is the kinetic energy matrix.

The dynamic model is given by the n Lagrange equations :

$$\Gamma^i = \sum_{j=1}^{n} a^{ij}\ddot{q}_j + \sum_{j,k=1}^{n} b^{i,jk}\dot{q}_j\dot{q}_k - G^i \qquad i = 1, 2, \dots, n.$$

in which :

a^{ij} are the elements of matrix A

$b^{i,jk}$, the Christofell symbols given by :

$$b^{i,jk} = 1/2\left(\partial a^{ij}/\partial q_k + \partial a^{ik}/\partial q_j - \partial a^{jk}/\partial q_i\right)$$

G^i are the gravity forces given by :

$$G^i = -\partial U/\partial q_i$$

It suffices to compute a^{ij} for $i \leqslant j$ as matrix A is symmetrical. Therefore the number of these elements which have to be calculated is equal to n(n+1)/2. Such elements are a function of q_k only if $k > i$. Thus the number of partial derivatives of the elements of the matrix A, with respect to the generalized coordinates q, which have to be calculated is equal to (n-1)n (n+1)/3. Then it is sufficient to compute $b^{i,jk}$ for $j \leqslant k$ and $i < k$. Thus, the number of these symbols which have to be calculated is equal to (n-1)n(n+1)/3. Stated more accurately :

$b^{i,jk} = 1/2\,(\partial a^{ij}/\partial q_k + \partial a^{ik}/\partial q_j)$ if $i<j\leq k$, $b^{i,jk} = 1/2\,(\partial a^{ji}/\partial q_k - \partial a^{jk}/\partial q_i)$ if $j\leq i<k$

The n Lagrange equations can be grouped under the matrix form :

$$\Gamma = A\ddot{q} + 2B\dot{q}\dot{q} + C\dot{q}^2 - G$$

in which the matrices $\ddot{q}$, $\dot{q}^2$, G and Γ are column matrices of dimension n x 1. The matrix $\dot{q}\dot{q}$ is a column matrix of dimension $\frac{n(n-1)}{2} \times 1$. The matrices A and C are square matrices of order n and B a rectangular matrix of dimension $n \times \frac{n(n-1)}{2}$.

Different procedures can be used to obtain practically the dynamic model.

In the following, two procedures are proposed and compared.

2.2.1. <u>A non iterative analytical procedure for obtaining a dynamic model of a robot manipulator</u>

This non iterative analytical procedure is based upon the work of Uicker-Kahn [13-5].

Its drawback is that it necessitates redundant calculations, that is calculations which are unnecessary as the notion of trace of a matrix is used. Various attempts have been made to simplify its organization [1]. A new organization of the calculations simplifying the Uicker-Kahn procedure has also been proposed by S. Megahed and this author : this solution requires 1677 multiplications and 1324 additions to solve the problem presented in the introduction for a general 6 revolute joint robot manipulator [9].

Recently S. Megahed has still simplified further the procedure by avoiding the utilisation of the notion of the trace of a matrix : his solution necessitates 1257 multiplications and 988 additions. Applied to the example (see § 4) this procedure necessitates only 602 multiplications and 465 additions. However this non iterative procedure appears to be redundant essentially because of the non utilization of the notion of augmented link (defined hereafter).

This drawback appears also in the iterative procedure presented in [3].

This is why a new procedure based on an iterative organization of the analytical calculation and using the notion of augmented link is at present proposed.

3. ITERATIVE ANALYTICAL PROCEDURE FOR REVOLUTE JOINTS ROBOTS MANIPULATORS [11]

3.1. Definition of vectors and tensors

To present the procedure vectors and tensors of the usual 3 dimensional Euclidian vectorial space are used. In accordance with Gibb's notation vectors are underlined once (e.g. $\underline{a}$) and tensors twice (e.g. $\underline{\underline{t}}$).

The following operations are used :

(a) outer product (designated by "symbol-free") :

(b) inner product (designated by.)

(c) cross product (designated by x)

it can be proved that :

$\underline{a} \times \underline{b} = \underline{\underline{\hat{a}}} \cdot \underline{b}$

where $\underline{\underline{\hat{a}}}$ is the skew-symmetric tensor associated with vector $\underline{a}$

(d) parenthesis product (designated by { }) such that :

$\{ \underline{\underline{t}}, \underline{\underline{u}} \} = \underline{\underline{t}} \cdot \underline{\underline{u}} + \underline{\underline{u}} \cdot \underline{\underline{t}}$

(e) bracket product (designated by [ ]) such that :

$[\underline{\underline{t}}, \underline{\underline{u}}] = \underline{\underline{t}} \cdot \underline{\underline{u}} - \underline{\underline{u}} \cdot \underline{\underline{t}}$

The transposed operation of a tensor (designated') is also used.

3.2. Definition of the robot manipulator structure

Links and joints are numbered from 1 to n starting from the base (link 0). The axis of the ith joint, situated between links i- 1 and i is defined by vector $\underline{z}_i$. Let O_i be a fixed point on this axis, whose definition is specified later, and G_i the mass centre of link i . Let be :

$$\underline{r}_{ij} = \overrightarrow{O_iO_j} \quad \text{and} \quad \underline{d}_{ij} = \overrightarrow{O_iG_j}$$

3.3. Calculation of the theoretical expression of the kinetic energy

The rotational motion around the ith axis - defined by the generalized coordinate q_i - gives link k angular velocity $\dot{q}_i \underline{z}_i$ and linear velocity of mass centre $\dot{q}_i(\underline{z}_i \times \underline{d}_{ik})$. Thus, if the superposition theorem is used the angular and linear velocities of link k due to all the elementary motions can be calculated :

$$\underline{\omega}_k = \sum_{i=1}^{k} \dot{q}_i \underline{z}_i \quad \text{and} \quad \underline{v}_k = \sum_{i=1}^{k} \dot{q}_i (\underline{z}_i \times \underline{d}_{ik})$$

Using the two previous equations, the calculation of rotational kinetic energy Tr and of translational kinetic energy Tt of the robot manipulator can be effected :

$$T_r = 1/2 \sum_{k=1}^{n} \underline{\omega}_k \cdot \underline{\underline{\varphi}}^k \cdot \underline{\omega}_k \quad \text{and} \quad T_t = 1/2 \sum_{k=1}^{n} m^k \underline{v}_k \cdot \underline{v}_k$$

These are expressions in which $\underline{\underline{\varphi}}^k$ is the link k inertia tensor corresponding to mass centre G_k and m^k is the link k mass. (Note that a superscript indicates that the corresponding quantity involves a mass in its dimensional equation). Kinetic energy T of the robot manipulator is simply : $T=T_r+T_t$ and then :

$$T=1/2\sum_{k=1}^{n}\sum_{i,j=1}^{n}\left(\underline{z}_i\cdot\left(\underline{\underline{\varphi}}^k-m^k\hat{\underline{\underline{d}}}_{ik}\cdot\hat{\underline{\underline{d}}}_{jk}\right)\cdot\underline{z}_j\right)\dot{q}_i\dot{q}_j$$

The permutation of the two symbols $\sum$ in the above expression gives :

$$T=1/2\sum_{i,j=1}^{n}a^{ij}\dot{q}_i\dot{q}_j \text{ with } a^{ij}=\underline{z}_i\cdot\underline{\underline{J}}^{ij}\cdot\underline{z}_j \text{ and } \underline{\underline{J}}^{ij}=\sum_{k=Max(i,j)}^{n}\left(\underline{\underline{\varphi}}^k-m^k\hat{\underline{\underline{d}}}_{ik}\cdot\hat{\underline{\underline{d}}}_{jk}\right)$$

The calculation of a^{ij} is now reduced to that of tensor $\underline{\underline{J}}^{ij}$. If it is noted that : $(\underline{\underline{J}}^{ij})'=\underline{\underline{J}}^{ij}$ it can be proved that : $a^{ji}=a^{ij}$ and thus the calculation of $\underline{\underline{J}}^{ij}$ is necessary only for i $\leqslant$ j.

3.4. Calculation of the theoretical expression of the potential energy

Assume that $\underline{z}_1$ is a unit vertical ascendant vector. The potential energy of the robot manipulator due to the Earth's gravity is simply :

$$U=g\sum_{i=1}^{n}m^i\underline{z}_1\cdot\underline{d}_{1i} \qquad \text{(g being the acceleration of gravity).}$$

By defining the notions of augmented link and generalized link, the calculation of the theoretical expressions of kinetic and potential energies will be greatly simplified.

3.5. Notions of augmented link and generalized link

Augmented link i is defined as the fictive link composed of link i and of the mass of links i+1, i+2,...,n attached at point Oi+1 (see Figure 1) [2 - 16].

Generalized link i is defined as the real link composed of links i, i+1,...,n (see Figure 2).

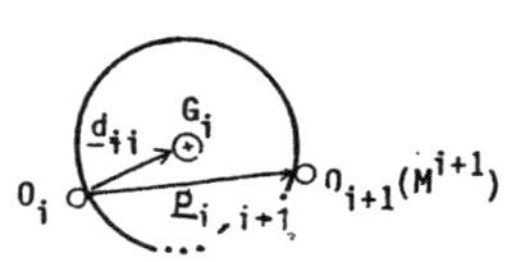

FIGURE 1. Augmented link i

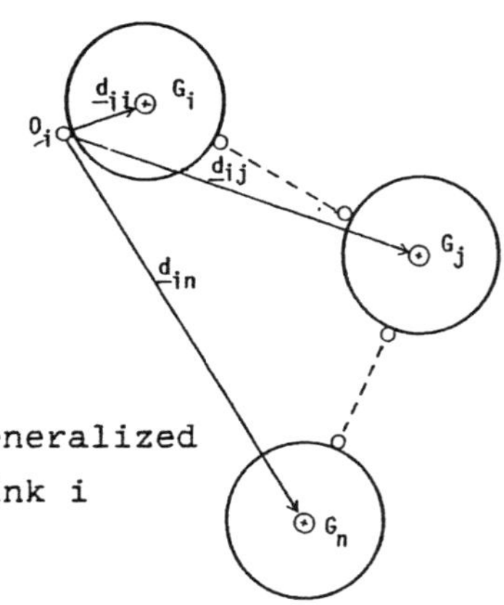

FIGURE 2. Generalized link i

Let be :

Mi the common mass of augmented and generalized links i, $\underline{u}i$ and $\underline{\underline{k}}i$ the first and the second order momenta of augmented link i corresponding to point Oi, respectively, $\underline{U}i$ and $\underline{\underline{K}}i$ the first and the second order momenta of generalized link i corresponding to point Oi, respectively : $M^i = \Sigma_{j=i}^{n} m^j$ and :

$$\underline{u}^i = m^i \underline{d}_{ii} + M^{i+1} \underline{r}_{i,i+1} \qquad \underline{\underline{k}}^i = \underline{\underline{\varphi}}^i - m^i \hat{\underline{\underline{d}}}_{ii} \cdot \hat{\underline{\underline{d}}}_{ii} - M^{i+1} \hat{\underline{\underline{r}}}_{i,i+1} \cdot \hat{\underline{\underline{r}}}_{i,i+1}$$

$$\underline{U}^i = \Sigma_{j=i}^{n} m^j \underline{d}_{ij} \qquad \underline{\underline{K}}^i = \Sigma_{j=i}^{n} \left(\underline{\underline{\varphi}}^j - m^j \hat{\underline{\underline{d}}}^{ij} \cdot \hat{\underline{\underline{d}}}^{ij} \right)$$

3.6. Simplified expressions of the kinetic and the potential energies

The previous expression of the kinetic energy matrix elements may be simplified as :

$$\underline{\underline{J}}^{ij} = \underline{\underline{K}}^j - \hat{\underline{\underline{r}}}_{ij} \cdot \hat{\underline{\underline{U}}}^j \quad (i \leq j)$$

Then it is possible to write :

$$a^{ij} = \underline{z}_i \cdot \underline{\underline{K}}^j \cdot \underline{z}_j + (\underline{z}_i \times \underline{r}_{ij}) \cdot (\underline{z}_j \times \underline{U}^j) \quad (i \leq j)$$

The previous expression of the potential energy U may also be simplified :

$$U = g \underline{z}_1 \cdot \underline{U}^1$$

3.7. Iterative analytical procedure of the vector $\underline{u}^j$ and the tensor $\underline{\underline{k}}^j$ calculations

$\underline{u}^j$ and $\underline{\underline{k}}^j$ are calculated iteratively as a function of $\underline{u}^{j+1}$ and $\underline{\underline{k}}^{j+1}$, respectively as follows :

$$\underline{U}^j = \underline{U}^{j+1} + \underline{u}^j$$

$$\underline{\underline{K}}^j = \underline{\underline{K}}^{j+1} + \underline{\underline{k}}^j - \left\{ \hat{\underline{\underline{r}}}_{j,j+1} \cdot \hat{\underline{\underline{U}}}^{j+1} \right\}$$

3.8. Calculation of the Christofell symbols

This calculation is based upon the expression of the partial derivatives of the vectors and tensors which are included in the expression of kinetic energy matrix elements. One obtains:

$$\underset{(i \leq j)}{\partial a^{ij} / \partial q_k} = \begin{cases} (\underline{z}_i \times \underline{z}_k) \cdot \underline{\underline{K}}^j \cdot \underline{z}_j + ((\underline{z}_i \times \underline{z}_k) \times \underline{r}_{kj} + (\underline{z}_i \times \underline{r}_{ik}) \times \underline{z}_k) \cdot (\underline{z}_j \times \underline{U}^j) & i \leq k \leq j \\ \underline{z}_i \cdot [\hat{\underline{\underline{z}}}_k, \underline{\underline{K}}^k] \cdot \underline{z}_j + ((\underline{z}_j \times \underline{r}_{jk}) \times \underline{z}_i + (\underline{z}_i \times \underline{r}_{ik}) \times \underline{z}_j) \cdot (\underline{z}_k \times \underline{U}^k) & i \leq j \leq k \end{cases}$$

and then :

$$b^{ijk} = \underline{z}_i \cdot \frac{1}{2} \left([\hat{\underline{\underline{z}}}_k, \underline{\underline{K}}^k] - \underline{\underline{(\underline{\underline{K}}^k \cdot \underline{z}_k)}}^{\wedge} \right) \cdot \underline{z}_j + (\underline{z}_i \times \underline{r}_{ik}) \cdot (\underline{z}_j \times (\underline{z}_k \times \underline{U}^k))$$

$\underline{\underline{(\underline{\underline{K}}^k \cdot \underline{z}_k)}}^{\wedge}$ is the skew-symmetric tensor associated with vector $\underline{\underline{K}}^k \cdot \underline{z}_k$

3.9. Calculation of the gravity forces

$$G^j = -g \underline{z}_1 \cdot (\underline{z}_j \times \underline{U}^j)$$

3.10. Practical calculation of the dynamic model

According to the Denavit-Hartenberg notation an orthonormal frame (O_{i+1}, $\underline{x}_{i+1}$, $\underline{y}_{i+1}$, $\underline{z}_{i+1}$) is attached to link i. (See Figure 3). The position and orientation of this frame with respect to its antecedent (O_i, $\underline{x}_i$, $\underline{y}_i$, $\underline{z}_i$) can be stated analytically in terms of the 4 x 4 elementary homogeneous transformation matrix :

$$T_{i,i+1} = \begin{pmatrix} C_i & -S_i\cos\alpha_i & S_i\sin\alpha_i & a_iC_i \\ S_i & C_i\cos\alpha_i & -C_i\sin\alpha_i & a_iS_i \\ 0 & \sin\alpha_i & \cos\alpha_i & r_i \\ 0 & 0 & 0 & 1 \end{pmatrix}$$

with $C_i = \cos q_i$
$S_i = \sin q_i$

The classical 3 x 3 elementary transformation matrix is designated by $R_{i,i+1}$.

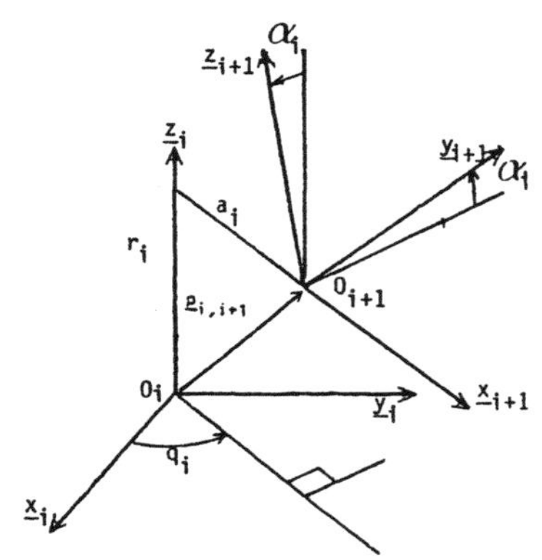

FIGURE 3. Definition of Denavit-Hartenberg parameters

The elementary homogeneous transformation matrices $T_{i,i+1}$ allow the calculation of homogeneous transformation matrix T_{ij} between frames (O_i, $\underline{x}_i$, $\underline{y}_i$, $\underline{z}_i$) and (O_j, $\underline{x}_j$, $\underline{y}_j$, $\underline{z}_j$) ($i < j$) :

$$T_{ij} = T_{i,i+1} T_{i+1,i+2} \cdots T_{j-1,j} \tag{3.34}$$

This matrix is of the form :

$$T_{ij} = \begin{pmatrix} R_{ij} & P_{ij} \\ 0 & 1 \end{pmatrix} \tag{3.35}$$

in which Rij is the classical transformation matrix between frames $(O_i, \underline{x}_i, \underline{y}_i, \underline{z}_i)$ and $(O_j, \underline{x}_j, \underline{y}_j, \underline{z}_j)$. Designate by $\underline{a}(j)$ and $\underline{\underline{t}}(j)$, the column matrix (3 x 1) and the square matrix (3 x 3) of vector $\underline{a}$ and tensor $\underline{\underline{t}}$ components, respectively, with regard to frame $(O_j, \underline{x}_j, \underline{y}_j, \underline{z}_j)$.

Let :

$$z_i(j) = (\alpha_{ij} \;\; \beta_{ij} \;\; \gamma_{ij})^t \qquad p_{ij}(j) = (u_{ij} \;\; v_{ij} \;\; w_{ij})^t$$

$$z_i \times p_{ij}(j) = (\ell_{ij} \;\; m_{ij} \;\; n_{ij})^t \qquad \underline{u}^j(j+1) = (x^j \;\; y^j \;\; z^j)^t$$

$$\underline{U}^j(j+1) = (X^j \;\; Y^j \;\; Z^j)^t \qquad \underline{U}^j(j) = (X^j_j \;\; Y^j_j \;\; Z^j_j)^t$$

$$\underline{\underline{k}}^j(j+1) = \begin{pmatrix} a^j & -f^j & -e^j \\ -f^j & b^j & -d^j \\ -e^j & -d^j & c^j \end{pmatrix} \qquad \underline{\underline{K}}^j(j+1) = \begin{pmatrix} A^j & -F^j & -E^j \\ -F^j & B^j & -D^j \\ -E^j & -D^j & C^j \end{pmatrix}$$

$$\underline{\underline{K}}^j(j) = \begin{pmatrix} A^j_j & -F^j_j & -E^j_j \\ -F^j_j & B^j_j & -D^j_j \\ -E^j_j & -D^j_j & C^j_j \end{pmatrix}$$

Moreover by defining:

$$A^{jk} = F^k_k \alpha_{jk} + 1/2(A^k_k - B^k_k + C^k_k)\beta_{jk} + D^k_k \gamma_{jk} \qquad B^{jk} = \tfrac{1}{2}(A^k_k - B^k_k - C^k_k)\alpha_{jk} - F^k_k \beta_{jk} - E^k_k \gamma_{jk}$$

$$L^{jk} = -X^k_k \gamma_{jk} \qquad N^{jk} = -Y^k_k \gamma_{jk} \qquad N^{jk} = X^k_k \alpha_{jk} + Y^k_k \beta_{jk}$$

We get :

$$a^{ij} = -E^j_j \alpha_{ij} - D^j_j \beta_{ij} + C^j_j \gamma_{ij} - Y^j_j \ell_{ij} + X^j_j m_{ij} \quad (i \leq j)$$

$$b^{i,jk} = A^{jk}\alpha_{ik} + B^{jk}\beta_{ik} + L^{jk}\ell_{ik} + M^{jk} m_{ik} + N^{jk} n_{ik} \quad (i<k \text{ and } j \leq k)$$

and :

$$G^i = g(\alpha_{1i} Y^i_i - \beta_{1i} X^i_i)$$

The elements $A^i_i, B^i_i, C^i_i, D^i_i, E^i_i, F^i_i, X^i_i, Y^i_i,$

are obtained iteratively using steps (a) and (b) :

$$\text{(a)} \quad \underline{\underline{K}}^i(i+1) = \underline{\underline{K}}^{i+1}(i+1) + \underline{\underline{k}}^i(i+1) - \{\hat{p}_{i,i+1}, \underline{\underline{\hat{U}}}^{i+1}\}(i+1)$$

$$\underline{U}^i(i+1) = \underline{U}^{i+1}(i+1) + \underline{u}^i(i+1)$$

$$\text{(b)} \quad \underline{\underline{K}}^i(i) = R_{i,i+1} \underline{\underline{K}}^i(i+1) R_{i+1,i}$$

$$\underline{U}^i(i) = R_{i,i+1} \underline{U}^i(i+1)$$

steps (a) and (b) are repeated for i = n, n-1,...,1 with the initial conditions :

$$\underline{\underline{K}}^n(n+1) = \underline{\underline{k}}^n(n+1) \, , \quad \underline{U}^n(n+1) = \underline{u}^n(n+1)$$

4. EXAMPLE

Consider the 6 revolute joint manipulator represented in Figure 4.

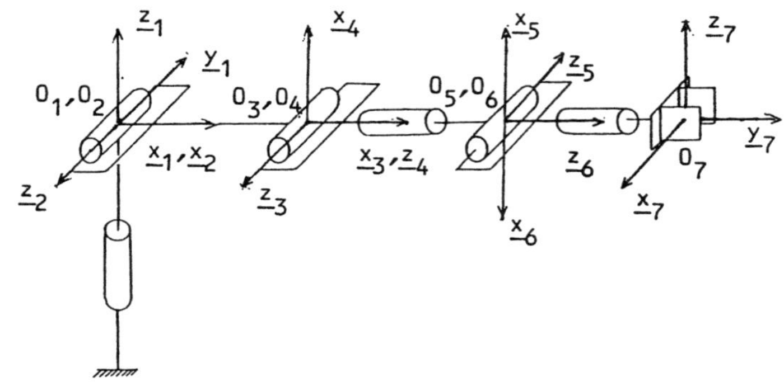

FIGURE 4. 6 revolute joints robot-manipulator

This manipulator is characterized by the constant parameters :

parameter \ i	1	2	3	4	5	6
α_i	$\pi/2$	0	$\pi/2$	$\pi/2$	$\pi/2$	$\pi/2$
a_i	0	a_2	0	0	0	0
r_i	0	0	0	r_4	0	r_6
x^i	0	x^2	0	0	0	0
y^i	y^1	0	0	y^4	0	y^6
z^i	0	0	z^3	0	z^5	0

The other parameters are masses m^i and inertial momenta a^i, b^i, c^i ($i = 1,2,\ldots,6$). Due to the symmetry of the links the inertial products d^i, e^i, f^i are zero.

The number of mathematical operations necessary for obtaining the dynamic model is indicated in the following. (Also indicated is the necessary number of sine and cosine).

Calcul. of \ number of	Sine and Cosine	*	+	* by 2	÷ by 2	Calcul. of \ number of	Sine and Cosine	*	+	* by 2	÷ by 2
$\alpha_{ij}, \beta_{ij}, \ldots$	10	25	7			$A\ddot{q}$		34	28		
$A^i, B^i, \ldots$		47	63	5	8	$B\dot{q}\dot{q}$		73	67		
a^{ij}		21	13			$C\dot{q}^2$		25	19		
$A^{jk}, B^{jk}, \ldots$		43	16			G		9	2		
$b^{i,jk}$		70	40			Γ			16		
$\dot{q}^2, \dot{q}\dot{q}$		21				Total	10	368	271	11	8

5. CONCLUSION

In this paper it has been demonstrated that the iterative analytical procedure using a Lagrange formulation is an efficient means of obtaining the dynamic model of a robot manipulator which has a simple kinematic chain structure with revolute and/or prismatic joints. The author is of the opinion that the minimization of the number of multiplications and additions involved in the dynamic model is due to the fact that the calculation of some constant quantities -concerning different links of the robot manipulator- is only possible once using analytical methods ; when employing numerical methods, these quantities must be calculated at each step. These constant quantities can be predicted thanks to the notion of augmented link. In fact, the proposed procedure is not exactly the minimal one as the choice of the set of generalized coordinates is of major importance for minimization and as a few simplifications can always a fortiori be effected for a given set. (The example shows that the set q1, q2, q2+q3, q4, q5, q6 instead of q1, q2, q3, q4, q5, q6 is better and also that for the given set a few simplifications are possible).

It is also important to note that the number of parameters involved in this procedure which is reduced in comparison with other procedures, is not necessarily the minimum one. By gathering together some parameters it will be possible to achieve a greater simplification and to identify then. It is now intended to obtain the dynamic model automatically in a symbolic form, with this procedure and for any given robot manipulator.

ACKNOWLEDGMENTS : The author wishes to express his thanks to Edson Ferreira, Bruno Gorla and Saïd Megahed for their critical comments concerning this paper.

REFERENCES

[1] A. K. Bejczy, 1974, "Technical memorandum 33-669", Pasadena J.P.L. California Institute of Technology.

[2] O. Fischer, 1906, "Einführung in die mechanik lebender Mechanismen", Leipzig.

[3] J.M. Hollerbach, 1980, "A.I. Memo n°533", Boston, MIT Artificial Intelligence Laboratory.

[4] J.M. Hollerbach and G. Sahar, 1983, "Partitioned inverse kinematic accelerations and manipulator dynamics", The International Journal of Robotics Research, vol.2, n°4.

[5] M.E. Kahn, "The near minimum time control of open loop articulated kinematic chains", Ph.D. thesis, Stanford University. 1970.

[6] A.G. Leskov and V.S. Medvedev, 1974, "Analysis of dynamics and synthesis of movement control of robot manipulator functional organs", Engineering Cybernetics, vol.12, n°6, pp.56-65.

[7] P.W. Likins, 1971, October, "Passive and semi-active attitude stabilizations-flexible space-craft", AGARD.

[8] J.Y.S. Luh, M.W. Walker and R.P.C. Paul, 1980, "On-line computational scheme for mechanical manipulators", ASME Journal of Dynamic Systems, Measurement and Control, vol.102, pp.69-76.

[9] S. Megahed and M. Renaud, 1982, "Minimization of the computation time necessary for the dynamic control of robot manipulators", 12th ISIR, Paris.

[10] M. Renaud, 1980, "Contribution à la modélisation et à la commande dynamique des robots manipulateurs", Thèse d'Etat, Université Paul Sabatier, Toulouse, France.

[11] M. Renaud, 1983, "An efficient iterative analytical procedure for obtaining a robot manipulator dynamic model", 1st Int. Symposium of Robotics Research, B.Woods.

[12] W.M. Silver, 1982, "On the equivalence of Lagrangian and Newton-Euler dynamics for manipulators", The International Journal of Robotics Research, vol.1, n°2.

[13] J.J. Uicker, 1968, "Dynamic behaviour of spatial linkages", ASME, Mech. 5, n°68, pp.1-15.

[14] M. Vukobratović and V. Potkonjak, 1982, "Dynamics of manipulation robots", Berlin, Heidelberg, New-York, Springer-Verlag.

[15] M.W. Walker and D.E. Orin, 1981, "Efficient dynamic computer simulation of robotic mechanisms", JACC, Charlotteville.

[16] J. Wittenburg, 1977, "Dynamics of systems of rigid bodies", Stuttgart, B.G. Teubner.

Global Treatment of Energy Dissipation Effects for Multibody Satellites

F.P.J. RIMROTT
University of Toronto

Summary

For actual satellites, consisting of several bodies, principal axes and principal inertia moments can usually be defined in some fashion, e.g. either by indentifying the undeformed shape of the bodies and the system, or by referring to an attitude of maximum or minimum kinetic energy for a given angular momentum, i.e. collinearity of angular momentum and angular velocity, which in turn is associated with no deformation, this time enforced by the kinetics of the system. If furthermore elasticity is generally neglected, in the sense that there shall be neither storage nor release of recoverable mechanical energy nor changes in the inertia moments, except where needed to account for energy dissipation, then the effects of internal energy dissipation can be described in a global fashion. In the present paper the attitude drift of a two-body gyrostat is studied in some detail. It is shown what deformations of each body are required to bring about an attitude drift. The deformed gyrostat configuration is found to be a function of the (slowly changing) attitude angle only, and thus it is (nearly) constant. As a consequence the gyrostat still consists of bodies that are essentially rigid, though deformed.

Nomenclature

A = $A^{R} + A^{P}$

A^{P} = Principal inertia moment of platform, m²kg

A^{R} = Principal inertia moment of rotor, m²kg

C = $C^{P} + C^{R}$

C^{P} = Principal inertia moment of platform, m²kg, about symmetry axis

C^{R} = Principal inertia moment of rotor, m²kg, about symmetry axis

C' = $C^{P} + C^{R}x$

C = Mass centre

Dynamics of Multibody Systems
IUTAM/IFToMM Symposium Udine 1985
Editors: G. Bianchi and W. Schiehlen

$\dot{D}$ = $\dot{D}^{P} + \dot{D}^{R}$

$\dot{D}^{P}$ = Energy dissipation rate of platform, W

$\dot{D}^{R}$ = Energy dissipation rate of rotor, W

$\dot{E}^{P}$ = Energy transmission rate to platform, W

$\mathbf{H}$ = Angular momentum, m^2kg/s

I^{P}_{uz} = Inertia product of platform, m^2kg

I^{R}_{uz} = Inertia product of rotor, m^2kg

I'_{uz} = $I^{P}_{uz} + I^{R}_{uz}\,x$

Q = Effective dissipation ratio rotor/platform

$\dot{T}$ = Kinetic energy rate, W

W_{d} = Dissipation energy, J

X = Amplitude, m

c = Damping coefficient, N s/m

$\dot{\mathrm{d}}$ = Dissipation rate

$Cuvz$ = Coordinate system rotating at $\mathbf{\Omega}$

$Cxyz$ = Body-fixed coordinate system rotating at $\boldsymbol{\omega}$

x = $\omega^{R}_{z}/\omega^{P}_{z} = \omega^{R}_{z}/\omega_{z}$ = Speed ratio

$\mathbf{\Omega}$ = Angular velocity of $Cuvz$ coordinate system, rad/s

β^{P} = Energy dissipation coefficient of platform, $m^2kg\,s^3$

β^{R} = Energy dissipation coefficient of rotor, $m^2kg\,s^3$

β_{e} = Effective dissipation coefficient, $m^2kg\,s^3$

η = Spin change allotment factor

ν = Nutation (cone) angle ("Attitude")

$\boldsymbol{\omega}$ = Angular velocity (of platform and gyrostat), rad/s

$\boldsymbol{\omega}^{R}$ = Angular velocity of rotor, rad/s

ω_{v} = v-component of angular velocity $\boldsymbol{\omega}$

ω^{R}_{z} = z-component of rotor's angular velocity $\boldsymbol{\omega}^{R}$, rad/s

ω_{z} = z-component of platform's angular velocity $\boldsymbol{\omega}$, rad/s

ω'_{z} = $\dfrac{C' - A}{A}\,\omega_{z}$

ω''_{z} = $\dfrac{Ax - C'}{A}\,\omega_{z}$

$<$ = Smaller than

$<<$ = Much smaller than

$<<<$ = Very much smaller than

Introduction

A very condensed presentation of the global dynamics of gyros with viscous internal energy dissipation is given, for a dual spinner. In principle, the theory can also be extended to multiple spinners.

In the theory the very slow drift rate observed in practice is introduced as a basic assumption, i.e.

$$|\dot{\nu}| \ll \frac{H \cos \nu}{C'}$$

and

$$|\dot{\nu}| \ll \frac{H \sin \nu}{A}$$

apply throughout.

The Platform of a Dual Spinner

We investigate first a dual spinner where only the platform P is dissipative. Energy dissipation is represented by numerous elemental viscous dampers, each of which has a damping coefficient c, and has an energy dissipation per cycle of

$$W_{\mathrm{d}} = \pi c \omega X^2 \tag{1}$$

where X is the amplitude of vibration. The dissipation rate per time is then

$$\dot{W}_{\mathrm{d}} = \frac{1}{2} c \omega^2 X^2 \tag{2}$$

The displacement amplitude X is proportional to the loading amplitude, which in turn is directly proportional to the acceleration amplitude. The accelerations can be shown [1] to produce alternating terms at a frequency ω_z' which contribute amplitude squares containing the factor

$$\omega_v^2 \, \omega_z^2 \tag{3}$$

Thus the dissipation rate becomes proportional to

$$\dot{d}^{P} = c_1 \omega^2 \omega_v^2 \omega_z^2 \tag{4}$$

The frequency ω appearing in expression (1) is

$$\omega = \omega_z' = \frac{C' - A}{A} \omega_z \tag{5}$$

such that the dissipation rate is

$$\dot{d} = c_1 \frac{(C' - A)^2}{A^2} \omega_v^2 \omega_z^4 \tag{6}$$

Now from gyrodynamics

$$\omega_v = \frac{H \sin \nu}{A} \tag{7}$$

$$\omega_z = \frac{H \cos \nu}{C'} \tag{8}$$

such that

$$\dot{d} = c_1 \frac{(C' - A)^2 H^6}{2A^4 C'^4} \cos^2 \nu \sin^2 2\nu \tag{9}$$

Each of the numerous viscous dampers homogeneously and isotropically distributed over the whole platform body contributes an appropriate $\dot{d}$. The dissipation rate integrated over the whole body consequently will assume an appearance

$$\dot{D}^{P} = \frac{(C' - A)^2 H^6}{2A^4 C'^4} \beta^{P} \cos^2 \nu \sin^2 2\nu \tag{10}$$

where β^{P} is the platform's energy dissipation coefficient.

If there is energy dissipation in the platform only, the kinetic energy change T and attitude drift rate $\dot{\nu}$ for the whole gyrostat can be shown [3] to be related by

$$\dot{T} = \frac{(C' - A)H^2}{2AC'} \sin 2\nu \; \dot{\nu} \tag{11}$$

If $\dot{T} = -\dot{D}^P$ then

$$(\dot{\nu})_{\dot{D}^R=0} = -\frac{(C' - A)H^4}{A^2C'^3} \beta^P \cos^2 \nu \sin 2\nu \tag{12}$$

Thus a dual spinner with energy dissipation in the platform only, will have a *negative* attitude drift rate, if $C' > A$ and $0 < \nu < 90°$.

The Rotor of a Dual Spinner

For a dual spinner, where only the rotor R is dissipative, we begin our analysis also with equation (1). The displacement amplitude X is again proportional to the loading amplitude, which in turn is directly proportional to the acceleration amplitude. The accelerations produce alternating terms at a frequency ω''_z which contribute amplitude squares containing the factor

$$\omega_v^2 \, \omega_z^{R^2} = x^2 \, \omega_v^2 \, \omega_z^2 \tag{13}$$

thus the dissipation rate becomes proportional to

$$\dot{d}^R = c_2 \, \omega^2 \, x^2 \, \omega_v^2 \, \omega_z^2 \tag{14}$$

The frequency appearing in equation (1) is, for the rotor,

$$\omega = \omega''_z = \frac{Ax - C'}{A} \omega_z \tag{15}$$

such that the dissipation rate integrated over the whole rotor becomes

$$\dot{D}^R = \frac{(Ax - C')^2 H^6}{2A^4 C'^4} \beta^R \cos^2 \nu \sin^2 2\nu \tag{16}$$

where β^R is the rotor's energy dissipation coefficient.

If there is energy dissipation in the rotor only, then the relation between kinetic energy change $\dot{T}$ and attitude drift $\dot{\nu}$ for the whole dual spinner can be shown [3] to be

$$\dot{T} = -\frac{(Ax - C')H^2}{2AC'} \sin 2\nu \;\; \dot{\nu} \tag{17}$$

Furthermore

$$\dot{T} = -\dot{D}^{R} \tag{18}$$

and consequently

$$(\dot{\nu})_{\dot{D}^{P}=0} = \frac{(Ax - C')H^4}{A^3C'^3} x^2 \beta^{R} \cos^2 \nu \sin 2\nu \tag{19}$$

Thus a dual spinner with energy dissipation in the rotor only will have a *positive* drift rate, if $Ax > C'$ and $0 < \nu < 90°$.

Equation (19), as well as equation (12), by the way satisfy the Second Collinearity Theorem [2], according to which there is no energy dissipation when angular momentum and angular velocity are collinear (i.e. $\nu = 0°, 90°, 180°$, etc.)

The Dual Spinner

If both platform P and rotor R of a dual spinner have energy dissipation then the resulting attitude drift will be the sum of equations (12) and (19), i.e.

$$\dot{\nu} = -\frac{(C' - A)H^4}{A^3C'^3}\left(\beta^{P} - \frac{Ax - C'}{C' - A} x^2 \beta^{R}\right) \cos^2 \nu \sin 2\nu \tag{20}$$

The terms in the brackets can be combined into an effective energy dissipation coefficient

$$\beta_e = \beta^{P} - \frac{Ax - C'}{C' - A} x^2 \beta^{R} \tag{21}$$

such that

$$\dot{\nu} = -\frac{(C' - A)H}{A^3C'^3}\beta_e \cos^2\nu \sin 2\nu \tag{22}$$

where β_e may now also be negative.

The deformation of rotor and platform kinetically necessary to allow for an attitude drift when there is no external torque, makes itself felt as an inertia product. Since it can be shown that axisymmetry militates against I_{uv}, only the I_{uz}'s will occur. For the rotor the required inertia product can be shown to be

$$I^R_{uz} = -\frac{(Ax - C')H^3}{A^2C'^2} x\beta^R \cos\nu \sin 2\nu \tag{23}$$

and for the platform

$$I^P_{uz} = \frac{(C' - A)H^3}{A^2C'^2}\beta^P \cos\nu \sin 2\nu \tag{24}$$

It is convenient to form a reduced inertia product

$$I'_{uz} = I^P_{uz} + I^R_{uz}x = \frac{(C' - A)H^3}{A^2C'^2}\left(\beta^P - \frac{Ax - C'}{C' - A}x^2\beta^R\right)\cos\nu \sin 2\nu \tag{25}$$

Equation (20) by the way, together with an assurance that the total kinetic energy rate is negative, satisfies the First Collinearity Theorem [3], according to which an attitude where angular momentum and angular velocity are not collinear is unstable, and the system tends towards collinearity at the lowest kinetic energy level.

Spin Change Allotment

During attitude drift $\dot{\nu}$, there is an exchange of magnitude between the components H_v and H_z of the (constant) angular momentum $\mathbf{H}$. From

$$H_v^2 + H_z^2 = H^2 = \text{constant} \tag{26}$$

one obtains

$$\dot{H}_v = -\frac{H_z}{H_v}\dot{H}_z \tag{27}$$

and since $H_v = A\,\omega_v = H\sin\nu$ and $H_z = C'\omega_z = H\cos\nu$, one obtains

$$\dot{H}_v = A\,\dot{\omega}_v = H_z\,\dot{\nu} = C'\omega_z\dot{\nu}$$

which is straightforward, and

$$\dot{H}_z = (C'\omega_z)^{\bullet} = C^{P}\dot{\omega}_z + C^{R}\dot{\omega}_z^{R} = -H_v\dot{\nu} = -A\,\omega_v\dot{\nu}$$

which is not. Somehow the total spin component change $\dot{H}_z = -A\,\omega_v\dot{\nu}$ is apportioned between

$$C^{R}\,\dot{\omega}_z^{R} = -\eta\,A\,\omega_v\,\dot{\nu} \tag{28}$$

and

$$C^{P}\,\dot{\omega}_z = -(1-\eta)A\,\omega_v\,\dot{\nu} \tag{29}$$

where η is the *spin change allotment factor* [3]. For all cases of practical interest (i.e. for $C' - A > 0$ and $Ax - C' > 0$) a gyrostat is stable (i.e. $\dot{\nu} < 0$) if

$$\eta \leq 0 \tag{30}$$

as is shown in reference [3].

From reference [3], the spin change allotment factor can be shown to be

$$\eta = -\frac{Q}{1-Q} \tag{31}$$

where $Q = (C' - A)\dot{D}^{R}/(Ax - C')\dot{D}^{P}$ (32)

is the effective dissipation ratio rotor/platform.

By comparing equations (10) and (16), it can be shown that

$$Q = \frac{(Ax - C')x^2 \beta^R}{(C' - A)\beta^P} \tag{33}$$

and since for stability, from equation (20),

$$\beta^P > \frac{(Ax - C')x^2 \beta^R}{(C' - A)} \tag{34}$$

it is concluded that for a stable system

$$0 \leq Q < 1 \tag{35}$$

which, when entered into equation (31), ensures that $\eta \leq 0$, i.e. exactly the requirement (30) for stability.

With the help of the spin change allotment factor η, the dissipation inertia products (23) and (24) become

$$I^R_{uz} = -\eta \frac{A\dot{\nu}}{\omega^R_z} = -\eta \frac{A\dot{\nu}}{x\,\omega_z} \tag{36}$$

$$I^P_{uz} = -(1 - \eta)\frac{A\dot{\nu}}{\omega_z} \tag{37}$$

The kinetic energy rates can be shown [3] to be

$$\dot{T}^R = \left(\frac{A^R}{A} C' - \eta Ax\right) \frac{H^2 \sin 2\nu}{2AC'} \dot{\nu} \tag{38}$$

$$\dot{T}^P = \left(\frac{A^P}{A} C' - (1 - \eta)A\right) \frac{H^2 \sin 2\nu}{2AC'} \dot{\nu} \tag{39}$$

and the energy dissipation rates

$$\dot{D}^R = \eta(Ax - C') \frac{H^2 \sin 2\nu}{2AC'} \dot{\nu} \tag{40}$$

$$\dot{D}^P = (\eta - 1)(C' - A) \frac{H^2 \sin 2\nu}{2AC'} \dot{\nu} \tag{41}$$

Kinetics

Euler's law requires for a torquefree gyrostat that

$$\dot{\mathbf{H}} = \mathbf{0} \tag{42}$$

a condition that leads to $\dot{\nu} = 0$ for a nondissipative gyrostat. For a dissipative gyrostat (ignoring a singularity at $Q = 0$) $\dot{\nu} \neq 0$, a condition that involves a deformation expressable by inertia products.

The angular velocity of the platform is

$$\boldsymbol{\omega}^P = \boldsymbol{\omega} = [\mathbf{e}_u \quad \mathbf{e}_v \quad \mathbf{e}_z] \begin{bmatrix} \dot{\nu} \\ \omega_v \\ \omega_z \end{bmatrix} \tag{43}$$

The angular velocity of the rotor is

$$\boldsymbol{\omega}^R = [\mathbf{e}_u \quad \mathbf{e}_v \quad \mathbf{e}_z] \begin{bmatrix} \dot{\nu} \\ \omega_v \\ x\,\omega_z \end{bmatrix} \tag{44}$$

The angular velocity of the $Cuvz$ coordinate system is

$$\boldsymbol{\Omega} = [\mathbf{e}_u \quad \mathbf{e}_v \quad \mathbf{e}_z] \begin{bmatrix} \dot{\nu} \\ \omega_v \\ \frac{C'}{A}\,\omega_z \end{bmatrix} \tag{45}$$

Let the deformed platform's inertia tensor be

$$[I]^{P} = \begin{bmatrix} A^{P} & 0 & I^{P}_{uz} \\ 0 & A^{P} & 0 \\ I_{uz} & 0 & C^{P} \end{bmatrix} \tag{46}$$

and the deformed rotor's

$$[I]^{R} = \begin{bmatrix} A^{R} & 0 & I^{R}_{uz} \\ 0 & A^{R} & 0 \\ I^{R}_{uz} & 0 & C^{R} \end{bmatrix} \tag{47}$$

The angular momentum of the platform is then

$$\mathbf{H}^{P} = [\mathbf{e}_{u} \quad \mathbf{e}_{v} \quad \mathbf{e}_{z}] \begin{bmatrix} A^{P}\dot{\nu} + I^{P}_{uz}\omega_{z} \\ A^{P}\omega_{z} \\ C^{P}\omega_{z} \end{bmatrix} \tag{48}$$

where the term $I^{P}_{uz}\dot{\nu}$ in the third component has been neglected because $I^{P}_{uz}\dot{\nu} \lll C^{P}\omega_{z}$.

The angular momentum of the rotor is then

$$\mathbf{H}^{R} = [\mathbf{e}_{u} \quad \mathbf{e}_{v} \quad \mathbf{e}_{z}] \begin{bmatrix} A^{R}\dot{\nu} + I^{R}_{uz}\omega_{z} \\ A^{R}\omega_{z} \\ C^{R}\omega^{R}_{z} \end{bmatrix} \tag{49}$$

with the term I^{R}_{uz} neglected in the third component, because $I^{R}_{uz}\dot{\nu} \lll C^{R}\omega^{R}_{z}$.

The angular momentum of the whole gyrostat is

$$\mathbf{H} = \mathbf{H}^P + \mathbf{H}^R = [\mathbf{e}_u \quad \mathbf{e}_v \quad \mathbf{e}_z] \begin{bmatrix} 0 \\ A\,\omega_v \\ C'\omega_z \end{bmatrix} = [\mathbf{e}_u \quad \mathbf{e}_v \quad \mathbf{e}_z] \begin{bmatrix} 0 \\ H \sin\nu \\ H \cos\nu \end{bmatrix} \tag{50}$$

where $H_u = 0$ can only be achieved if

$$H_u = A\,\dot{\nu} + I'_{uz}\,\omega_z = 0 \tag{51}$$

As a consequence

$$I'_{uz} = -\frac{\dot{\nu}}{\omega_z} A \tag{52}$$

with $\dot{\nu}$ as given by equation (20), the reduced inertia product I'_{uz} as given by equation (25) is obtained.

Conclusion

A theory for the attitude drift rate, and attitude stability, for gyrostats with viscous internal energy dissipation is given. The standard approach used by space engineers to this point in time has been to make the rotor *nondissipative* and the platform *dissipative*, to ensure attitude stability. The incorporation of a nondissipative solid into a design is a chancy undertaking, in particular if one takes into consideration that in real cases the rotor's energy dissipation is often found to be much higher than the platform's. The present theory sheds light into how much energy dissipation is actually occurring in the rotor and on how much energy dissipation in the rotor can be tolerated without affecting the gyrostat's attitude stability. New design rules should call now for an energy dissipation ratio Q between rotor and platform that is smaller than one (and greater than zero), to ensure attitude stability. A small Q includes the previous condition of rotor nondissipativity, now in the form of a rotor dissipativity as small as possible, which means a rotor as stiff *and small* as possible.

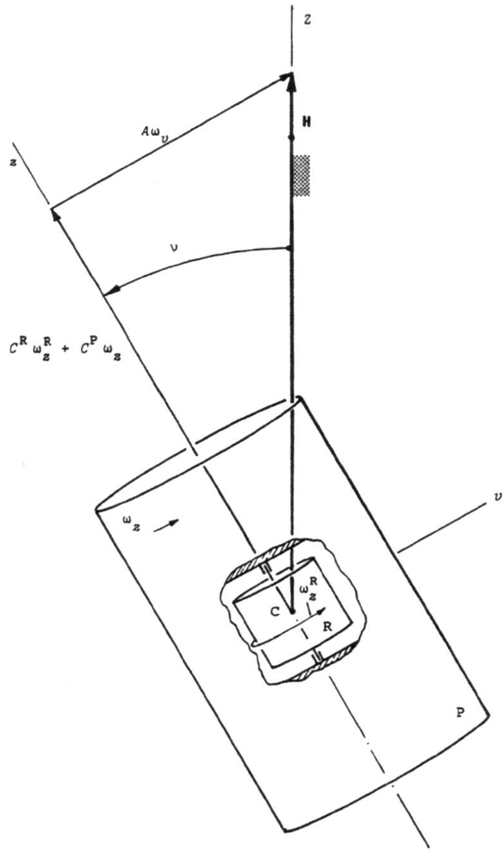

Fig. 7. A Torquefree Nondissipative Dual Spinner

References

[1]. Rimrott, F.P.J.
"Nutation Accretion or Decay due to Internal Energy Dissipation"
Transactions of the CSME, paper 85-CSME-34

[2]. Rimrott, F.P.J.
"Dissipative Rigids in Gyrodynamics"
ZAMM, 66, 7, (1985), 287-301

[3]. Rimrott, F.P.J.; You-sun Li
"Attitude Stability of Torquefree Axisymmetric Dual Spinners with Energy Sinks, Considering Spin Change Allotment between Rotor and Platform"
Mechanics Research Communications, 9, 6, (1982), 403-410

Equilibria and Stability of Satellite Gyrostats: A Comprehensive Review

R.E.ROBERSON

Department of Applied Mechanics and Engineering Science,
University of California San Diego,
La Jolla, CA 92037, USA

V.A.SARYCHEV

Institute of Applied Mathematics
Academy of Sciences of the USSR
Moscow, USSR

ABSTRACT

The paper summarizes the basic results for the types and number of equilibria of an orbiting gyrostat, the stability of these equilibria and the history of their development.

1. Introduction

Finding the equilibrium orientations of a satellite gyrostat traveling in a circular orbit in an inverse square field and establishing the stability properties of these orientations have been subjects for investigation for a number of years. Many of this puzzle have been discovered and fit together during this time, and investigators have gotten an idea of how the complete picture will look. However, until recently the final pieces remained well-hidden and the puzzle could not quite be finished. Now these missing pieces have been discovered. It seems to be an appropriate time to stand back and look at the whole picture as it now stands. That is the primary purpose of this paper.

A second purpose is to prepare a solid foundation for future work. Inevitably, other investigators will discover that the picture can be constructed from fewer and more elegant pieces, perhaps in a simpler way, now that one version of the puzzle has been fit together. This comprehensive review is intended to given them a useful starting point.

Equilibrium problems for the satellite gyrostat divide themselves naturally into two main classes, the *direct problems* and the *inverse problem.* The former can be subdivided further into major cases whose treatments have differing degrees of difficulty. These cases must be identified clearly at the outset, for they are reflected in previously published work on the satellite gyrostat problem. We can do this with precision only after the equilibrium conditions have been cast in symbolic terms, so that is our first step. Thereafter, each of the cases is reviewed separately.

2. Equilibrium Equations

All equations are given initially in matrix form. The following definitions are used:

O	a point, the center of mass of the gyrostat, supposed to follow a circular orbit in an inverse square field
ω_o	angular velocity of point O about its attracting center
I_{ij}	inertia components of the gyrostat about O, in body axes
I	3x3 matrix $[I_{ij}]$
A, B, C	principal moments of inertia, eigenvalues of I
h_i	components of internal angular momentum, in body axes
h	3x1 matrix $[h_i]$
ω_i	components of absolute angular velocity of the gyrostat rotating about O, resolved in body axes

Dynamics of Multibody Systems
IUTAM/IFToMM Symposium Udine 1985
Editors: G. Bianchi and W. Schiehlen

ω	3x1 matrix $[\omega_i]$
E	3x3 identity matrix
$\mathbf{e}_i^o$	triad of orthonormal, dextral unit vectors associated with the orbit: $\mathbf{e}_3^o$ is the outward-pointing geocentric vertical through O; $\mathbf{e}_2^o$ normal to the orbit plane such that $\mathbf{e}_2^o$ is in the direction of forward motion
$\mathbf{e}^o$	the 3×1 array $[\mathbf{e}_i^o]$
$\mathbf{e}_i^B$	triad of orthonormal, dextral unit vectors embedded in the gyrostat
$\mathbf{e}^B$	the 3×1 array $[\mathbf{e}_i^B]$
Z	a 3×3 direction cosine matrix defined by $\mathbf{e}^B = Z\mathbf{e}^o$
α,β,γ	3×1 matrices, columns of Z, thus matrices of components of $\mathbf{e}_1^o,\mathbf{e}_2^o,\mathbf{e}_3^o$ in body axes
$\tilde{v}$	for any 3×1 matrix v, $\tilde{v}$ stands for the 3×3 skew-symmetric matrix $[\epsilon_{\lambda\mu\nu}v_\mu]$

The rotational dynamical equations for the satellite gyrostat are simply Euler's rotational equations in which an expression is used for the applied torque which corresponds to an inverse square gravitational field. In matrix form it is

$$I d\omega/dt + \tilde{\omega}(I\omega + h) = 3\omega_o^2\,\tilde{\gamma} I \gamma \tag{1}$$

The constant coefficient on the right side results from the assumption that the orbit is circular. This expression for gravitational torque was given in scalar components in the classical literature. d'Alembert [1] had such an expression for the special case of a spheroid in an inverse square gravitational field, as did Euler [2]. It was used by Lagrange in his prize work on the libration of the moon in 1763 [3], and in his subsequent 1780 [4] analysis of this problem. LaPlace's *Mécanique céleste* probably helped put these torque expressions into the mainstream of astronomy. Since the advent of artificial satellites they have been written in various vector and tensor forms by a number of authors. The matrix form used here was first expressed, to the best of our knowledge, in [5].

The absolute angular velocity of the orbiting vector triad $\{\mathbf{e}_\lambda^o\}$ is $\omega_o\mathbf{e}_2^o$. Thus if the gyrostat is to remain in an equilibrium orientation in this frame it must have that same constant angular velocity, whose components in body axes are $\omega_o\beta$. Using this for matrix ω in Eq.1 and setting $\dot{\omega}=0$ for equilibrium gives as a necessary condition for equilibrium

$$\omega_o\tilde{\beta}(\omega_o I\beta + h) = 3\omega_o^2\tilde{\gamma} I\gamma \tag{2}$$

Provided the body is given initial conditions corresponding to equilibrium, this condition also is sufficient.

It is convenient to get rid of the constant ω_o by renormalizing h. Thus we define matrix

$$H = h/\omega_o \tag{3}$$

whose physical dimensions are those of moment of inertia, and Eq.2 becomes

$$\tilde{\beta}(I\beta + H) = 3\tilde{\gamma} I\gamma \tag{4}$$

Henceforth we work with the equilibrium condition in this form.

3. The Inverse Problem

Two fundamental classes of problems based on Eq.4 can be distinguished. *Direct problems* are those in which I and H are given and all solutions for α,β,γ are to be found. In the *inverse problem* I and α,β,γ are given and H is to be found. The latter problem can be rephrased as: given the inertia matrix and the desired orientation for the gyrostat, choose an internal angular momentum which makes this orientation one of equilibrium. Only the inverse problem has a simple and straightforward solution.

Rewrite Eq.4 as

$$\tilde{\beta}H = 3\tilde{\gamma}I\gamma - \tilde{\beta}I\beta \tag{5}$$

Premultiply Eq.5 successively by γ^T and α^T to get

$$\alpha^T H = -\alpha^T I\beta \tag{6a}$$

$$\gamma^T H = -4\gamma^T I\beta \tag{6b}$$

Then, because α,β,γ form a basis for 3-space, we can expand H as a linear combination of these vectors to get

$$H = -(\alpha^T I\beta)\alpha + \lambda\beta - (4\gamma^T I\beta)\gamma \tag{7}$$

where all quantities on the right are given except λ. Its value is arbitrary as regards the establishment of equilibrium, though not as regards stability.

4. Special Direct Problems

4.1 Types of Problems

The "general case" of the direct problem is the one in which I has no repeated eigenvalues and all $H_i \neq 0$. Only cases for which one or more $H_i = 0$ are easily solved, so we begin with situations in which H has one of these specialized forms. If I has two or more equal eigenvalues, without loss of generality one can choose body axes so that one of these special cases of H is realized. Although the rigid body is a special subcase of Case 1 below, this paper concerns only the true gyrostat for which $\mathbf{H} \neq 0$.

To classify the direct problems we follow [5] and express $I\alpha, I\beta, I\gamma$ in the spanning frame α,β,γ. Using conditions of symmetry (such as $\alpha^T I\gamma = \gamma^T I\alpha$) and the equilibrium conditions, Eq.4, we obtain

$$(I - \lambda E)\alpha + J_1\beta = 0 \tag{8a}$$

$$(I - \mu E)\beta + J_1\alpha + (J_3/4)\gamma = 0 \tag{8b}$$

$$(I - \nu E)\gamma + (J_3/4)\beta = 0 \tag{8c}$$

in which λ,μ,ν are unknown constants. Also, J_1 and J_3 are the projections of H on α and γ respectively. Four exhaustive and mutually exclusive cases are characterized by $(J_1 = J_3 = 0)$, $(J_1 = 0, J_3 \neq 0)$, $(J_1 \neq 0, J_3 = 0)$ and $(J_1 J_3 \neq 0)$.

4.2 Number and Location of Equilibria

Case 1 $(J_1 = J_3 = 0)$

The conditions defining Case 1 are necessary and sufficient conditions that β coincide with a principal axis of inertia. Equations 8 show that if the condition is satisfied then both α and γ are eigenvalues of I. Hence, regardless of the magnitude of J_2, β also must be an eigenvalue. This shows sufficiency. If β is an eigenvalue, Eq.8b reduces to a linear combination of α,β,γ. These vectors are linearly independent, so their coefficients must vanish, i.e. $J_1 = J_3 = 0$ and necessity is established. An argument based on Eqs.8, using linear algebra, was given in [6] to settle definitively an erroneous claim in the literature that this set of equilibria is not exhaustive.

Case 1 cannot exist for $\mathbf{H} \neq 0$ except when $\mathbf{H}$ is aligned with a principal axis of inertia and that axis is in the direction of $\pm\,\beta$. There then remain 4 choices for one of the two remaining principal axes, whereupon the direction of the third axis is determined. Thus there is a total of 8 possible equilibria.

Case 2 $(J_1 = 0, J_3 \neq 0)$

The conditions of this case are necessary and sufficient conditions that α be a principal axis of inertia. The arguments for both necessity and sufficiency are based on Eqs.8 and parallel those for Case 1. If **H** has no component on α, it follows that **H** must lie in the plane of the two remaining axes, i.e. in the β,γ-plane normal to the orbit path. Without loss in generality, we may suppose that $\pm\,\alpha$ coincides with the principal body axis $\mathbf{e}_1^B$, so **H** lies in the plane of body axes $\mathbf{e}_2^B$,$\mathbf{e}_3^B$.

Putting J_1=0 in Eqs.8 and eliminating β from the result, one is led to

$$\left[(I-\mu E)(I-\nu E)-(J_3/4)^2E\right]\gamma = 0 \tag{9}$$

The body component γ_1 of γ must vanish because γ is orthogonal to α which coincides with $\mathbf{e}_1^B$. Not both α and γ can be eigenvectors of I except in Case 1, so it follows that both γ_2 and γ_3 must be non-zero. Hence Eq.9 yields

$$(B-\mu)(B-\nu)-(J_3/4)^2 = 0 \tag{10a}$$

$$(C-\mu)(C-\nu)-(J_3/4)^2 = 0 \tag{10b}$$

Eliminating ν from these equations, we find that

$$2\mu = B+C\pm\sqrt{(B-C)^2-(J_3/2)^2} \tag{11}$$

An identical expression is gotten for ν.

Let J be the magnitude of the **H**-vector and suppose that **H** lies in the plane of $\mathbf{e}_2^B$, $\mathbf{e}_3^B$ at an angle ϕ to the former, such that

$$H_2 = J\cos\phi, \qquad H_3 = J\sin\phi \tag{13}$$

In the direct problem, both J and ϕ are known quantities. In Case 2, body axes can be gotten by a rotation of the orbit frame through some angle θ about the forward direction α. Define θ so that

$$\alpha = \begin{bmatrix}1\\0\\0\end{bmatrix}, \qquad \beta = \begin{bmatrix}0\\\cos\theta\\-\sin\theta\end{bmatrix}, \qquad \gamma = \begin{bmatrix}0\\\sin\theta\\\cos\theta\end{bmatrix} \tag{14}$$

Then

$$J_3 = \begin{bmatrix}0 & J\cos\phi & J\sin\phi\end{bmatrix}\gamma = J\sin(\theta+\phi) \tag{15}$$

Substituting β and γ from Eq.14 and μ from Eq.11 into Eq.8b, then eliminating the square root in the result, leads to the equilibrium condition

$$\sin(\theta+\phi) = \kappa\,\sin2\theta \tag{16}$$

in which

$$\kappa = 2(C-B)/J \tag{17}$$

Given ϕ and the parameters that define κ, Eq.16 is a transcendental equation to solve for θ, to give the equilibrium orientation of the body frame in the orbit frame.

Introduce a variable t defined by

$$\sin\theta = 2t/(1+t^2), \qquad \cos\theta = (1-t^2)/(1+t^2) \tag{18}$$

into Eq.16 to get the quartic in t

$$t^4-t^3\left(\frac{4\kappa+2\cos\phi}{\sin\phi}\right)+t\left(\frac{4\kappa-2\cos\phi}{\sin\phi}\right)-1 = 0 \tag{19}$$

Equation 19 has at least one solution for t regardless of the values of κ and ϕ. (Solve the equation for κ, writing the result as $\kappa=g(t)$. Put $t=-1+\epsilon\,(\epsilon>0)$ and note that $g(t)\to-\infty$ *as* $\epsilon\to0$. Next, put $t=-\epsilon$ and note that $g(t)\to+\infty$ *as* $\epsilon\to0$. Thus, for any κ, there must be a root of Eq.19 in the range $-\pi\leqslant\theta\leqslant\pi$). It follows that there must be 2 or 4 real roots of Eq.19 in this range of θ.

Classical algebra teaches that the discriminant $\Delta=0$ separates these cases, $\Delta>0$ corresponding to 4 real roots (or zero, which has been shown to be untrue) and $\Delta<0$ corresponding to 2 real roots. To explore the division between the two cases, one writes out the discriminant

$$\Delta = \frac{64}{\sin^6\phi}\left[\kappa^2-1)^3-27(\kappa^2)\sin^2 2\phi\right] \tag{20}$$

The value of κ which separates the cases of 2 and 4 equilibria is given by the solution of

$$(4\kappa^2)^3-3(4\kappa^2)^2+3(1-9\sin^2\phi\cos^2\phi)(4\kappa^2)-1 = 0 \tag{21}$$

which, by Tartaglia's method for solving the cubic, is

$$4\kappa^2 = 1+3(\sin\phi\cos\phi)^{2/3}(\cos^{2/3}\phi+\sin^{2/3}\phi) \tag{22}$$

Larger $|\kappa|$ (smaller values of J) give 4 equilibria, smaller $|\kappa|$ (larger values of J) give 2 equilibria.

In summary, the body axis to which **H** is normal can be aligned with α in either sense. Then there are 2 or 4 real values of θ in the range $-\pi \leqslant \theta \leqslant \pi$ which correspond to equilibrium positions. Thus there is a total of 4 or 8 equilibria for Case 2.

Case 3 ($J_1\neq 0, J_3=0$)

The conditions of this case are necessary and sufficient conditions that γ be a principal axis of inertia. The development for Case 3 is entirely parallel to that for Case 2. We may suppose withoug loss in generality that γ lies along the principal body axis $\mathbf{e}_3^B$, so H lies in the norizontal β,γ-plane and in the plane of body axes $\mathbf{e}_1^B, \mathbf{e}_2^B$. Putting $J_3=0$ in Eqs.8 and eliminating β leads to

$$[(I-\mu E)(I-\lambda E)-(J_1)^2 E]\alpha = 0 \tag{23}$$

Neither α_1 and α_2 can be zero, so

$$(A-\mu)(A-\lambda)-(J_1)^2 = 0 \tag{24a}$$

$$(B-\mu)(B-\lambda)-(J_1)^2 = 0 \tag{24b}$$

and one finds that

$$2\lambda = A+B\pm\sqrt{(A-B)^2-(2J_1)^2} \tag{25}$$

with an identical expression for μ.

We put

$$H_2 = J\cos\phi, \qquad H_1 = -J\sin\phi \tag{26}$$

and

$$\alpha = \begin{bmatrix}\cos\theta\\ -\sin\theta\\ 0\end{bmatrix} \quad \beta = \begin{bmatrix}\sin\theta\\ \cos\theta\\ 0\end{bmatrix} \quad \gamma = \begin{bmatrix}0\\ 0\\ 1\end{bmatrix} \tag{27}$$

so that

$$J_1 = \begin{bmatrix}-J\sin\phi & J\cos\phi & 0\end{bmatrix}\alpha = -J\sin(\theta+\phi) \tag{28}$$

Substituting α and β from Eq.25 and λ from Eq.23 into Eq.8a, one is led to precisely the same equilibrium condition as in Case 2, namely Eq.16, except that now

$$\kappa = (B-A)/2J \tag{29}$$

Equations 19 and 20, and the argument that accompanies them in Case 2, are unchanged in Case 3 except that κ may have a different value.

Historical Remarks on the Special Cases

The three special cases above were discovered independently, at about the same time, by Anchev [7] and Roberson [5]. The latter observed explicitly that the cases are mutually exclusive and exhaustive, and made free use of matrix formulations of equations. An unnecessary complication of the analysis in [7] is caused by the use of an inertial reference frame rather than $\mathbf{e}^o$. The equilibrium conditions gotten in the two works are the same. Actually, there had been forerunners of these cases in the form of spinning rigid bodies, which are effectively gyrostats whose outer bodies have been stripped away, leaving only the spinning symmetric rotor. We shall not dwell on these subcases, but refer the reader to [8] for the pertinent references. We mention only that Likins [9] found the three special cases above for such systems. A status summary of orbiting gyrostat equilibria was given in 1968 [8], which also showed that the unrestricted problem (taking into account the coupling between orbital and attitude motion) can be replaced by a restricted problem (considering attitude motion alone) simply by changing the value of parameter κ slightly.

4.3 Stability of Equilibria

The first treatment of stability of orbiting gyrostats was that of Anchev [7]. He constructed Liapunov functions by Chetaev's method and found sufficient conditions for the stability of all three cases. However, in Case 2 he uses only a restricted perturbation from the equilibrium, so he is led to a sufficient condition for stability that is slightly, but unnecessarily, restricted. Rumiantsev [10],[11] treats stability by another method involving a "dynamic potential energy function". (The method is fully described by Likins in [12].) At one point the variables used to establish positive definiteness of the function are treated as independent, whereas they are in fact not. The consequence is that, while the sufficient conditions he finds are correct, they are a little more stringent than necessary.

Roberson [13] finds stability conditions using a different Liapunov function, but also uses a third approach based on the Thompson-Tait-Chetaev theorem for the linearized second order matrix differential equation describing small rotations about equilibrium. The equilibrium is stable if the stiffness matrix K is positive definite. Logically, the second method is equivalent to the method of the dynamic potential energy function. Both methods used in [13] lead to precisely the same sufficient conditions for stability, and to slightly larger stability regions than those of the earlier works.

Case 1

Sufficient conditions for stability are as follows:

1. $\alpha,+\beta,\gamma$ aligned with $\mathbf{e}^B$

C< A< B	stable for all J
C< B< A	stable if $J > \mid B-A \mid$
B< C< A	stable if $J > max(4\mid B-C\mid, \mid B-A\mid)$

2. $-\alpha,-\beta,\gamma$ aligned with $\mathbf{e}^B$

C< A< B	stable if $J < \mid B-A \mid$

Case 2

Sufficient conditions for stability when α is aligned with $\mathbf{e}_1{}^B$ are:

$$J\cos\phi + (B-C)\cos 2\theta > 0 \tag{30a}$$

$$J\cos\phi + 4(B-C)\cos 2\theta > 0 \tag{30b}$$

$$(J\cos\phi - P)Q > (3J\sin\phi/4)^2 \tag{30c}$$

where

$$P = A - B\cos^2\theta - C\sin^2\theta, \qquad Q = A - B\sin^2\theta - C\cos^2\theta \tag{30d}$$

Case 3

Sufficient conditions for stability when γ is aligned with $\mathbf{e}_3{}^B$ are:

$$J\cos\phi+(B-A)\cos 2\theta > 0 \tag{31a}$$

$$3(A+B-2C)+P > 0 \tag{31b}$$

$$9(A-C)(B-C)+3(J\cos\phi+P)Q > 0 \tag{31c}$$

where

$$P = A\sin^2\theta+B\cos^2\theta-C \qquad Q = A\cos^2\theta+B\sin^2\theta-C \tag{31d}$$

5. The General Case ($J_1J_3\neq 0$)

In this case, none of the vectors α,β,γ can be a principal axis of inertia. To prove this one need only assume that one of them is a principal axis, whence Eqs.8 show that either $J_1=0$ or $J_3=0$, contrary to hypothesis. Moreover, as shown in [14], no Case 4 solutions exist for a body having an axis of symmetry.

It was recognized both in [13] and [14] that Case 4 has two major subcases. We distinguish Case 4A, in which a principal axis is normal to one of α,β *or* γ but is not aligned with any of these axes, from Case 4B in which no principal axis is either along or normal to α,β *or* γ.

5.1 Case 4A

This subcase is solved completely in [14]. It is shown there that it can exist only when one principal axis is in the α,γ-plane. Let body axes be given relative to orbit axes by

$$\mathbf{e}^B = {}^3A(\psi)\,{}^2A(\theta)\mathbf{e}^o \tag{32}$$

where ${}^3A(\psi)$ and ${}^2A(\theta)$ are the usual canonical rotation matrices about the indicated axes. Then

$$\alpha = \begin{bmatrix}\cos\psi\cos\theta\\ -\sin\psi\cos\theta\\ \sin\theta\end{bmatrix},\quad \beta = \begin{bmatrix}\sin\psi\\ \cos\psi\\ 0\end{bmatrix},\quad \gamma = \begin{bmatrix}-\cos\psi\sin\theta\\ \sin\psi\sin\theta\\ \cos\theta\end{bmatrix} \tag{33}$$

Using these in the equilibrium conditions, Eq.4, results in

$$H_3\cos\psi = 3(C-B)\sin\psi\sin\theta\cos\theta \tag{34a}$$

$$-H_3 = 3(C-A)\cos\psi\sin\theta\cos\theta \tag{34b}$$

$$(B-A)\sin\theta\cos\theta+H_2\sin\psi-H_1\cos\psi = 3(A-B)\sin\psi\cos\psi\sin^2\theta \tag{34c}$$

Divide Eq.34a by 34b to get

$$\tan^2\psi = \frac{A-C}{C-B} \tag{35a}$$

The requirement that $\tan^2\psi$ exceed zero implies that

$$A < C < B \quad \text{or } B < C < A \tag{36}$$

That is, the moment of inertia about the axis lying in the α,γ-plane must be the *intermediate* moment of inertia. Equation 35a has four solutions. Let ψ be that solution for which

$$\sin\psi = \sqrt{\frac{A-C}{A-B}} \qquad \cos\psi = \sqrt{\frac{C-B}{A-B}} \tag{35b}$$

We shall refer to this as the primary solution. The remaining three solutions are $-\psi,\pi\pm\psi$.

Next, multiply the left side and right side of Eq.34a by the left and right sides respectively of Eq.34b. This yields

$$\sin^2 2\theta = \frac{4H_3^2}{9(C-B)(A-C)} \tag{37a}$$

From this it is clear that H_3 is subject to the inequality

$$|H_3| < \frac{3}{2}\sqrt{(C-B)(A-C)} \tag{38}$$

In lifting the square root of Eq.37, the choice of algebraic sign must be consistent with the algebraic signs of $\sin\psi$ and $\cos\psi$. A solution for θ is

$$\theta = \frac{1}{2}\ principal\ arcsin\left(\frac{2H_3}{3\sqrt{(C-B)(A-C)}}\right) \tag{37b}$$

lying in the range $-\pi/2 < \theta < \pi/2$. Clearly, $\theta \pm \pi$ also are solutions. As a convention, when $H_3 > 0$ two independent solutions are chosen to be $\theta > 0$, $\theta - \pi$. When $H_3 < 0$ they are chosen to be $\theta < 0$, $\theta + \pi$. Thus there is a total of 8 equilibria for Case 4A.

5.2 Case 4B

To treat the general case, premultiply the equilibrium equation, Eq.4, by the transposes of α,β,γ respectively. The resulting three scalar equations are

$$\gamma^T I \beta = -\frac{1}{4} H^T \gamma \tag{39a}$$

$$\alpha^T I \gamma = 0 \tag{39b}$$

$$\alpha^T I \beta = -H^T \alpha \tag{39c}$$

Although the development can be carried further in matrix form, so it does not matter how body axes are defined, it now is convenient to specify that body axes are principal axes of inertia as was done in the previous cases. Define differences of principal moments of inertia by

$$S_1 = B - C\ , \qquad S_2 = C - A\ , \qquad S_3 = A - B \tag{40}$$

Equation 39b, together with $\alpha^T \gamma = 0$, can be used to get α in terms of γ. In particular, in matrix form $\alpha = (I\gamma)\tilde{}\,\gamma/D$, where D is chosen so that $\alpha^T \alpha = 1$. In scalar form this relationship can be written

$$\alpha_i = S_i \gamma_j \gamma_k / D \tag{41a}$$

in which -- here and below -- i,j,k is a cyclic permutation of 1,2,3 and D is

$$D = \pm\sqrt{(S_1\gamma_2\gamma_3)^2 + (S_2\gamma_3\gamma_1)^2 + (S_3\gamma_1\gamma_2)^2} \tag{41b}$$

The orthogonality relationships among α,β,γ include $\beta = \tilde{\gamma}\alpha$, from which

$$\beta_i = \gamma_i (S_k \gamma_j^2 - S_j \gamma_k^2)/D \tag{41c}$$

In view of Eqs.41ac, Eqs.39ac take the form

$$D = -\frac{1}{4} H^T \gamma \tag{42a}$$

$$\begin{aligned}(S_1S_2S_3)\gamma_1\gamma_2\gamma_3 &= D(H_1S_1\gamma_2\gamma_3 + H_2S_2\gamma_3\gamma_1 + H_3S_3\gamma_1\gamma_2) \\ &= -\frac{1}{4}(H_1S_1\gamma_2\gamma_3 + H_2S_2\gamma_3\gamma_1 + H_3S_3\gamma_1\gamma_2)(H_1\gamma_1 + H_2\gamma_2 + H_3\gamma_3)\end{aligned} \tag{42b}$$

Equation 42a can be squared and Eq.41b used to eliminate D, yielding

$$(S_1\gamma_2\gamma_3)^2 + (S_2\gamma_3\gamma_1)^2 + (S_3\gamma_1\gamma_2)^2 = \frac{1}{16}(H_1\gamma_1 + H_2\gamma_2 + H_3\gamma_3)^2 \tag{42c}$$

Equations 42c and 42b, together with the normality relation $\gamma^T\gamma=1$, suffice to determine all the γ_i, hence, in view of Eqs.41a,c, the complete equilibrium orientation. Note that we need only seek the equilibria for which $\gamma_3>0$, for it is evident from the equations that if γ is a solution, so is $-\gamma$. The six parameters S_i, H_i can be replaced by five ratios because both equations are homogeneous in these parameters, but there is no special benefit in doing so at this point.

The solution method is to represent γ in terms of two parameters such that γ satisfies the orthogonality condition. Thereafter, Eqs.42c and 42b become two polynomial equations in two unknowns. In the pioneering solution given in [15], the authors choose

$$\gamma^T = [\ \sin\theta\sin\phi \quad \sin\theta\cos\phi \quad \cos\theta\ ] \tag{43a}$$

However, purely algebraic expressions can serve just as well, e.g.

$$\gamma^T = [\ xy \quad x \quad 1\ ]/\sqrt{x^2(y^2+1)+1} \tag{43b}$$

$$\gamma^T = [\ 4x \quad 4y \quad 4-x^2-y^2\ ]/(4+x^2+y^2) \tag{43c}$$

(The latter is a stereographic projection of γ on the x,y-plane.) In the development below we use Eq.43b.

Replacement of γ in terms of x and y in Eqs. 42c and 42b yields the simultaneous equations

$$\sigma_2 x^2+\sigma_1 x+\sigma_0 = 0 \tag{44a}$$

$$\tau_4 x^4+\tau_3 x^3+\tau_2 x^2+\tau_1 x+\tau_0 = 0 \tag{44b}$$

in which

$$\sigma_i = \sum_{j=0}^{k(i)} p_{ij} y^j \quad (i=0,2;\ k(0)=1, k(1)=2, k(2)=2) \tag{45}$$

$$p_{00}=S_1H_1H_3 \qquad p_{01}=S_2H_2H_3 \tag{46a}$$

$$p_{10}=S_1H_1H_2 \qquad p_{11}=4S_1S_2S_3+S_1H_1^2+S_2H_2^2+S_3H_3^2 \tag{46b}$$

$$p_{12}=S_2H_1H_2 \tag{46c}$$

$$p_{21}=S_3H_2H_3 \qquad p_{22}=S_3H_1H_3 \tag{46d}$$

and

$$\tau_i = \sum_{j=0}^{i} q_{ij} y^j \quad (i=0,4) \tag{47}$$

$$q_{00}=H_3^2 \qquad q_{10}=3H_1H_3 \tag{48a}$$

$$q_{11}=2H_1H_3 \qquad q_{20}=H_2^2+H_3^2-16S_1^2 \tag{48b}$$

$$q_{21}=2H_1H_2 \qquad q_{22}=H_3^2+H_1^2-16S_2^2 \tag{48c}$$

$$q_{30}=2H_2H_3 \qquad q_{31}=2H_1H_3 \tag{48d}$$

$$q_{32}=2H_2H_3 \qquad q_{33}=2H_1H_3 \tag{48e}$$

$$q_{40}=H_2^2 \qquad q_{41}=2H_1H_2 \tag{48f}$$

$$q_{42}=H_1^2+H_2^2-16S_3^2 \qquad q_{43}=2H_1H_2 \tag{48g}$$

$$q_{44}=H_1^2 \tag{48h}$$

The theory of elimination in higher algebra teaches that Eqs.44ab have a common root when the "resultant"

$$\begin{vmatrix} \sigma_0 & \sigma_1 & \sigma_2 & 0 & 0 & 0 \\ 0 & \sigma_0 & \sigma_1 & \sigma_2 & 0 & 0 \\ 0 & 0 & \sigma_0 & \sigma_1 & \sigma_2 & 0 \\ 0 & 0 & 0 & \sigma_0 & \sigma_1 & \sigma_2 \\ \tau_0 & \tau_1 & \tau_2 & \tau_3 & \tau_4 & 0 \\ 0 & \tau_0 & \tau_1 & \tau_2 & \tau_3 & \tau_4 \end{vmatrix} \tag{49}$$

equals zero. Each element is a polynomial in y; thus, in principle, the determinant can be expanded into a single polynomial in y. Working out explicit values for the coefficients of this polynomial is a monumentally tedious task, fraught with the danger of algebraic error, unless one turns to a suitable symbolic manipulation computer program. Some of the coefficients consist of so many terms that it is not feasible to write out literal expressions for them here. However, a systematic method of solution exists that is quite practicable with the aid of a computer, once numerical values of the p and q are known. Explicit representation of the coefficients in terms of the original parameters of the problem simply is not needed, and is so complicated in any case that it is not enlightening. Details of this method will be described elsewhere.

REFERENCES

1. d'Alembert, J.le R., **Récherches sur la précession des equinoxes et sur la nutation de l'axe de la terre, dans le système Newtonien,** David, Paris, 1749.

2. Euler, L., "Recherches sur la précession des equinoxes et sur la nutation de l'axe de la terre", **Mém. Acad. Sci. Berlin 1749** (1753), 289-325.

3. Lagrange, J.L., "Recherches sur la libration de la lune", **Prix de l'Acad. Roy. Sci. Paris 9** (1764).

4. Lagrange, J.L., "Théorie de la libration de la lune, et des autres phénomenes qui dépendent de la figure non sphérique de cette planete", **Nouv. Mém. Acad. Sci. Berlin 1780** (1782), 203-309.

5. Roberson, R.E., Hooker, W.W., "Gravitational equilibria of a rigid body containing symmetric rotors", **Astrodynamics, Guiodance and Control**, (Proc. 17th Int. Astronaut. Congr., Madrid, 1966), Vol.4, pp.203-210, Dunod, Paris, 1967.

6. Likins, P.W., Roberson, R.E., "Uniqueness of equilibrium attitudes for Earth-pointing satellites", **J. Astronaut. Sci. 13** (1966), 87-88.

7. Anchev, A.A., "Flywheel stabilization of relative equilibrium of a satellite", **Cosmic Research 4** (1966),173-183.

8. Roberson, R.E., "Equilibria of orbiting gyrostats", **J. Astronaut. Sci. 15**(1968),242-248.

9. Likins, P.W., "Stability of a symmetrical satellite in attitudes fixed in an orbiting reference frame", **J. Astronomical. Sci. 12**(1965), 18-24.

10. Rumiantsev, V.V., "On the stability of stationary motions of a satellite with a rotor and a cavity containing fluid", **Cosmic Res. 5**(1967),163-169.

11. Rumiantsev, V.V., "On stability of stationary motions of the gyrostat satellite",**Proc 18th Int. Astronaut. Congr.** (Beograd, 1967).

12. Likins, P.W., "Stability analysis of mechanical systems with constraint damping",**AIAA J 5**(1967),2091-2094.

13. Anchev, A.A., "Ob otnositel'nom ravnovesii sputnika c rotorami", **Knjige Zbornik Radova 9 Jugoslovenskog Kongresa za Racionalni i Primenjenu Mehaniku**(Split, 1963), pp.17-24.

14. Longman, R.W., Roberson, R.E., "General solution for the equilibria of orbiting gyrostats subject to gravitational torques", **J. Astronaut. Sci. 16**(1969),48-58.

15. Sarychev, V.A., Gutnik, S.A., "Relative equilibria of a gyrostat satellite", **Cosmic Research 22**# 3 (1984), 257-260.

Multibody Formalism Applied to Non-Conventional Railway Systems

Jean-Claude SAMIN et Pierre Y. WILLEMS
Université Catholique de Louvain
Louvain-La-Neuve, Belgium.

Abstract
The paper presents the main features of the AUTODYN programme developed in Louvain-La-Neuve . This multi-purpose programme, based on the virtual work principle, can be used for vehicle dynamics as well as in robotics and for other mechanical systems. The generalized coordinates describe the relative motion in the joints between bodies. A library of joints, including rail-wheel connections, with their corresponding interaction models is available. Moreover the joints can be restricted by means of user-defined constraints and some of their coordinates can have a prescribed motion. Constraints resulting from closed loops are automatically generated and the final form of the equations is obtained by using the Lagrange multipliers technique and a coordinate partioning method.

The AUTODYN Multibody Programme*

In this paper, a multibody system is assumed to be composed of a certain number of interconnected rigid bodies. These bodies are characterized by their geometry and their mass distribution and are connected by some mechanical devices, such as springs, dampers, hinges, universal joints, or have more sophisticated connections such as the sliding or the rolling of their surfaces. The connection will be defined as a **joint**. Relative motion can take place in the joint and mutual interaction is produced. The corresponding forces and torques can be either unknown constraint quantities or can be described by dynamical relations as functions of the relative motion. The variables which describe this relative motion will

----

(*) The software implementation was realized with the collaboration of Ph. Chatelle, P. Maes and G. Schmitz, and the finantial support of B.N. - Constructions Ferroviaires et Métalliques and of the Belgian "Institut pour l'encouragement de la recherche scientifique dans l'industrie et l'agriculture"IRSIA

Dynamics of Multibody Systems
IUTAM/IFToMM Symposium Udine 1985
Editors: G. Bianchi and W. Schiehlen

be the variables of the problem.

In order to establish the equations of motion, the configuration of the system with respect to inertial space is described in terms of these variables and this description is greatly simplified by the use of the graph of the system.

As an example let us consider the system described by Fig. 1 and schematically represented in Fig. 2.

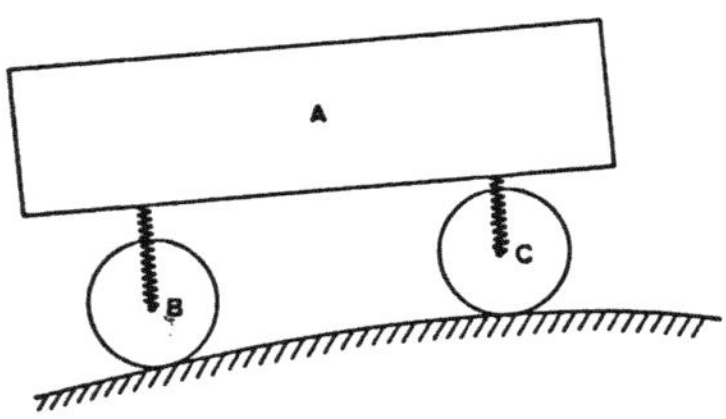

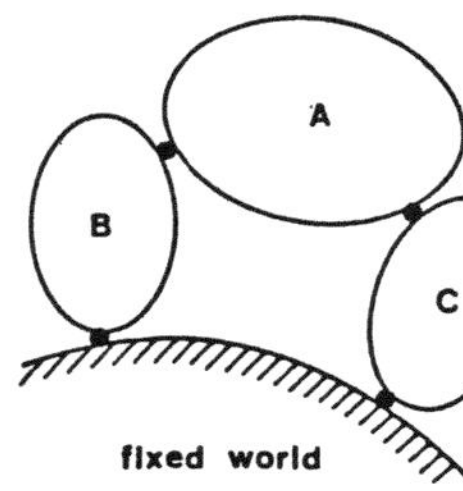

Fig.1 Example of planar vehicle Fig.2 Multibody Model

The absolute motion of the frame A may be expressed as the sum of its relative motion with respect to the wheel B and the relative motion of this body with respect to the ground. It is clear that this description is not unique; we could have used a path through the other wheel.

In general, the fact that the motion of a body can be expressed by different kinematical relations between joint variables, is equivalent to the existence of a kinematical **loop** and implies that the corresponding joint variables are not independant. In the above-mentioned example the closure of the loop is made through the ground.

The ground is present in the description of many mechanical systems and we took the option of considering the ground as a particular fixed body of all our systems.

Moreover, the loops are broken in order to remove the kinematical ambiguity due to their presence. This can be done by cutting either joints or bodies. For reasons which will appear more clearly later on, our deliberate choice was to cut bodies.

Description of the graph

The conventions on the numbering of bodies and joints and on the localization of cuts will now be introduced. In this procedure graph matrices - incidence matrix S, path matrix T and circulation loop matrix R [1].[2] - have been abandoned in favour of a more readily usable notion of filiation [3]. The **filiation** simply expresses that a joint connects two neighbouring bodies and can be formalized under the form of a simple matrix function which is also well suited for numerical investigations.

First of all, a **reference body** and a **reference joint** have to be defined. The index 0 is associated to the reference body considered as the founder of the system.

For tree structures the graph numbering is very simple. All the descendants will be numbered according to the filiation order, starting from the reference body; the joints will be given the number of the body they precede in this order.

It is clear that the procedure favours some branches of the system. In particular one may define a **main body** which is connected to the reference body by the reference joint; the main body and the reference joint then have the index one.

If the same procedure is applied in the presence of loops, it will soon appear (Fig. 3a) that a given body is the parent of two (or more) other bodies which have common descendants. This indicates that the given body is the origin of a loop and it will be cut as illustrated is Fig. 3b.

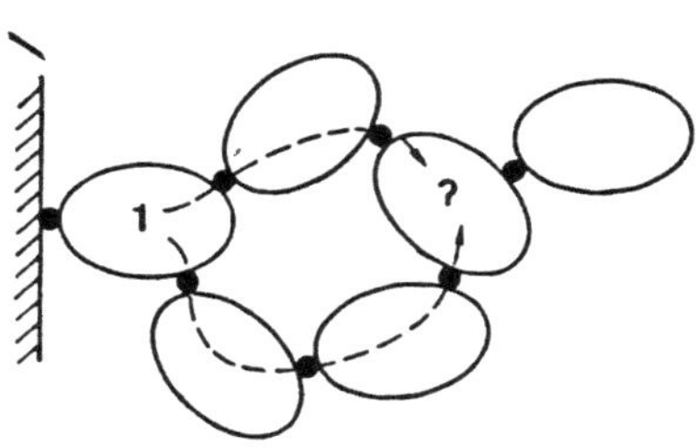

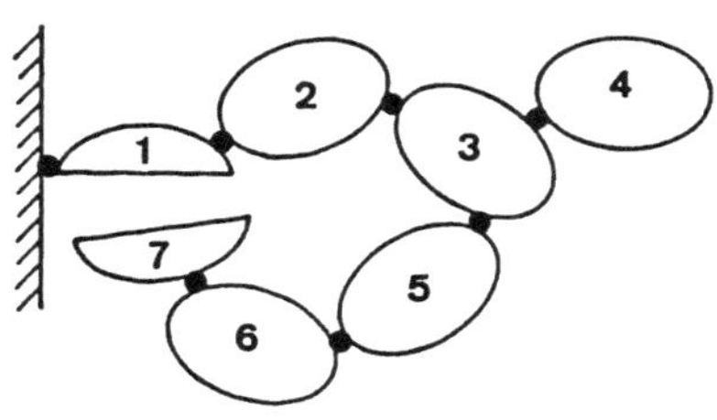

Fig.3a Loop and augmented tree Fig.3b

The cut fragment will then be a descendant of the original body and this fictitious terminal body will be designed as the

closure body of the loop. It should be noted that all the bodies of an original loop are in a direct line of filiation in the so-defined **augmented tree** structure. Once the cutting procedure is completed NLOOP independant loops have been defined and the numbering proceeds as before for the augmented tree.

The filiation vector **(INBODY)** has as many elements as there are joints in the system (or equivalently as there are descendants of the reference body in the augmented system) and these elements are defined as follows : INBODY (j) = i where i is the parent of the body j; this also means that the joint a, with a=j, connects the body j to its parent i.(*)

The path between a body and the reference body is obtained by a recursive use of the INBODY function until the index zero is reached. Summation on elements of a path is simply obtained by this recursive procedure.

Loop information can be obtained by the use of this vector and of a 2 × NLOOP matrix **INLOOP**. The α column of this matrix corresponds to the loop α and contains the indices of the origin and the closure of the loop. As the closure body is the descendant of the origin of the loop, the recursive use of the function INBODY, initiated at j=INLOOP(2,α) and interrupted when INBODY(i)=INLOOP(1,α) provides the path in the loop. For the example of Fig.3 the matrix INLOOP is simply INLOOP=$\binom{1}{7}$.

Choice of the reference and the main bodies

For ground-based applications (mechanisms, robots, vehicles), the ground is a normal choice as reference body. This body will be made inertial by being assigned infinite mass and inertia (say 10^{30}, for computational purposes). An arbitrary point of the ground can be chosen as its "centre of mass" and this joint will then coincide with the centre of mass of the complete system.

The joint between the ground and the main body of the system,

----

(*) The indices of bodies are designed by the letters i,j,k the indices of joints by the letters a,b,c and the indices of loops by the letters α,β,γ.

the reference joint, will generally have six degrees of freedom but some of the corresponding variables may be prescribed as functions of time. This could also be the case for any joint in the system. This allows us to disregard the usual distinction between fixed and moving (inertial and kinematical) reference frames [4] [5].

This procedure is illustrated in Fig.4 which represents the augmented body used to investigate the motion of the system represented in Figs. 1 and 2. The system includes two loops; there are two cut fragments. The ground is considered as the reference body and the frame is the main body.

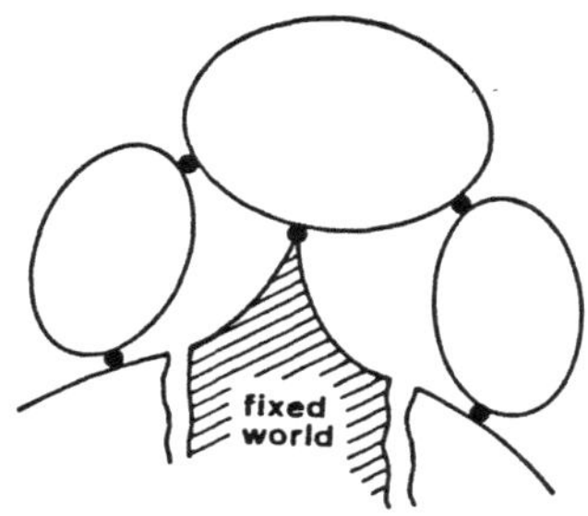

Fig.4 Augmented Multibody model

If the constant speed behaviour is investigated, one can consider that the longitudinal motion of the reference joint is imposed.

Dynamical reference frames centered at the centre of mass of the system (apart from the ground) could also be easily introduced by adding (generally between the reference and the main body) a fictitious massless body contrained to be located at this centre of mass and having an appropriate rotational motion.

Kinematical Description of bodies and joints

A reference frame, $\{\hat{\underset{\sim}{X}}^i\}$, is associated to the body i. The mass of this body is denoted m^i and its inertia tensor with respect to its mass centre G^i, is expressed by its components in $\{\hat{\underset{\sim}{X}}^i\}$, i.e. by the inertia matrix I^i.

For fictitious bodies, mass and inertia are chosen equal to zero and the mass centre is thus arbitrary. However in order to simplify to systematical derivation of constraint relations of the corresponding kinematical loops, we assume that the mass centres and the reference frames of the origin and closure bodies coincide when the loops are closed.

A joint connecting two bodies - the parent and child - is represented by Fig. 5. The end points of the joint are fixed

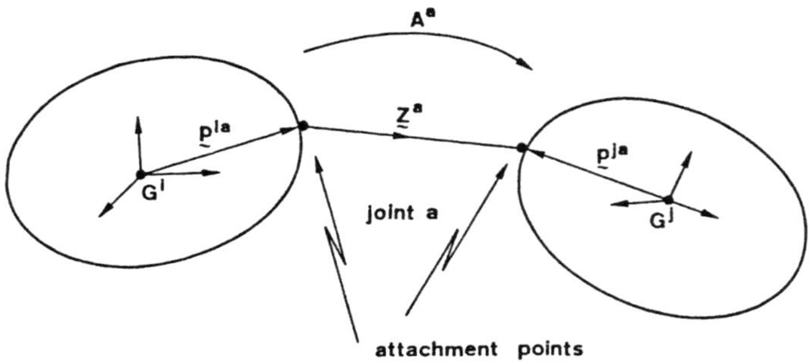

Fig.5 Joint Vectors

with respect to the two bodies. Expressed in the corresponding frames, their positions vectors with respect to the mass centres thus have constant components, which are part of the geometrical data. Whatever the physical nature of the joint, the relative motion of the child with respect to its parent can be described by means of the relative position vector $\underset{\sim}{z}^a$ and the relative rotational matrix A^a given by

$$\underset{\sim}{z}^a=[\hat{\underset{\sim}{X}}^j]^T z^a \qquad [\hat{\underset{\sim}{X}}^j]=A^a[\hat{\underset{\sim}{X}}^i]$$

These quantities are then described in terms of the specific joint coordinates. The user will have to provide the corresponding analytical expression of the functions

$$z^a=z^a(q^a,t) \qquad A^a=A^a(q^a,t)$$

where time appears explicitly when (part of) the motion is kinematically imposed.

At this stage, the variables q^a are not necessarily independent: the user may include holonomic constraints under the form

$$g^a(q^a,t)=0 \qquad (1)$$

The first and second derivatives of the functions z^a, A^a and g^a are required for the derivation of the equations; they have to be provided together with the dynamical relations for the generalized forces corresponding to the degrees of freedom.

The development of corresponding subroutines may be arduous and the programme is planned so as to call a library of joint subroutines. Presently, six standard joints are included in the package; their parameters may be interactively adapted to the user application. They have been chosen in order to cover a wide range of application, and are as follow :

- ROTRA :sequence of three Tait Bryan rotations (3,2,1) followed by three translations;
- TRARO :sequence of three translations followed by three rotations in reverse order (1,2,3);
- AXE :rotation around one axis (1,2 or 3);
- GLIS :translation along one axis (1,2 or 3);
- ROUERAIL:wheel on a straight track - motion of a revolute surface on a cylinder;
- RORACOUR:wheel on a track with constant curvature - motion of a surface of revolution on a torus.

Joints ROTRA and TRARO are opposed in sequence. This is useful as our programme does not consider joint orientation. The relative motion always consists in the motion of the child with respect to its parent. For example, if the child is a frame connected to a rotating wheel a ROTRA joint will be used. In this case the first encountered variable corresponds to the rotation of the wheel and moreover the suspension deformation and the corresponding damping and stiffness forces and torques will be easily expressed in terms of the remaining variables. On the other hand, if the wheel is the child a TRARO joint will be used.

For these joints an interactive subprogramme permits to indicate the kinematically constrained or imposed variables. They offer standard kinematical and dynamical laws. If the standard options are not sufficient, the user may introduce other relations and joint constraints. As an example, for connecting rods, one can express that the distance between the end points of the joint is constant.

Joints AXE and GLIS are simplified versions of the preceding ones. They permit the composition of any sequence of degrees of freedom together with the addition of intermediate mass-less bodies.

Joints ROUERAIL and RORACOUR were specifically developed for railway applications. These joints are described in [6].

Automatic derivation of kinematical relations and loop constraints

The orientation and position of the various bodies of the system can be computed by the recursive use of the relations

$$[\hat{\underset{\sim}{X}}^{j}]=A^{a}[\hat{\underset{\sim}{X}}^{i}] \qquad (2)$$

with a=j i=INBODY(j)

$$\underset{\sim}{x}^{j}=\underset{\sim}{x}^{i}+\underset{\sim}{p}^{ia}+\underset{\sim}{z}^{a}-\underset{\sim}{p}^{ja} \qquad (3)$$

The velocities $\dot{\underset{\sim}{x}}^{i}$, angular velocities $\underset{\sim}{\omega}^{i}$ as well as the compatible virtual displacements $\delta \underset{\sim}{x}^{i}$ and rotations $\delta \underset{\sim}{\pi}^{i}$ can be generated from these relations and the functions A^{a} and z^{a} (and their time derivatives) provided by the joint subroutines. The generation of the linear and angular accelerations requires the second derivatives of A^{a} and z^{a} which are also provided by the joint subroutines.

Finally all the vectors are expressed in the reference body frame and the linear and angular momentum vectors of the various bodies are computed as functions of the total set q of joint variables

$q=(..,q^{a},...)^{T}$ (and their time derivatives).

Without considering all the details, we should point out that the barycentric properties [7] and the notion of augmented bodies [8] are used in a particular way. Indeed, as the reference body has a (quasi) infinite mass the barycentres (almost) coincide with the joint end (on the parent side) and half of the barycentric vectors are (practically) equal to zero. These vectors are not included in the various terms of the equations except in those which directly depend on the mass of the total system (see for instance [7] 5.56).

The programme will derive appropriate expressions for the loop constraints. If i is the index of the origin of the loop α and k is the index of the closure body, loop constraints are obtained by expressing the fact that bodies i and k remain in coincidence during the motion; this can be written as

$$\underset{\sim}{x}^{k}=\underset{\sim}{x}^{i} \qquad [\hat{\underset{\sim}{X}}^{k}]=[\hat{\underset{\sim}{X}}^{i}]$$

The recursive use of (2) and (3) along the loop also permits to write these relations under the form

$$\underset{\sim}{x}^{k}=\underset{\sim}{x}^{i}+\underset{\sim}{z}^{\alpha}(q,t) \qquad [\hat{\underset{\sim}{X}}^{k}]=A^{\alpha}(q,t)[\hat{\underset{\sim}{X}}^{i}]$$

and the loop constraints will be given by

$$\underset{\sim}{z}^{\alpha}(q,t)=0 \text{ and } A^{\alpha}(q,t)=E \qquad (4)$$

From these relations, six independant algebraic constraints can be derived (the matrix A , being the product of orthogonal matrices, is orthogonal).
It should be noted that, as the loop contraints are always satisfied, the time derivative of the relation (4), $\dot{A}$ =0, implies that A $\dot{A}^T$=0. The latter relation simply expresses the fact that the sum of relative angular velocities of the corresponding loop is equal to zero.
One must point out the advantages of cutting the loops through bodies. If the loops are broken through a joint, it is clear that the corresponding joint coordinates, say q^c, will not appear in tree kinematical relations (which are functions of the other joint variables q^b). On the other hand the interactions Q^c in the cut joint are normally described by the user in terms of the variables q^c, but they have to be provided to the programme in terms of the q^b. For elementary geometry this can be done analytically by the user, but this is a much more fastidious task for more sophisticated systems (see for instance the difficulty of the inverse kinematical problem in robotics [9]). The loop constraints now have the form

$$\begin{aligned} \underset{\sim}{z}^{\alpha}(q^b,t) &= \underset{\sim}{z}^c(q^c,t)+\underset{\sim}{p}^{ic}-\underset{\sim}{p}^{kc} \\ \tilde{A}^{\alpha}(q^b,t).\tilde{A}^c(q^c,t) &= E \end{aligned} \tag{5}$$

and even if we assume that there are no user contraints between the q^c and that none of the q^c are prescribed functions of time, the numerical algorithms will depend on the number n^c of joint variables q^c. For dynamical purposes, the programme has to solve (5) for the variables q^c; this can be done by projecting (5) along the corresponding degrees of freedom [10]. However, if $n^c<6$ some remaining constraints between the q^b variables are still included in (5) and have to be solved independently.
In our procedure, both the loop and user defined contraints can be written under the form

$$S(q,t)=0 \tag{6}$$

The time derivatives of the contraints can be written under the form

$$A\dot{q} + \frac{\partial S}{\partial t} = 0 \tag{7}$$

where the elements of the matrix A corresponding to the loops are automatically generated by the programme from the various matrices A^a and the components of the relative angular velocity vectors; the elements corresponding to the user constraints have to be provided by the user subroutine.

<u>Automatic derivation of equations of motion</u>

The equations of motion are obtained from the virtual work principle, the generalized variables being the joint variables q. The virtual work can be expressed as

$$\Sigma_i [(\dot{\underset{\sim}{N}}^i - \underset{\sim}{F}^i) \cdot \delta\underset{\sim}{x}^i + (\dot{\underset{\sim}{H}}^i - \underset{\sim}{L}^i) \cdot \delta\underset{\sim}{\pi}^i] = \Sigma_a \delta q^a Q^a + \delta W$$

where $\underset{\sim}{N}^i$ and $\underset{\sim}{H}^i$ are the linear and angular momentum vectors of the body (with respect to the centre of mass);

$\underset{\sim}{F}^i$ and $\underset{\sim}{L}^i$ are the resultants of forces and torques (around the centre of mass) applied to the body i (joint interactions excluded);

$\delta\underset{\sim}{x}^i$ and $\delta\underset{\sim}{\pi}^i$, the virtual displacement and rotation vectors;

Q^a the generalized forces in the joints;

δW the virtual work of constraint forces.

The expressions of $\underset{\sim}{F}^i$ and $\underset{\sim}{L}^i$ have to be provided by the user (the effect of gravity can be obtained by an existing subroutine). The expressions of the generalized joint forces Q^a are available through the outputs of joint subroutines. The contribution to δW of the loop constraints are equal to zero for compatible virtual displacement; indeed, the loops remain closed for these displacements as the origin and closure body have no virtual relative motion. In most cases, the user constraints also do not work virtually - this has to be checked - and δW will then be equal to zero for virtual displacements which are compatible, i.e. satisfying : $A\delta q=0$. This implies that δW has the form : $\delta W=\lambda^T A\delta q$ where λ is a vector of Lagrangian multipiers.

Finally the equations of motion are obtained under the form

$$M(q,t)\ddot{q} = F(q,\dot{q},t) + Q + A^T(q,t)\lambda \qquad (8)$$

where $Q = (.. \; Q^a \; ..)^T$

This equation together with the constraints (6) provide the complete set of equations of motion. The numerical values of the matrices M, F, Q and A having been obtained, the programme reduces the equations by a coordinate partitioning method [11]

A regular submatrix of A has to be defined. This regular matrix can be suggested by the user in order to choose the independant variables, otherwise the programme will define an optimal choice from the numerical point of view.The partitioning of the variables is written $q=(u^T,v^T)^T$ where u are the independant coordinates. The matrix A is then written as $A=[A_u;A_v]$ where A_v is regular.

For each value of u, the nonlinear constraints (6) are solved by means of a Newton-Raphson method based on the recursive use of the linear equation

$$S(u,\hat{v},t) = A_v(u,\hat{v},t)(\hat{v}-v) \qquad (9)$$

The time derivatives of variables v are then computed from (7) written under the form

$$A_v(u,v,t)\dot{v} + A_u(u,v,t)\dot{u} + \frac{\partial S}{\partial t} = 0 \qquad (10)$$

The equation of motion (8) and the second derivative of the constraints are then rewritten as

$$\begin{bmatrix} M_{uu} & M_{uv} & A_u \\ M_{uv} & M_{vv} & A_v \\ A_u^T & A_v^T & 0 \end{bmatrix} \begin{bmatrix} \ddot{u} \\ \ddot{v} \\ \lambda \end{bmatrix} = \begin{bmatrix} F_u + Q_u \\ F_v + Q_v \\ C\lambda \end{bmatrix} \qquad (11)$$

From these equations, λ and v are eliminated and the equations of motion reduce to a system of the form

$$\mathbf{M}(u,t)\ddot{u} + \mathbf{C}(u,\dot{u},t) = 0 \qquad (12)$$

The system (12) can be used for various types of applications: equilibrium investigation, modal analysis after appropriate numerical linearization, control design and simulation...

For rail vehicle dynamics, an original five degrees of freedom joint has been developed for the wheel/rail contact. In our approach, the track belongs to the multibody model so that the geometrical problem arising from multiple wheel/rail contacts is solved by means of closed loops. This allows to deal with non classical vehicles (loose wheels, articulated bogie frame) for which the wheel/rail kinematics cannot be decoupled from the kinematics of the vehicle itself. The case of persistent flange contact can also be considered by splitting the wheel which has a double contact with the rail and by considering an appropriate closed-loop. The generalized coordinates of the contact joint have been chosen in order to simplify the description of the wheel and rail profiles. They are suitable for straight and curved track and allow to obtain the interactions in rather simple form from Kalker's linear theory and/or Coulomb friction law since the normal component of the contact force is available from the dynamics (11).

The nonlinear output of the programme has been used to investigate the steady-state configuration of vehicles on straight and curved tracks and to perform modal analysis for constant speed motions along the tracks. In the latter case, the equations of motion were linearized numerically.

References

1. R.E. ROBERSON and J. WITTENBURG : A Dynamical Formalism for and Arbitrary Number of Interconnected Rigid Bodies, with reference to the Problem of Satellite Attitude Control. Proc. of the 3rd International Congress of Automatic Control (London, 1966) Butterworth and Co, London (1967).
2. Ph. BOLAND, J.C. SAMIN an P.Y. WILLEMS : Stability Analysis of Interconnected Deformable Bodies with Closed-loop Configuration. AIAA Journal 13 (1975), 864-867.
3. S. MEGAHED and M. RENAUD : Dynamic Modelling of Robot Manipulators Containing Closed Kinematic Chains. Proc. of Advanced Software in Robotics, Liège, Belgium 4-6 may 1983.
4. W. KORTUM and W. SCHIELEN : General Purpose Vehicle System Dynamics Software Based on Multibody Formalism. Vehicle System Dynamics 14 (1985) 229-263.
5. R. SCHWERTASSEK and R.E. ROBERSON : A State-Space Dynamical Representation for Multibody Mechanical Systems. Part I : Systems with Tree Configuration. Acta Mechanica 50 (1984).
6. Ph. CHATELLE, J. DUPONCHEEL and J.C. SAMIN : Investigation on Non Conventional Railway Systems through a generalized Multibody Approach, 8th IAVSD-IUTAM Symposium on the Dynamics of Vehicle on Road and Tracks, Cambridge Mass, August 1983. Lisse : Swets ans Zeitlinger (1984).
7. J. WITTENBURG : Dynamics of Systems of Rigid Bodies, Stuttgart : B.G. TEUBNER 1977.
8. Ph. BOLAND, J.C. SAMIN and P.Y. WILLEMS : Stability Analysis of Interconneted Deformable Bodies in a Topological Tree. AIAA Journal 12 (1974).
9. B. GORLA and M. RENAUD : Modèles des Robots Manipulateurs : application à leur commande. Toulouse Ed. CEPADUES 1984.
10. R. SCHWERTASSEK and R.E. ROBERSON : A State-Space Dynamical Representation for Multibody Mechanical Systems. Part II : Systems with Closed Loops. Acta Mechanica 51 (1984).
11. R.A. WEHAGE and E.J. HAUG : Generalized Coordinate Partitioning for Dimension Reduction in Analysis of Constrained Dynamic Systems. Journal of Mechanical Design 104 (1982).

Strength Estimations in Multibody Systems

W. SCHIEHLEN and E. KREUZER

Institute B of Mechanics
University of Stuttgart, Stuttgart, F.R.G.

Summary

The internal stresses in a rigid body cannot be calculated because the problem is statically undetermined. A method for the estimation of internal stresses in multibody systems is proposed using the assumption of slow rigid body motion with respect to structural vibrations. The method includes all external forces as well as all inertia forces acting on the system under consideration. Thus, additional information for the strength estimation is available supporting a proper design of the system.

Introduction

The main engineering problems in applied dynamics are the analysis of motion and the computation of strength. For the analysis of large working motions as well as small structural vibrations the method of multibody systems is very well qualified as was demonstrated by many authors. The state of the art has been presented by Magnus [1], Haug [2], Kortüm and Schiehlen [3]. For the computation of strength the method of multibody systems is not applicable, in principle. The rigid bodies assembled in a multibody system are statically and/or dynamically undetermined with respect to the interior stress distribution. Therefore, it is a well-known procedure to use the method of finite element systems for the computation of stresses and strength in engineering. The static loads on the body under consideration are replaced by dynamic loads found from an earlier motion analysis. This means, however, a quasistatic procedure neglecting the dynamical interactions.

Finite element systems and lumped mass systems are characterized by a large number of degrees of freedom and a

Dynamics of Multibody Systems
IUTAM/IFToMM Symposium Udine 1985
Editors: G. Bianchi and W. Schiehlen

unique stress distribution within the material bodies. Multibody systems result in a low number of degrees of freedom, but the stress distribution is nonunique. On the one hand, for a large number of degrees of freedom a complete dynamical analysis is very expensive. On the other hand, the rigid body model does not allow a strength computation at all. As a compromise, in this paper a strength estimation for rigid bodies is presented. This means that in addition to the analysis of motion some basic information on stresses useful for strength estimations can be obtained. For this purpose the original system has to be partitioned resulting in additional reaction forces. The proposed method will be applied to the connecting rod of an engine.

Strength Estimations

For a simplified strength estimation in multibody systems the following assumptions will be used:

(i) a rigid body model,

(ii) a continuous distribution of inertia and applied forces within the body, and

(iii) a linear stress distribution within the main volume of the body.

These assumptions are hypothetic from a theoretical point of view, but they can be justified in an engineering sense.

A rigid body under consideration may be subject to continuous and discrete forces. The discrete forces are assumed to act in nodal points, Fig. 1. Consequently, in theory, the neighborhood of these nodal points will be characterized by stress peaks. In engineering practice, however, the stresses near the nodal points are reduced by an optimal design using the geometrical shape or a material reinforcement. Furthermore, due to Saint Venant's principle, stress peaks decrease strongly with distance in elastic bodies. And finally, the stress peaks are limited in plastic bodies by viscous behavior. Thus, the volume of a rigid body can be divided in a main volume of linear stress distribution and an insignificant volume near the nodal points where the stress is reduced due to the phenomena discussed.

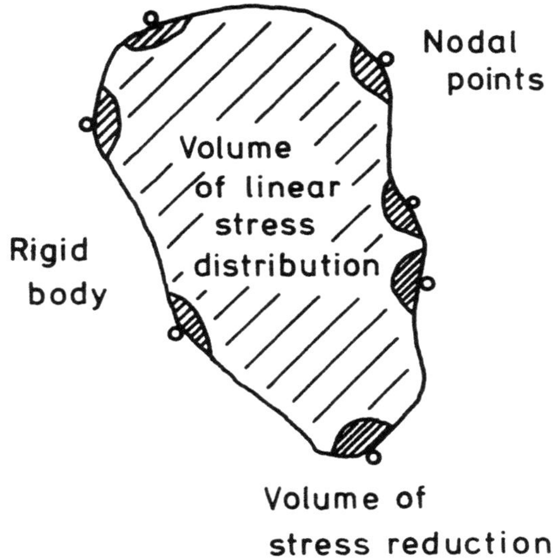

Fig. 1. Assumptions to the stress reduction and stress distribution

In a first step the natural axis of a rigid body is defined as an axis within the body connecting the nodal points of largest distance L , Fig. 2. The natural axis may be a straight or curved line and it is characterized by one material coordinate u_1 . Perpendicular to the natural axis a cross-section is introduced cutting the body in two parts, the left and the right body. The intersection point of the cross-section and the natural axis is OS , Fig. 3. For the cross-section the center of area CS and the principle axes e_2, e_3 can be found. Then, the material particles in the cross-section can be defined by coordinates u_2, u_3 and the material particles of the body are identified by u_1, u_2, u_3

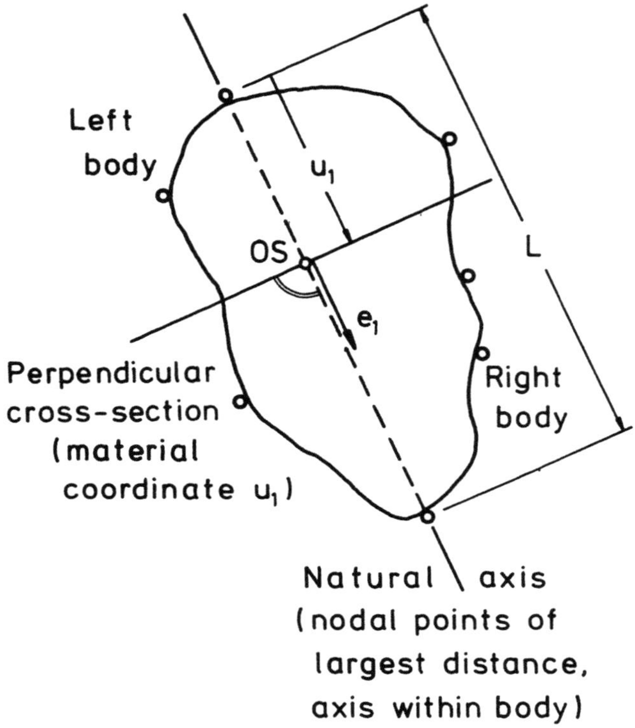

Fig. 2. Definition of the natural axis of a rigid body

Secondly, the reactions on the left body are computed with respect to the center of area CS . The reaction force vector $f^r(u_1,t)$ and the reaction torque vector $l^r(u_1,t)$ are both a function of the material coordinate u_1 and time t , Fig. 3. These reactions can be computed uniquely even for a multibody system.

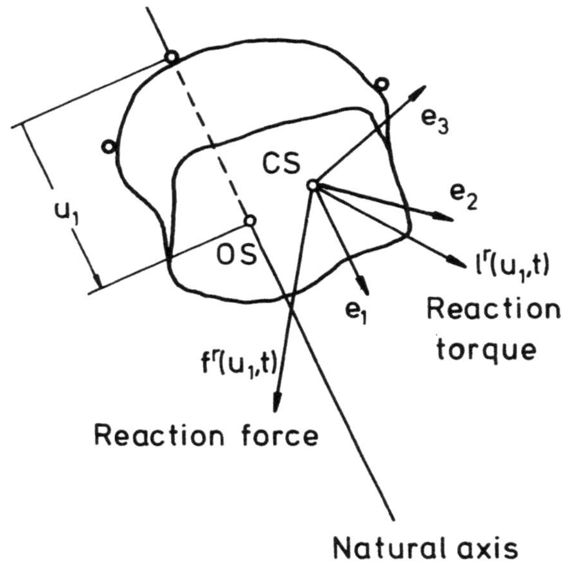

Fig. 3. Reactions on the left part of body

Thirdly, with the assumption of the linear stress distribution, the stresses for each material point can be obtained as shown in Ref. [4]. It turns out for the normal stress

$$\sigma_{11}(u_1,u_2,u_3,t) = \frac{f_1^r}{A} + \frac{l_2^r}{J_2}\,u_3 - \frac{l_3^r}{J_3}\,u_2 \qquad (1)$$

and the shear stresses read as

$$\tau_{12}(u_1,u_3,t) = \frac{f_2^r}{A} - \frac{l_1^r}{J_p}\,u_3 \quad , \qquad (2)$$

$$\tau_{13}(u_1,u_2,t) = \frac{f_3^r}{A} + \frac{l_1^r}{J_p}\,u_2 \quad , \qquad (3)$$

where A is the area and J_2, J_3, J_p are the corresponding second moments of area and polar second moment of area, respectively, of the cross-section characterized by the material coordinate u_1 .

Finally, according to strength theory a scalar characteristic stress σ_c is found from the stresses defined in (1) to (3). As an example, for the maximum distortion-energy hypothesis it remains

$$\sigma_c(u_1,u_2,u_3,t) = \sqrt{\sigma_{11}^2 + 3(\tau_{12}^2 + \tau_{13}^2)} \tag{4}$$

In conclusion, the scalar charateristic stress depends on space and time.

System Analysis

The original multibody system may consist of p rigid bodies interconnected by q constraints resulting in f = 6p-q degrees of freedom. The partitioned system, Fig. 4, has p + 1 bodies and p + 6 constraints, but an unchanged number of degrees of freedom, f = 6p-q .

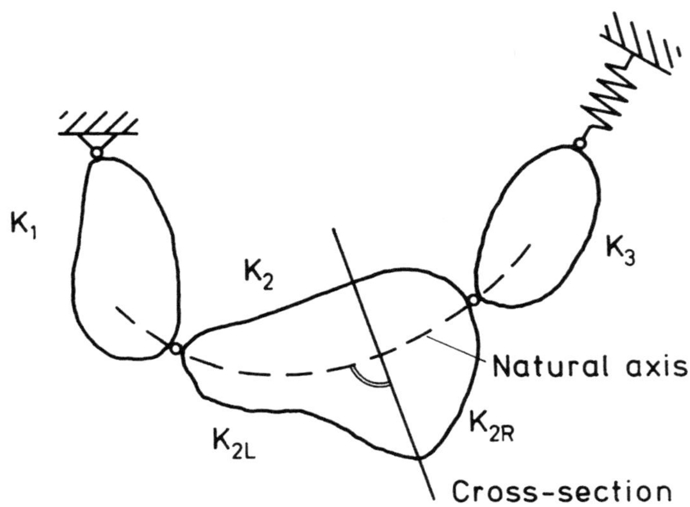

Fig. 4. Partitioned multibody system

The Newton-Euler equations, see Ref. [4], can now be written down for the original system

$$\bar{\bar{M}}\ \bar{J}\ \ddot{y}(t) + \bar{q}^c = \bar{q}^a + \bar{Q}\ g_o(t) \tag{5}$$

or the partitioned system, respectively,

$$\bar{\bar{M}}(u_1)\ \bar{J}(u_1)\ \ddot{y}(t) + \bar{q}^c(u_1) = \bar{q}^a(u_1) + \bar{Q}(u_1)\ g_p(t) \quad . \tag{6}$$

Here, $y(t)$ is the fx1-vector of generalized coordinates, $g_o(t)$ is a qx1-reaction vector and $g_p(t)$ is a (q+6)x1-reaction vector. Furthermore, $\bar{\bar{M}}$ are block diagonal inertia matrices, $\bar{J}$ are Jacobian matrices, $\bar{q}^c$, $\bar{q}^a$ represent coriolis and applied forces and $\bar{Q}$ are distribution matrices of adequate dimension. In particular it yields

$$g_p(t) = [g_o^T(t) \quad f^{rT}(u_1,t) \quad l^{rT}(u_1,t)]^T \quad , \tag{7}$$

where f^r, l^r represent the vectors of the reactions used in (1) to (3).

The equations of motion

$$M\ \ddot{y}(t) + k = q \tag{8}$$

may be obtained from (5) or (6), respectively. The fxf-inertia matrix reads as

$$M = \bar{J}^T\ \bar{\bar{M}}\ J = \bar{J}^T(u_1)\ \bar{\bar{M}}(u_1)\ \bar{J}^T(u_1) \tag{9}$$

Corresponding relations can be written for the fx1-vectors k, q of the generalized gyroscopic and applied forces. However, for the original system only a reduced number of parameters exists.

The equations of reaction

$$N_p\ g_p(t) + \hat{q}_p = \hat{k}_p \tag{10}$$

can be obtained from (6) only. The (q+6)x(q+6)-reaction matrix of the partitioned system is given as

$$N_p = \bar{Q}^T(u_1)\ \bar{\bar{M}}^{-1}(u_1)\ \bar{Q}(u_1) \tag{11}$$

Furthermore, $\hat{q}_p$, $\hat{k}_p$ are (q+6)x1-vectors of the applied and gyroscopic forces affecting the reactions. In particular, for the partitioned system the total number of parameters has to be used. Efficient numerical methods for the computation of the generalized constraint forces are described by Schramm [5].

Application to Engine

The described procedure for strength estimations is now illustrated by the application to the connecting rod of a one-cylinder engine, Fig. 5.

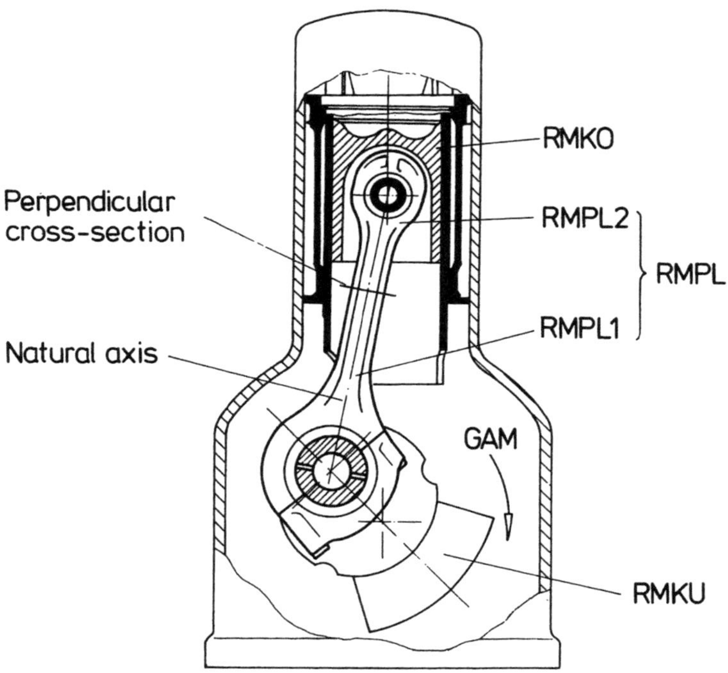

Fig. 5. One-cylinder engine

Modeling the real system as a multibody system consisting of $p = 3$ rigid bodies with $q = 17$ constraints results in $f = 1$ degree of freedom. Partitioning the system as described above, one arrives at $p = 4$ bodies with $q = 23$ constraints and hence again $f = 1$ degree of freedom. Consequently, the motion of the system is uniquely described by the generalized coordinate GAM . The equations of motion of the original and partitioned system are generated symbolically by the NEWEUL program [6]. In Fig. 6 are shown only the inertia quantities of both systems. In particular, there holds the relation RMPL=RMPL1+RMPL2 for the masses of the original and partitioned connecting rod.

For the determination of the stresses within the connection rod the natural axis is defined to be the geometrical symmetry axis. The reactions in the cross-section are calculated by a program developed by Settelmeyer [7]. This program simulates the motion and uses the equations of reaction also provided by the NEWEUL program.

Original system

```
M = S**2*RMKU
    +R**2*RMKO*SINGAM**2
    +2.*RL*R*RMKO*DEO1*SINDEL*SINGAM
    +RL**2*RMKO*DEO1**2*SINDEL**2+
    +RMPL*R**2
    +2.*B*RMPL*R*DEO1*SINDEL*SINGAM
    +B**2*RMPL*DEO1**2
    -2.*B*RMPL*R*DEO1*COSDEL*COSGAM
    +RIKUZ
    +RIPLZ*DEO1**2
```

Partitioned system

```
M = S**2*RMKU
    +R**2*RMKO*SINGAM**2
    +2.*RL*R*RMKO*DEO1*SINDEL*SINGAM
    +RL**2*RMKO*DEO1**2*SINDEL**2
    +RMPL1*R**2
    +2.*B1*RMPL1*R*DEO1*SINDEL*SINGAM
    +B1**2*RMPL1*DEO1**2
    -2.*B1*RMPL1*R*DEO1*COSDEL*COSGAM
    +RMPL2*R**2
    +2.*B*RMPL2*R*DEO1*SINDEL*SINGAM
    +2.*B2*RMPL2*R*DEO1*SINDEL*SINGAM
    +B**2*RMPL2*DEO1**2
    +2.*B2*B*RMPL2*DEO1**2
    +B2**2*RMPL2*DEO1**2
    -2.*B*RMPL2*R*DEO1*COSDEL*COSGAM
    -2.*B2*RMPL2*R*DEO1*COSDEL*COSGAM
    +RIKUZ
    +RIPLZ*DEO1**2
```

Fig. 6. Scalar inertia of original and partitioned systems

Figure 7 illustrates the simulation of the transient into the steady state motion. In Fig. 8 are shown the strength estimations along the rod resulting from the maximum

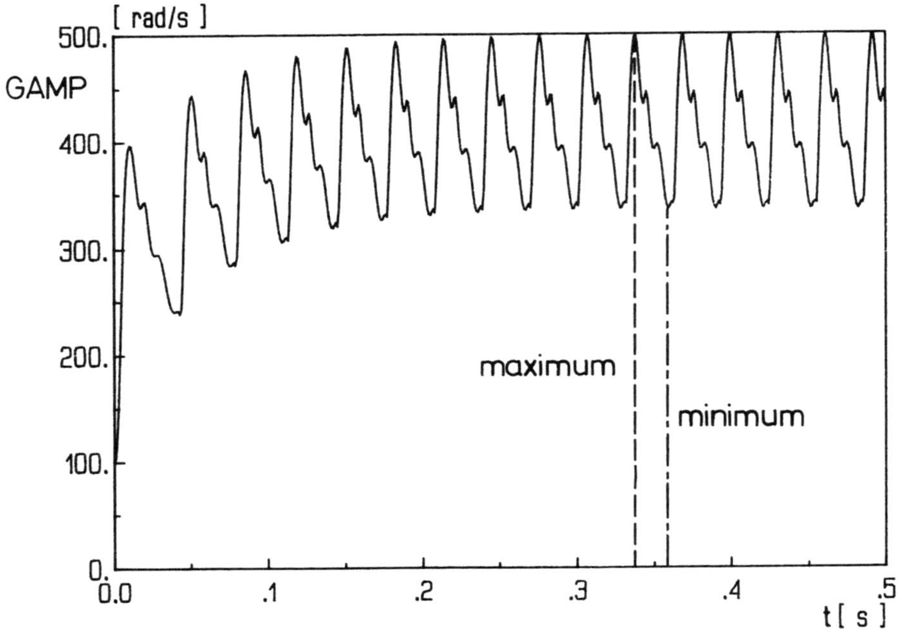

Fig. 7. Time dependence of anglular velocity GAMP

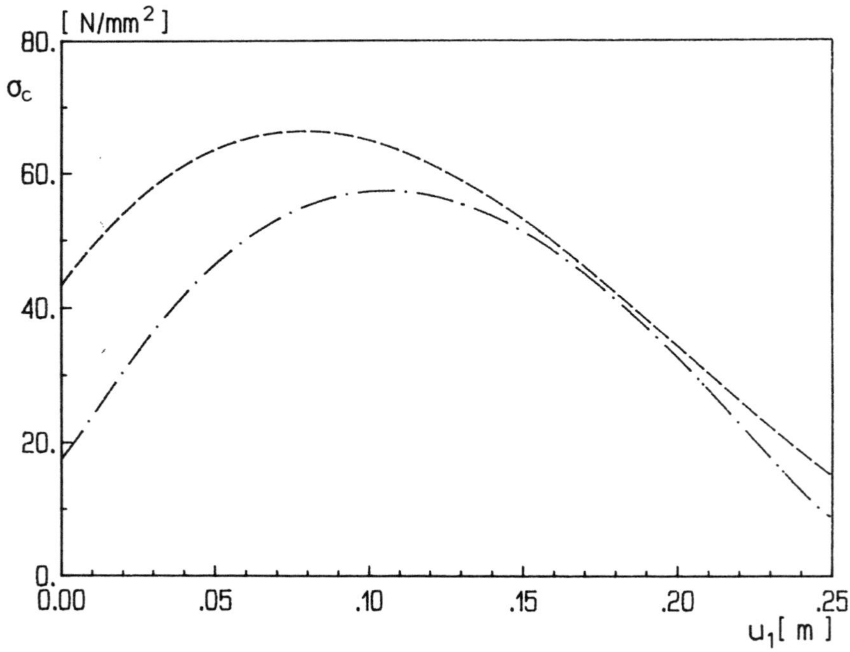

Fig. 8. Strength estimations along the rod, (maximum velocity ---, minimum velocity —·—)

distortion-energy hypothesis for the time instants of the maximum and minimum velocity in the steady state motion and the critical points in each cross-section depending on u_1.

Conclusion

It has been shown that the method of multibody systems can also be used for strength estimations. To determine the stresses within a body the equations of motion have to be supplemented with the equations of reaction. Efficient numerical algorithms are necessary to solve both sets of equations, because according to strength theory, for each particle of a body and all times a characteristic scalar stress has to be calculated to determine the critical point of a body. Of course, in general, the critical point of a body within a multibody system will be time-dependent.

References

1. Magnus, K. (ed.): Dynamics of multibody systems. Berlin/...: Springer 1978.
2. Haug, E. (ed.): Computer Aided Analysis and Optimization of Mechanical System Dynamics. Berlin/...: Springer 1984.
3. Kortüm, W.; Schiehlen, W.: General purpose vehicle system dynamics software based on multibody formalisms. Vehicle System Dynamics 14 (1985) pp. 229-263.
4. Schiehlen, W.: Technische Dynamik. Stuttgart: Teubner 1985.
5. Schramm, D.: Ein effizienter Algorithmus zur numerischen Berechnung der Zwangskräfte mechanischer Systeme. To appear in ZAMM 66 (1986)
6. Kreuzer, E.; Schmoll, K.-P.; Schramm, D.: Programmpaket NEWEUL'84. AN-10, Stuttgart: Institut B für Mechanik, Universität Stuttgart 1984.
7. Settelmeyer, E.: Spannungsanalyse an Roboterarmen. STUD-12, Stuttgart: Institut B für Mechanik, Universität Stuttgart 1984.

A Perspective on Computer-Oriented Mutlibody Dynamical Formalisms and their Implementations

R. SCHWERTASSEK

DFVLR, Oberpfaffenhofen, FRG

ROBERT E. ROBERSON

University of California San Diego
La Jolla CA, USA

ABSTRACT

Past work on computer-oriented formalisms for rigid/gyrostatic bodies is surveyed, together with their numerical computer program implementations.

1. Introduction

Computer-oriented treatments of multibody dynamical systems can be categorized in various ways. One basic division is into numerical digital simulation or into symbolic programs, the latter constructing equations in literal symbolic form. Another is into systems of rigid, possibly gyrostatic, bodies on one hand and elastic or partly elastic bodies on the other. Space does not permit our surveying all of these, so our scope extends only to formalisms for rigid/gyrostatic bodies and to numerical programs.

This work concerns developments during the last two decades: earlier history is given in [44]. Previous surveys include [10], mainly on machine dynamics; [38] and [41], mainly on formalisms of Lagrangian type; [24], on multibody spacecraft dynamics; [44] on Eulerian formalisms; [29] on Lagrangian and Eulerian methods for systems with tree structure, but not on their computer implementations. Various methods are compared as analytical methods in [27], again without reference to computation. More general surveys are found in [49] and [60].

Development problems for multibody formalisms have been largely solved, except perhaps for certain details, and computer codes based on many formalisms are available. In some cases, however, these codes are not fully polished for user convenience, efficiency and style. The most recent developments in coding relate to convenient options for specifying input data and to the use of preprocessing and postprocessing to promote efficiency in computation and storage. Computer graphics to show the motions of the system visually are increasingly available.

One classification of approaches is into synthetic methods, which start by applying the dynamical laws of Newton and Euler to the individual bodies of the system, and analytical methods based on principles of mechanics as formulated by d'Alembert, Jourdain, Lagrange, Hamilton and others. Generically, the synthetic methods are called Eulerian and the analytical methods Lagrangian. Computer-oriented multibody dynamical formalisms, and especially the computer implementation of such formalisms, also can be compared on the basis of other factors:

1. The generality of the class of systems that can be modeled.
   - a. tree vs. general system structure
   - b. planar vs. spatial motion
   - c. modes of interbody motion (rotation, translation, coupled)
2. Choice of dependent variables.
   - a. This choice can be made by the formalism itself or can be left to the user. Choice by the user makes for flexibility, but can be complicated and burdensome

Dynamics of Multibody Systems
IUTAM/IFToMM Symposium Udine 1985
Editors: G. Bianchi and W. Schiehlen

in specific cases. More generally, is the philosophy of computer implementation one of minimum user involvement with details?

b. Variables may be absolute, describing all motion with respect to a common reference frame, or relative, describing motions between interacting bodies. In some cases they may be mixed. Absolute variables lead to equations having simpler form, whereas relative variables may yield higher accuracy and make it easier to represent interactions between bodies.

3. Number of equations used. A minimal set of equations may be used or redundant equations may be admitted. Minimal sets of equations are more difficult to derive, more complicated in form, possibly faster to solve, but might not yield directly all the desired information about the systems (e.g. constraint forces and torques). Descriptions with redundant equations are more easily derived, generally simpler in form, slower to solve, but provide additional information.

2. Lagrangian Approaches

General

Formalisms based on Lagrange's equations have been preferred by authors dealing with mechanisms and robots, systems which often contain closed loops of links. (See [42], for example.) Most early applications involved only planar motion. Graph-theoretic concepts have been widely used to represent the system topology.

Program DYMAC (Dynamics of Machinery)

Program development is described in [1], [39] and [40], among others. The program is limited to planar motion of a system having general topology. In the system graph branches represent links (bodies) and nodes represent joints connecting links. Dependent variables are planar inertial displacements of the centers of mass of the bodies, resolved in an inertial frame, and rotation angles of the bodies in inertial space. The starting point is d'Alembert's equation, written in terms of virtual displacements of the two components of displacement and the angle of rotation for each body. M user-chosen coordinates forming matrix q, some redundant, are defined.

The system satisfies $2L$ loop constraints expressing the closing of displacement polygons around closed loops; N_S spatial (scleronomic) constraints and N_T temporal (rheonomic) constraints. The constraint equations in the three cases are respectively

$$f_L(q) = 0, \quad f_S(q) = 0, \quad f_T(q,t) = 0 \tag{1}$$

From Eqs.1, $\dot{q}$ must satisfy

$$\begin{bmatrix} \partial f_L/\partial q \\ \partial f_S/\partial q \end{bmatrix} \dot{q} = 0 \tag{2}$$

A suitable subset q_I of q, consisting of N_F+N_T variables (N_F is the number of degrees of freedom) is chosen and velocity variables q_{II} are defined as $\dot{q}_I$. The rheonomic constraints are written at the velocity level in the form

$$w_T{}^T q_{II} = \partial f_T/\partial t \tag{3}$$

Without giving details of the manipulation, the final result is

$$\begin{bmatrix} A & w_T \\ w_T{}^T & 0 \end{bmatrix} \begin{bmatrix} \dot{q}_{II} \\ \lambda \end{bmatrix} = \begin{bmatrix} Q \\ -\dot{w}_T{}^T q_{II} \end{bmatrix} \tag{4}$$

which can be solved for all of the unknowns, $\dot{q}_{II}$ and the Lagrange multipliers λ. Paul speaks of the solution which includes Lagrange multipliers as the method of excess differential equations. The alternative is a solution based upon the minimal set of dependent variables q_I. Computational experience is that the use of excess differential equations typically requires only about 40% of the computer time required by the method of minimal equations.

Programs DAMN and DRAM

The original program, DAMN (dynamic analysis of mechanical networks), was completed at the University of Michigan in 1969, having its roots in work by M.A.Chace [6]. It was improved and expanded by D.A.Smith [11], was renamed DRAM (dynamic response of articulated machinery) and has continued to develop in response to realistic industrial problems. A review of the program's status as of 1977 is given in [8]. Like DYMAC, this is a program for the planar motion of a system having arbitrary topology, founded on the same principles but with different coding and solution procedures.

The system is structured by a graph whose nodes represent bodies and whose branches represent translational or rotational contacts between them, hence the dual of the graph used for DYMAC. The dynamical description is Lagrange's equation with Lagrange multipliers, written (with a change in notation to resemble Eq.4)

$$\frac{d}{dt}\left(\frac{\partial T}{\partial \dot{q}_I}\right) - \frac{\partial T}{\partial q_I} = Q + w_L \lambda \tag{5}$$

The actual generation of the left side is done in a way that avoids giving rise to two separate terms which cancel [5], [9]. The equation for loop constraints is as in the first item of Eq.1. From these follows a single partitioned matrix equation having the form of Eq.4, except that $\dot{q}_{II}$ is replaced by $\ddot{q}_I$. All constraints are handled automatically by the choice of a minimal set of generalized coordinates.

Program ADAMS

ADAMS originated from Orleandea's Ph.D. dissertation [36]. (See also [32], [37], [7].) It applies to three-dimensional systems having arbitrary structure. The philosophy of programming is based on the tailoring of variables and equations to the numerical methods used. In particular, the system matrices are made very sparse by these choices, sparse matrix techniques are used to solve linear algebraic equations and Gear's method is used for integrating the resulting stiff differential equations of motion.

Let $x_{Ij}{}^i$, $i=1,N$ be three inertial coordinates of the center of mass of body i when j=1,3 and three Euler angles representing the orientation of body axes relative to an inertial frame when $j=4,6$. Let $x_{IIj}{}^i$ be the corresponding velocity variables, $Q_j{}^i$ be the applied forces and torques and λ be a matrix of K Lagrange multipliers. In what follows, index i ranges over the bodies, i.e. $1,N$. Index j ranges over 1,6. There are K constraint equations of all types,

$$f_k(x_{Ij}{}^i\ , t)=0 \qquad k=1,2...K \tag{6}$$

The dynamical differential equations are

$$\dot{p}_j{}^i - \frac{\partial T}{\partial x_{Ij}{}^i} - Q_j{}^i + \sum_{k=1}^{K} \frac{\partial f_k}{\partial x_{Ij}{}^i}\lambda_k = 0 \tag{7a}$$

in which the derivative of $p_j{}^i$ is replaced by the derivative of $m^i x_{IIj}{}^i$ for $j=1,3$ but in which the $p_j{}^i$ are left as dependent variables for $j=4,6$. The kinematic differential equations are

$$\dot{x}_{Ij}{}^i = x_{IIj}{}^i \tag{7b}$$

Finally, the angular momenta variables are connected to other variables of the problem by

$$p_j{}^i = \frac{\partial T}{\partial x_{IIj}{}^i}, \quad (j=4,6) \tag{7c}$$

Thus, there are K Eqs.6, $6N$ Eqs.7a, $6N$ Eqs.7b and $3N$ Eqs.7c, or a total of $(15N+K)$ equations containing the $6N$ variables $x_{Ij}{}^i$, the $6N$ variables $x_{IIj}{}^i$, the $3N$ variables $p_j{}^i$, and the K variables λ. Assuming that the constraints are independent, the number of equations and unknown variables is consistent. Even aside from the Lagrange multipliers, there is more than a minimal number of dependent variables. However, it is claimed that the equations still can be solved expeditiously because of the sparseness of the matrices to which one is led during the solution. In particular, such matrices can be written as LU decompositions literally and linear equations can be solved with corresponding ease.

Coordinate Partitioning Method

Generalized coordinate partitioning is described in [34], [35]. Its theory is based on Lagrange's equations with Lagrange multipliers, similar to the theory underlying the ADAMS computer program except that angular orientations are described by four Euler parameters rather than three Euler angles and the angular momenta are not used as separate dependent variables. A matrix generalized coordinate for each of the N bodies, consists of the three inertial components of absolute position relative to an inertial origin, together with four Euler normalized parameters characterizing the absolute rotation of the body. These seven coordinates are the elements of vectors q^i and the latter are assembled as the columns of matrix q.

The velocity form of the matrix constraint equation is used as a basis for partitioning the absolute coordinates into a set of dependent and a set of independent variables. Both holonomic and nonholonomic constraints are admitted. Because the holonomic constraint equations generally are nonlinear, they are solved numerically to express dependent variables in terms of the independent variables. The $7N$ Lagrange equations are reduced to a state space form in terms of the independent variables, then integrated numerically. These ideas have been implemented in the DADS program. Its pre- and postprocessing capabilities are similar to those of ADAMS.

Program IMP

IMP (integrated mechanism program) [54],[55] is unique in three particulars.

1. Hamilton's canonical equations are used to formulate the system equations.
2. Generalized coordinates are defined on the basis of a notational scheme using 4 by 4 matrices originated by Denavit and Hartenberg.
3. The program applies only for systems with closed loops.

Generalized coordinates for each joint between links, from 1 to 5 in number, are selected. One selects a subset of variables among the coordinates for the totality of links such that the subset of variables is sufficient to represent all system motion. It is the subset which provides the position variables in the dynamical equations. The remaining dependent variables are the corresponding generalized momenta. Program IMP comes in several versions.

Gauss' Method of Least Constraint

A different foundation was used by Lilov and Lorer in [30], namely the derivation of multibody equations by way of Gauss' principle of least "constraint". The derivation is reduced to the solution of a minimization problem with constraints. The main purpose of the work is stated to be the investigation of robot dynamics. The bookkeeping notation is essentially that of [46], [59] and [31].

3. Eulerian Approaches

General

Most of the early uses of formalisms based on the Newton-Euler equations were by authors who modeled spacecraft as multibody systems. In that context, three-dimensional motion was the rule and closed loops of links were not present. The rationale of different Eulerian methods usually is quite similar, the differences often being mainly in the way the "bookkeeping" problems for the variables and parameters are handled.

Method of Margulies and Hooker

One of the earliest approaches to self-organizing computer- oriented dynamical formalisms for arbitrarily interconnected systems of an arbitrary number of bodies was initially planned jointly by Margulies and Roberson, who then worked on the problem separately. Margulies, in collaboration with Hooker, published his approach first [19]. They state that their method is a generalization of that of [15], applicable to a two-body system. It describes the three-dimensional motion of a system having topological tree structure, admitting relative rotations but no translation in the joints. Graphs are not introduced explicitly, but several sets of indices are defined which serve the same purpose. An inertial origin is used and dependent variables are absolute angular velocity vectors.

An important contribution is the modern resurrection of the concepts of augmented body and barycenter, introduced by Fischer around the turn of the century. (See [44].) The dynamical equations are written in a form equivalent to

$$A\dot{\omega} = E + PT^c \tag{8}$$

Explicit forms are given for matrix A and the "extra terms" E. Array P is a matrix of locked-mode vectors for the joints while T^c is an array of constraint torque vectors. Constraint equations are

$$P^T\omega = 0 \tag{9}$$

Two methods for handling constrained modes in the joints are clearly described. In the "elimination method", Eq.8 is solved for $\dot{\omega}$ and the result is used in the time derivative of Eq.9. This gives an equation that can be solved for T^c in terms of known quantities. Then T^c is returned to Eq.8 to get an equation of motion free of unknown constraint torques. The same technique is used by later authors who have described it as using a "generalized inverse". In the "augmentation method", Eq.9 is differentiated and adjoined to Eq.8 to get the higher order system

$$\begin{bmatrix} A & -P \\ P^T & 0 \end{bmatrix}\begin{bmatrix} \dot{\omega} \\ T^c \end{bmatrix} = \begin{bmatrix} E \\ -\dot{P}^T\omega \end{bmatrix} \tag{10}$$

to be solved simultaneously for all unknowns, $\dot{\omega}$ and T^c. A simulation program based on this paper, but extended to gyrostats, is MLTBDY by G.E. Fleischer [13]. He refers to the method as the "barycentric formulation".

Method of Roberson and Wittenburg

In parallel with the work of Margulies and Hooker, Roberson began his approach to a matrix formulation of multibody dynamical equations, stressing the use of graph-theoretic concepts. The derivation of a number of formulae directly involves graph theory, so the graph serves as more than just a convenient way to list the vertices and branch numbers and their relationships. Unlike its use elsewhere, there is no requirement that the vertices or branches be numbered in any preferential way, or that a spanning tree be known *a priori*. He was joined shortly thereafter by Wittenburg and their first interim report on this work was [46]. It was a preliminary step, concerned with systems having tree topology, in motion with respect to a moving reference, without translation and rotational constraints in the interbody connections. Only a few general remarks are made there about the treatment of interaction constraints. Restrictions were gradually relaxed: e.g. relative translation is included in [43], but much of the work remained unpublished (as internal reports of DFVLR). An external report on the formalism and the computer program MULTIBODY was published in 1978 [52], and the first descriptions of the full formalism are [53] and [45]. The generality of the systems it can model is greater than found in most other rigid/gyrostatic multibody formalisms. An attempt to extend the formalism to elastic bodies has been made by Graul, resulting in computer code MAFES [62].

A few details about the present form created by Roberson and Schwertassek follow.

a. System structure is defined by a packed incidence matrix. A spanning tree, a "path matrix", and a submatrix of the fundamental circuit and cut-set matrices of the graph are generated algorithmically. In programming, this part of the problem is done off-line by a program SGRAPH.

b. A feasible configuration is defined and used to construct new body axes if desired, converting all body properties to these new axes. Free mode 6-vectors (rotation/translation) are defined for each joint, denoted ϕ , the basis of R^6 is completed with $\overline{\phi}$, and the dual basis $[\psi\ \ \overline{\psi}]$ is generated. These operations are done off-line by program PARAMS.

c. A third program, IVPROB, assembles the dynamical and kinematical equations and solves the initial value problem by a suitable integration routine.

Variables in the theory are initially direction cosine matrices $B^a = [B^a{}_{\alpha\beta}]$ across "primary" joints, parametrized by Euler normalized parameters, relative displacements $z^a = [z^a{}_\alpha]$ and relative translational velocities $V^a = [V^a{}_\alpha]$. Superscript a denotes one of the primary joints represented by the arcs of the system graph, and lower case Greek indices range over 1,2,3. The kinematical equations of motion are

$$\dot{B}^b = -\tilde{\Omega}^b B^b \tag{11a1}$$

$$\dot{z}^b = V^b \tag{11a2}$$

with index b ranging over all of the joints. The position variables must satisfy the kinematic consistency conditions on the motions in closed loops, namely

$$z^c{}_\alpha = z^c{}_\alpha(z^a{}_\alpha, B^a{}_{\alpha\beta}) \tag{11b1}$$

$$B^c{}_{\alpha\beta} = B^c{}_{\alpha\beta}(B^a{}_{\alpha\beta}) \tag{11b2}$$

developed in [53]. Index c ranges over the "secondary" joints closing the loops, represented by the chords of the system graph. The dynamical equations in terms of variables Ω^a and V^a are developed from the laws of Newton and Euler. The path matrix of the graph operates on the initial quations to yield the same unreduced system equations as in the nested body approach:

$$J\dot{G}_A = \Lambda_A + W\Lambda_C + Q \tag{12a}$$

Here J is a symmetric positive definite generalized mass matrix, G_A is a generalized velocity matrix $[[\Omega^a]^T \;\; [V^a]^T]^T$, Λ are generalized force matrices $[[L^b]^T \;\; [F^b]^T]^T$ in which branch index b is an arc index for Λ_A and a chord index for Λ_C. Matrix W describes the way in which the interactions Λ_C across the chords contribute to the interactions Λ_A across the arcs, while Q subsumes all other terms. For tree systems there are no chords and the W term is absent. In case of systems with closed loops, Eqs.12a are accompanied by the consistency conditions on the velocity variables

$$G_C = W^T G_A \tag{12b}$$

in which matrix G_C contains velocities $[[\Omega^c]^T \;\; [V^c]^T]^T$ across the secondary joints.

The kinematical equations, Eqs.11a, are rephrased in terms of minimal sets of "joint state variables" describing the motion across the joints. They must be integrated while satisfying Eqs.11b. Variables z^b and B^b needed in the latter equations are gotten from the joint state variables by the explicit forms of the position constraints across joint b. The dynamical equation, Eqs.12a, is reduced to one in fewer variables in the light of the constraints on the velocities.

Linear algebra teaches that velocity G^b across joint $b=a$ or $b=c$ can be represented in terms of a smaller set of velocity variables g.

$$G = \phi g + \overline{\phi}\overline{g} \tag{13a}$$

where $\phi^T\overline{\phi}=0$ and rows of $\overline{g}$ corresponding to rheonomic constraints are known time functions, whereas those corresponding to scleronomic constraints are zero. (Index b has been suppressed for notational simplicity.) In an analogous way, the Λ can be represented as

$$\Lambda = \psi\lambda + \overline{\psi\lambda} \tag{13b}$$

Here λ elements are known functions of relative motion and $\overline{\lambda}$ elements are unknown constraint torques, the projections of Λ on the constraint modes. Using Eqs.13ab (with added superscripts for arcs or chords) in Eq.12a, after writing the latter out in superscript form, one can premultiply the result by ϕ^{a^T} to rid the equation of the joint constraints $\overline{\lambda}^a$ in the arcs. In a non-tree structure, loop consistency conditions, Eqs.12b, permit the construction of system mode vectors which can be used much like the joint mode vectors, to eliminate the chord constraints $\overline{\lambda}^c$ still remaining in the equations. (See [53].) Alternatively, the chord constraints can be included as unknowns and the dynamical equations can be augmented by the loop closure conditions at the velocity level, to get a soluble augmented system.

Meanwhile, Wittenburg [57] published his development of the formalism begun in [46]. It applies to tree systems, point- connected (relative rotation only), but also considers impacts. Later he changed the starting point to the d'Alembert equation in [58], [31] and [59], but otherwise the development is much the same. Constraints are removed from the equations by the elimination method used by Hooker and Margulies. The loop constraint equations are gotten after defining a spanning tree by cutting bodies (see [64]) rather than separating joints into primary and secondary ones. By 1983 his interests had turned to symbolic programming of the multibody formalism, resulting in computer code MESA VERDE [61].

Information Flow Diagram

One of the few approaches to mechanism simulation using Eulerian methods resulted in the general-purpose program MEDUSA [12]. It simulates planar mechanisms whose components are limited to rigid bodies and flexible beams; translational and rotational springs and dampers; force and torque generators; motion generators. The user models the components to fit these requirements. Formal rules are used to create a network, a so-called "information flow diagram", in which blocks represent inertia characteristics, equilibrium conditions for free body diagrams and rigid body kinematics. This diagram, representing the primitive forms of the Newton-Euler equations of planar mechanisms, is described by a user-provided subroutine. The equations still contain unknown constraint forces and torques. Constraints on motion are formulated in terms of accelerations in the joints, leading to second order differential equations to be adjoined to the unreduced equations of motion. This approach was selected because the reduction to state space form was considered too unwieldy. A Lagrangian approach was considered unsatisfactory because geometrical meaning is lost and it is difficult to incorporate flexible bodies. The performance of MEDUSA did not seem very efficient for simulating mechanisms, which may be a reason for the apparent dislike of Eulerian methods by those involved in mechanism applications.

Nested Body Methods

In [20] Hooker addresses the same type systems as Hooker and Margulies, namely a tree system of bodies constrained to undergo only relative rotation. Dependent variables are absolute velocity of a main body and rotation angles about the free mode vectors in the joints. To get rid of constraint torque, say in joint j, the dynamical equations for bodies in a set S_j outboard (or inboard) of j are summed. All internal torques in this set sum to zero, leaving just one dynamical equation containing the single internal torque in joint j. This equation is projected on the null space of the subspace defined by the constraint mode vectors in the joint. This leads to one or more scalar equations in which constraint torques are absent. This "nested body" method for isolating a single joint constraint torque in one dynamical equation seems to have been first devised by Velman [56] in the context of a specific four-body system. Various authors also mention an unpublished 1969 work (an Aerospace Corp. report) by W.J. Russell in connection with the nesting of bodies. Russell also apparently focused on a specific rather than an arbitrary configuration.

Fleischer and Likins based a series of FORTRAN subroutines on Hooker's work [14]. It differs from [13] in the same way that [20] differs from [19]. Systems have tree topology, joints are hinged (no relative translation), bodies must be labeled preferentially. Absolute angular velocity vectors are used as dependent variables. Computation involves barycenters. The user is required to specify a nominal configuration and is permitted to prescribe rheonomic constraints.

Ness and Farrenkopf [33], and Farrenkopf in earlier (1969) unpublished work, describe an "inductive algorithm" for generating multibody system equations. Preferential labeling of the system is required. There is no explicit use of a graph, but a list serves the purpose.

Major attention is to systems having tree structure: otherwise, an artifice is suggested but not further explored. Their method is effectively a nested body method, but in doing the nesting one works inward from the terminal bodies, one at a time, and where chains of bodies join, the subsystems gotten thus far are combined. This joining of subsystems is in contrast to the usual nested body approach in which no more than one body at a time is adjoined to a previous subsystem of bodies.

More recently, program N-BOD2 has been based on the ideas of nested bodies in [16] and [17]. It applies to trees of rigid bodies and point masses, the latter only as terminal bodies. The extended bodies can only rotate relative to one another, whereas the point masses translate. The velocity state is described by absolute linear and angular velocities of a main body and by relative velocities of the permitted interbody motions. The kth nest is the set consisting of body k and all bodies outward (relative to the main body) of it.

In a posthumous work [18], Hooker returns to the multibody problem, again with a nested body approach and a system model like that of [19]. This time the rotational equations are written about inboard interbody attachment points so the "troublesome" (a subjective characterization) moment arms of interbody constraint forces do not appear. Again, constraint torques are removed by projection on the null space of the subspace defined by constraint torque modes.

Methods Utilizing Jourdain's Principle

The formalism of Schiehlen and Kreuzer [51], [50], [49], reduces the dynamical equations to state space form by applying d'Alembert's principle in the case of holonomic systems and Jourdain's principle in the case of nonholonomic systems. In [51] only holonomic systems are considered. The Newton/Euler dynamical equations are put into the form

$$M\overline{J}\dot{y}+\overline{k}(y,\dot{y},t) = \overline{q}^{e}(y,\dot{y},t)+\overline{Q}(y,t)g \tag{14}$$

in which y is a matrix of generalized coordinates, M is a generalized mass matrix, J is a Jacobian matrix, $\overline{k}$ subsumes a number of miscellaneous terms, $\overline{q}^{e}$ represents external generalized forces, g denotes constraint forces and $\overline{Q}$ describes the distribution of the constraint forces among the equations. d'Alembert's principle is used to show that $\overline{J}^{T}Q=0$. Thus, if Eq.14 be multiplied throughout by the transpose of $\overline{J}$, the constraint torques are removed. The case of nonholonomic systems is treated in [49] and the remaining references. A linear transformation from the velocities $\dot{y}$ to new velocity variables z is made using the nonholonomic constraint equations and an equation analogous to Eq.14 is written in $\dot{z}$, namely

$$M\overline{L}\dot{z}+\overline{k}(y,z,t) = \overline{q}^{e}(y,z,t)+\overline{Q}g \tag{15}$$

Here $\overline{L}$ is a transformed $\overline{J}$ matrix, and the definition of $\overline{k}$ is changed. This time, Jourdain's principle is used to show that $\overline{L}$ is orthogonal to the constraint term, so multiplication by its transpose removes the constraints. After solving the equation without constraints, one can return to Eq.14 or 15 and find g by using the generalized inverse $((\overline{Q}^{T}\overline{Q})^{-1}\overline{Q}^{T})$. This theory underlies the symbolic development of the equations by the computer program NEWEUL.

Earlier, Kane proposed a method for deriving equations of motion in [26], [27] whose underlying ideas and results are essentially those of the Schiehlen and Kreuzer procedure. The coefficients multiplying the generalized velocities $\dot{z}$ (in Schiehlen's notation) to get absolute linear and angular velocities of the bodies are called "partial velocities" by Kane: these play a central role in his formalism. A closely related work by other authors is [25]. This approach has been realized in a symbolic computer program SD-EXACT [63]. A numerical program UCIN-EULER based on Kane's equations has been developed by Huston and others and is described in [23], [21]. It applies to systems with tree configuration and with hinges and ball-and-socket joints. Motions across spherical joints are described by Euler parameters. [22] is an extension to translation in the joints. This program has been applied to a variety of practical problems.

Transformation Operator Method

This method was devised by Jerkovsky [24] for three-dimensional motion of systems whose bodies can undergo relative rotation and translation and whose topological structure is general. Two formulations are presented: a "velocity formulation" in which the dependent variables are absolute angular velocities and absolute translational velocities of the bodies; a "momentum formulation" in which the corresponding variables are angular momenta about the bodies' centers of mass and linear momenta of the bodies. The system graph's nodes represent bodies and branches represent interconnections between bodies. The graph requires preferential labeling. Body 1 is a reference body marking the root of a spanning tree, and other bodies are enumerated by monotonically increasing numbers moving outward in that tree from the root. An $n \times n$ "path matrix" π is defined such that $\pi_{ij}=1$ if body j is inboard of body i and is zero otherwise. The bulk of the discussion is for tree systems, with a somewhat abbreviated description toward the end concerning the treatment of systems with loops.

Primitive momentum equations are $G=K$, where vector G has an upper subarray of vector angular momenta of the bodies about their respective centers of mass and a lower subarray of vector linear momenta, and K has an upper subarray of vector torques and a lower subarray of vector forces, both internal and external. Primitive velocity equations are $\mu \cdot \dot{\sigma} + Y = K$, where $\dot{\sigma}$ has an upper subarray of absolute angular velocity vectors and a lower subarray of absolute translational velocity vectors, μ is a generalized inertia/mass dyadic, and Y contains the Euler cross-product terms as an upper array and a lower array of zeros. A "transformation operator", dyadic A, derived from path matrix π, transforms all arrays to get a new momentum formulation

$$\dot{\overline{G}} + \overline{X} = \overline{K} \tag{16}$$

and velocity formulation

$$\overline{\mu} \cdot \dot{\overline{\sigma}} + \overline{Y} = \overline{K} \tag{17}$$

The transformation converts the absolute angular and translational velocity vectors in σ into relative angular and translational velocity vectors between contiguous bodies in $\overline{\sigma}$. To this point, most difficulties would appear to be in getting $\overline{X}$ and $\overline{Y}$, although the process is described as straightforward.

Equations 16,17 apply directly only when all interconnections have six degrees of freedom. To introduce constraints into the joints, dependent variables are expanded in terms of unit mode vectors in the joints. Rotational mode vectors are not orthogonal, so their dual bases also are needed. Torque and force can be expanded the same way. Some of the modes represent free motions in which torque or force are known functions of relative motion. The other modes are constrained, so their coefficients represent torques or forces of constraint. The equations of motion are partitioned into free and constrained parts. The free equations are solved by numerical integration to get the free dependent variables. Constrained dependent variables can be gotten from the latter algebraically. This method of eliminating constraints is essentially that used by Roberson and Wittenburg. An alternative solution method is described which is the augmentation method of Hooker and Margulies. The unpublished formalism of Russell also is based on momentum equations, differing from Jerkovsky by referring momenta and velocities to the system center of mass rather than to the centers of mass of the individual bodies.

Vector-Network Method

This method, formulated by G.C.Andrews in his Ph.D. dissertation [2], applies the concepts of electrical network theory directly to mechanical systems. See also [3] and [4]. It is applicable to the three-dimensional motion of any network of systems for which there exists a "proper" spanning tree, i.e. a tree that contains only mass elements and displacement/velocity drivers, other elements being confined to the cotree. Absolute displacement, velocity and angular velocity vectors (relative to an inertial frame) are used as dependent variables and, at least in the case of translation, such vectors are resolved in the inertial frame. Nodes of a graph are defined at the points where mechanical elements (masses, springs, dampers, displacement and velocity drivers, force drivers) are interconnected. The directed branches of the graph are the across-variables of the system (displacement or velocity vectors). Actually, the graph need not be metrized this way: the method works if such physical quantities are simply put into correspondence with the branches. "Terminal" or constitutive equations are written for each element, relating across-variables to through variables for that element. These include the d'Alembert force for mass elements. The circuit equations are substituted into the cotree terminal equations to express the cotree forces in terms of the tree displacements and velocities. The cotree terminal equations then are used in the cutset equations to express the tree forces in terms of the tree displacements and velocities. The mechanical equations of motion follow.

This method is the basis for the "self-formulating" computer program VECNET at the University of Waterloo. It deals with three- dimensional particle systems which include passive elements and displacement/velocity and force drivers. The theory applies to rotation of non-point mass elements as well, but the computer program is confined to translational motions. Common linear terminal equations are built into the program: others, as well as terminal equations for drivers, are user-provided. An input record contains the element number, type, parameter value, and the numbers of the two nodes between which the element is attached. The three numbers in the record constitute, in effect, a packed form of the graph's incidence matrix. The incidence matrix is used to construct fundamental cutset and circuit matrices. The illustrative examples given in the literature on the method show a preferential assignment of element numbers in tree and cotree. VECNET has given rise to other programs. DYNIS (dynamic interactive simulation program) is essentially VECNET with improved, interactive I/O [48]. PLANET (plane-motion simulation using network techniques) and PLANET II are programs which treat only planar motion, but they also include plane motion of rigid bodies [47].

REFERENCES

1. Amin, A., "Automatic formulation and solution techniques in dynamics of machinery", Ph.D. dissertation, University of Pennsylvania, Philadelphia, 1979.

2. Andrews, G.C., "The vector-network model: a topological approach to mechanics", Ph.D. dissertation, University of Waterloo, 1971.

3. Andrews, G.C., Kesavan, H.K., "The vector-network model: a new approach to vector dynamics", **Mechanism and Machine Th. 10** (1975), 57-75.

4. Andrews, G.C., Kesavan, H.K., "Simulation of multibody systems using the vector-network model", **Dynamics of Multibody Systems, Proc. IUTAM Symposium.** (Munich, 1977), K. Magnus ed., Springer, Berlin, 1978, pp.1-13.

5. Bayazitoglu, Y.O., Methods for automated analysis of three-dimensional mechanical dynamic systems with application to nonlinear vehicle dynamics", Ph.D. dissertation, ;University of Michigan, 1972.

6. Chace, M.A., "Analysis of time-dependent of multi-freedom mechanical systems in relative coordinates", **J. Engrg. Ind. 89** (1967), 119-125.

7. Chace, M.A., "Methods and experience in computer aided design of large- displacement mechanical systems", **NATO Advanced Study Inst. on Computer Aided Analysis and Optimization of Mechanical System Dynamics**, J. Haug, ed., Springer, Berlin, 1984, pp.233-259.

8. Chace, M.A., Angell, J.C., "Interactive simulation of machinery with friction and impact using Dram", **Trans. SAE 86** (1977), 239-250. (Paper 770050)

9. Chace, M.A., Bayazitoglu, Y.O., "Development and application of a generalized d'Alembert force for multifreedom mechanical systems", **J. Engrg. Ind. 93** (1971), 317-327.

10. Chace, M.A., Sheth, P.N., "Adaptation of computer techniques to the design of mechanical dynamic machinery", ASME Paper 73-DET-58, ASME, NY, 1973.

11. Chace, M.A., Smith, D.A., "DAMN - digital computer programn for the dynamic analysis of generalized mechanical systems", **Trans. SAE 80** (1971), 969-983.

12. Dix, R.C., Lehman, T.J., "Simulation of the dynamics of machinery", **J. Engrg. Ind. 94** (1972), 433-438.

13. Fleischer, G.E., "Multi-rigid-body attitude dynamics simulation", JPL Tech. Report 32-1516, Pasadena, 1971.

14. Fleischer, G.C., Likins, P.W., "Attitude dynamics simulation subroutines for systems of hinge-connected rigid bodies", JPL Tech. Report 32-1592, Pasadena, 1974.

15. Fletcher, H.J., Rongved, L., Yu, E.Y., "Dynamics analysis of a two-body gravitationally oriented satellite", **Bell System Tech. J 42** # 5 (1963), 2239-2266.

16. Frisch, H.P., "A vector-dyadic development of the equations of motion for N-coupled rigid bodies and point masses", NASA TN D-7767, 1974.

17. Frisch, H.P., "The N-BOD2 User's and Programmer's Manual", NASA Tech. Paper 1145, 1978.

18. Hooker, W.W., "Equations of motion for interconnected rigid and elastic bodies: a derivation independent of angular momentum", **Celestial Mech. 2** (1975), 337-359.

19. Hooker, W.W., Margulies, G., "The dynamical attitude equations for n-body satellite", **J. Astronautical Sci. 12** (1965), 123-128.

20. Hooker, W.W., "A set of r dynamical attitude equations for an arbitrary n-body satellite having r rotational degrees of freedom", AIAA J 8 # 7 (1970), 1205-1207.

21. Huston, R.L., Passerello, C.E., "On multi-rigid-body system dynamics", **Computers and Structures 10** (1979), 439-446.

22. Huston, R.L., Passerello, C.E., "Multibody structural dynamics including translation between bodies", **Computers and Structures 12** (1980). 713-720.

23. Huston, R.L., Passerello, C.E., Harlow, M.W., "Dynamics of multirigid-body systems", **J. Appl. Mech 45** (1978), 889-894.

24. Jerkovsky, W., "The structure of multi-body dynamics equations", **Proc. Sympos. Dynamics and Control of Large Flexible Structures** (Blacksburg VA, 1977), pp.387-421. [Also **J. Guidance and Control 1** (1978), 173-182.]

25. Kamman, J.W., Huston, R.L., "Dynamics of constrained multibody systems", **J. Appl. Mech. 51** (1984), 899-903.

26. Kane, T.R., "Dynamics of nonholonomic systems", **J. Appl. Mech. 28** (1961), 574-578.

27. Kane, T.R., Levinson, A.D., "Formulation of equations of motion for complex spacecraft", **J. Guidance Control 3** (1980), 99-112.

28. Kane, T.R., Wang, C.F., "On the derivation of equations of motion", **J. Soc. Ind. Appl. Math. 13** (1965), 487-492.

29. Likins, P.W., "Point-connected rigid bodies in a topological tree", **Celestial Mechanics 11** (1975), 301-317.

30. Lilov, L., Lorer, M., "Dynamic analysis of multirigid-body system based on the Gauss principle", **ZAMM 62** (1982), 539-545.

31. Lilov, L., Wittenburg, J., "Bewegungsgleichungen für Systeme starrer Körper mit Gelenken beliebiger Eigenschaften", **ZAMM 57** (1977), 137-152.

32. Maros, D., Orlandea, N., "Contributions to the determination of the equations of motion for multidegree of freedom systems", **J. Engrg. Ind. 93** (1971), 191-195.

33. Ness, D.J., Farrenkopf, R.L., "Inductive methods for generating the dynamic equations of motion for multibodied flexible systems", ASME, NY, 1971.

34. Nikravesh, P.E., Chung, I.S., "Application of Euler parameters to the dynamic analysis of three-dimensional systems with nonholonomic constraints", **J. Mech. Design 104** (1982), 785-791.

35. Nikravesh, P.E., Haug, E.J., "Generalized coordinate partitioning for analysis of mechanical systems with nonholonomic constraints", **J. Mechanisms, Transmissions Automation in Design 105** (1983), 379-384.

36. Orlandea, N., "Node-analogous, sparsity-oriented methods for simulation of mechanical systems", Ph.D. dissertation, University of Michigan, 1973.

37. Orlandea, N., Chace, M.A., Calahan, D.A., "A sparsity-oriented approach to the dynamic analysis and design of mechanical systems - Part 1", **J. Engrg. Ind. 99** (1977), 773-784.

38. Paul, B., "Analytical dynamics of mechanisms - a computer oriented overview", **Mechanism and Machine Theory 10** (1975), 481-507.

39. Paul, B., "Dynamic analysis of machinery via program DYMAC", SAE Paper 770049, Soc. of Automative Engrs., 1977.

40. Paul, B., **Kinematics and Dynamics of Planar Machinery**, Prentice Hall, Englewood Cliffs, NJ, 1979, Part 2.

41. Paul, B., "Computer oriented analytical dynamics of machinery", **NATO Advanced Study Inst. on Computer Aided Analysis and Optimization of Mechanical System Dynamics**, (Iowa City, 1983), E.J.Haug, ed., Springer, Berlin, 1984, pp.41-87.

42. Paul, R.P., **Robot manipulators: Mathematics, programming and control**, MIT Press, Cambridge, 1981.

43. Roberson, R.E., "A form of the translational dynamical equations for relative motion in systems of many non-rigid bodies", **Acta Mechanica 14** (1972), 297-308.

44. Roberson, R.E., "Computer-oriented dynamical modelling of spacecraft: historical evolution of Eulerian multibody formalisms since 1750", **Proc. 28th Int. Astronaut. Congr.** (Prag, 1977), IAF Paper 77-A11.

45. Roberson, R.E., **Duo Ti Xi Tong Dong Li Xue (Dynamics of Multibody Systems)**, Lectures given at the Nanjing Aeronautical Institute, 5-16 September 1983, Nanjing Aeronautical Institute, Sep 1984. (Chinese)

46. Roberson, R.E., Wittenburg, J., "A dynamical formalisms for an arbitrary number of interconnected rigid bodies, with reference to the problem of satellite attitude control", **Proc. 3rd Congr. Int. Fed. Auto. Control** (London, 1966), Vol.1 Book 3 System Dynamics, Butterworth, London, undated, Paper 46D.

47. Rogers, R.J., Andrews, G.C., "Simulating planar systems using a simplified vector-network method", **Mechanism and Machine Th. 10** (1975), 509-519.

48. Savage, G.J., Andrews, G.C., "DYNIS: a dynamic interactive simulation program for three- dimensional mechanical systems", **Proc. 3rd man-computer communications seminar**, (Ottawa, 1973).

49. Schiehlen, W.O., "Dynamics of complex multibody systems", **SM Archives 9** (1984), 159-195.

50. Schiehlen, W.O., "Computer generation of equations of motion", **NATO Advanced Study Inst. on Computer Aided Analysis and Optimization of Mechanical System Dynamics**, E.J. Haug, ed., Springer, Berlin, 1984, pp.183-215.

51. Schiehlen, W.O., Kreuzer, E.J., "Rechnergestützes Aufstellen der Bewegungsgleichungen gewöhnlicher Mehrkörpersysteme", **Ing.-Archiv 46** (1977), 185-194.

52. Schwertassek, R., "Der Roberson/Wittenburg-Formalismus und das Programmsystem MULTIBODY zur Rechnersimulation von Mehrkörpersystemen." Report DFVLR-FB-78/08, DFVLR, Wissenschaftliches Berichtswesen, Köln, 1978.

53. Schwertassek, R., Roberson, R.E., "A state-space dynamical representation for multibody mechanical systems. Part I: Systems with tree configuration", **Acta Mechanica 50** (1983),141-161. "Part II: Systems with closed loops", **Acta Mechanica 51** (1984), 15-29.

54. Sheth, P.N., "A digital computer based simulation procedure for multiple degree of freedom mechanical systems with geometric constraints", Ph.D. dissertation, University of Wisconsin, 1972.

55. Sheth, P.N., Uicker, J.J., "IMP (integrated mechanism program), a computer-aided design analysis system for mechanisms and linkage", **J. Engrg. Ind. 94** (1972), 454-464.

56. Velman, J.R., "Simulation results for a dual-spin spacecraft", **Proc. Sympos. Attitude Stabilization and Control of Dual-Spin Spacecraft**, (El Segundo CA, Dec 1967), SAMSO-TR-68-191, Air Force Systems Command, Space and Missile Systems Organization, pp. 11-23.

57. Wittenburg, J., "The dynamics of systems of coupled rigid bodies. A new general formalism with applications" **Stereodynamics** Centro Internazionale Matematico Estivo, I Ciclo 1971 (Bressanone, 2-12 Jun 1971), Edizioni Cremonese, Roma, 1972.

58. Wittenburg, J., "Nonlinear equations of motion for arbitrary systems of interconnected rigid bodies", **Dynamics of Multibody Systems**, Proc. IUTAM Symposium (Munich, 1977), K.Magnus ed., Springer, Berlin, 1978, pp.357-376.

59. Wittenburg, J., **Dynamics of Systems of Rigid Bodies**, B.G.Teubner, Stuttgart, 1977.

60. Wittenburg, J. "Dynamics of multibody systems", **Proc. 15th Int. Congr. of Theoretical and Applied Mechanics, IUTAM/ICTAM** (Toronto, 1980), F.P.J. Rimrott, B. Tabarrok eds., pp.197-207.

61. Wittenburg, J., Wolz, U., "MESA VERDE: a symbolic program for nonlinear articulated-rigid- body dynamics", **Proc 10th ASME Design Engr. Conf. on Mech. Vibration and Noise** (Cincinnati, Sep 1985).

62. Graul, S.F., "Mehrkörpersystem-Algorithmus in Vector-Dyadisches Form für beliebig angeordnete elastischen Substrukturen", Ph.D. dissertation, T.H.Aachen, 1985.

63. Rosenthal, D.E., Sherman, M.A., "Symbolic multibody equations via Kane's method", AIAA paper 83-803, AAS/AIAA Astrodynamics Specialist Conf., Lake Placid NY, 1983.

64. Lilov, L., Chirikov, V.A., "On the dynamics equations of systems of interconnected bodies", PMM 45 (1981), 383-390.

Interactive Design for Flexible Multibody Control

R. SINGH, R. VANDERVOORT and P. LIKINS

Dynacs Engineering Company and Lehigh University

Summary

In response to the need for efficient design and analysis of such complex electromechanical systems as spacecraft, robots, transportation vehicles and manufacturing equipment, a generic formulation of equations of motion for a broad class of mathematical models is presented, and an associated interactive computer program is described.

Introduction

During the past twenty years, the task of formulating generic equations of motion for systems of rigid and deformable bodies in various topological configurations has received serious attention by many capable analysts, motivated sometimes by the intriguing character of the equations themselves and sometimes by the practical demands of electromechanical system design. As a research problem in dynamics this subject is now mature, although interesting derivative problems abound.

The emphasis currently is no longer on the method used to formulate the equations, but rather on their utility, particularly for control system design. That utility derives from the computer programs developed around the numerical simulation capability provided by the equations of motion, so the emphasis in this paper is on the description of such programs. The foundation rests on the equations of motion, however, and on the technique adopted to process constraint equations as required to reduce the system equations to their minimum kinematic dimensions, as required for control.

System Topology

A typical multibody system in an arbitrary configuration is

Dynamics of Multibody Systems
IUTAM/IFToMM Symposium Udine 1985
Editors: G. Bianchi and W. Schiehlen

shown in Figure 1A. Body 3 is here selected (arbitrarily) as the reference body and it is assumed (for convenience) that Body 3 is connected to an imaginary inertially fixed body (an inertial reference frame) labeled 0.

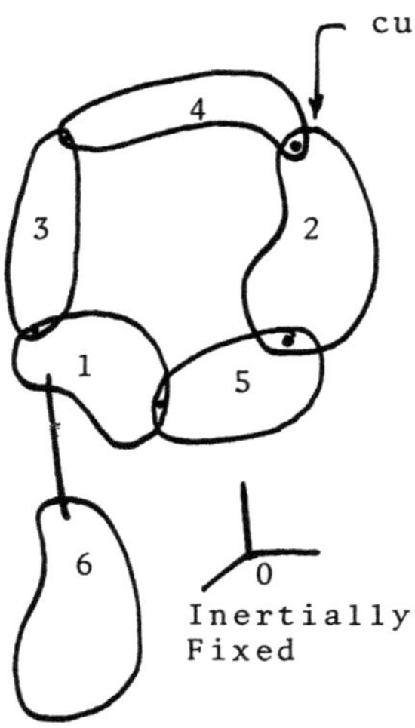

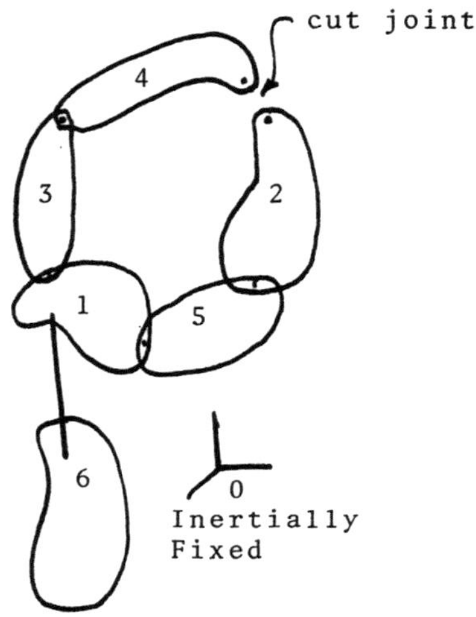

Fig. 1A. An Arbitrary Topology Fig. 1B. Equivalent Tree Topology

For the system shown in Figure 1A there does not exist a unique kinematical path from Body 0 (inertial reference) to every body in the system. To develop a unique kinematic description one can cut any set of joints that will make the resulting topology an open tree (for example, as shown in Figure 1B).

For the topology of Figure 1B a direct path array can be constructed readily by assigning to the jth body a number C(j) which is the body number of the adjacent body heading inboard to the reference body, and then to Body 0. Every j has a unique path set whose elements are 0 and all the bodies in the path from 0 to j including j. Thus given j and C(j) one could readily draw Figure 1B. The cut-joints can be handled by replacing them with the equations of kinematic constraints described in the next section.

The examples of Figures 1A and 1B are generalizable, with results applicable to any set of interconnected rigid or deformable bodies.

Figure 2 defines a hinge as a pair of two material points, one on each of two adjoining bodies of the equivalent tree. The

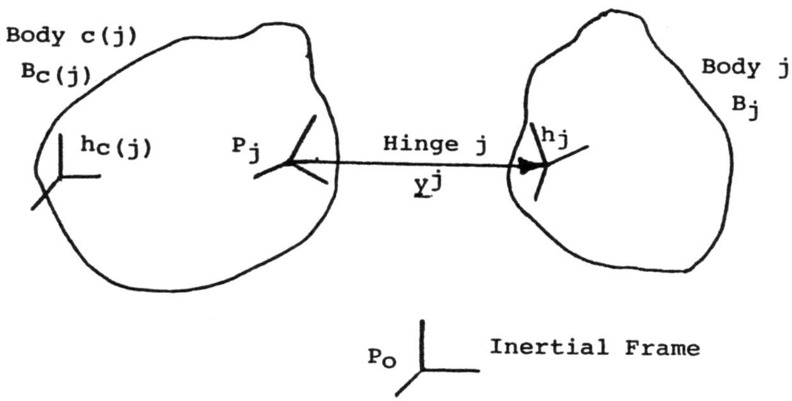

Figure 2. A Typical Hinge

discrete degrees of freedom of the jth body are described by the relative translations of h_j (a point defined by B_j, such as its mass center or a specific material point), with respect to P_j (a point defined by $B_{c(j)}$), and the relative rotations of the frames embedded in the bodies or associated reference frames at h_j and P_j. Points P_j, h_j and the reference frames at these two points constitute the jth hinge. For consistency, a fictitious hinge is assigned to the connection between a point of the reference body and the inertially fixed point P_o. Thus the number of hinges equals the number of bodies in the system.

The following symbols are used to describe the open tree configuration of the type illustrated by Figure 1B:

N_B = number of bodies
NT_j = translational degrees of freedom of the jth hinge
NR_j = rotational degrees of freedom of the jth hinge
NM_j = deformational degrees of freedom of the jth body
NS = total number of degrees of freedom of the equivalent tree system (typified by Figure 1B)

$$= \sum_{j=1}^{NB} (NT_j + NR_j + NM_j)$$

Equations of Motion

For a mechanical system of NB flexible bodies in a topological tree configuration, as typified by Figure 1B, the equations of motion are presented in Reference [1], combining Equations (14) and (15). The derivation of these equations is discussed in [1] and [2]. These equations can be cast in the second order matrix form

$$M\ddot{q} = F(\dot{q},q,t) \tag{1}$$

where the NS elements of q represent the generalized coordinates of the system, and M is a square, symmetric matrix.

For a more general configuration of the same NB bodies, constrained by the closure of loops in the topology (as in Figure 1A), the equations of motion may be obtained by combining Eq. (1) with kinematical constraint equations of the form

$$A\dot{q} = B \tag{2}$$

where for a system with m scalar constraints A is a matrix of dimension mxNS, and both A and B generally depend on q and t. Such constraints are classically called simple or Pfaffian and nonholonomic.

For the closed loop case, of course, the symbol F in Eq. (1) includes forces required to impose the constraints as well as other forces. It may be convenient to separate these forces symbolically, as

$$M\ddot{q} = F'(\dot{q},q,t) + F^c(\dot{q},q,t) \tag{3}$$

confining to F^c the unknown constraint forces. For control system design these unknowns are usually of no interest, and they will be eliminated by combining Eqs. (2) and (3) through the singular value decomposition of matrix A, as shown in [3].

Singular Value Decomposition

Orthogonal decomposition of an mxn matrix L by singular value decomposition is closely related to the eigenvalue-eigenvector decomposition of the symmetric positive semi-definite matrices L^TL and LL^T. Let $r \leq m$ be the rank of L. Then there are orthogonal matrices U and V of order mxm and nxn, respectively, such that

$$U^TLV = \begin{bmatrix} \Sigma & 0 \\ 0 & 0 \end{bmatrix} = S \tag{4}$$

where $\Sigma = \text{diag}\,(\lambda_1, \lambda_2, \ldots, \lambda_r)$ and $\lambda_1 \geq \lambda_2 \geq \ldots \geq \lambda_r > 0$.

The diagonal elements of the decomposition are called the singular values of the matrix L. The singular values are unique, although U and V are not.

It is easy to verify that

$$V^TL^TLV = \text{diag}\,(\Sigma^2, 0) \tag{5}$$

Thus $(\lambda_1^2, \ldots, \lambda_r^2)$ must be the nonzero eigenvalues of L^TL arranged in the descending order and the requirement that λ_i be nonnegative completely determines the λ_i. The eigenvectors of L^TL are the columns of V. If L^TL has a multiple eigenvalue $\lambda^2 > 0$, the corresponding columns of V may be chosen as an orthonormal basis for the space spanned by the eigenvectors corresponding to λ^2.

From Eq. (4)

$$L = USV^T \tag{6}$$

Now with proper partitioning of U and V Eq. (4) can be expressed as

$$L = \begin{bmatrix} U_1 & \vdots & U_2 \end{bmatrix} \begin{bmatrix} \Sigma & \vdots & 0 \\ \hline 0 & \vdots & 0 \end{bmatrix} \begin{bmatrix} V_1^T \\ V_2^T \end{bmatrix} = U_1 \Sigma V_1^T \tag{7}$$

From the above one obtains

$$U_1 = LV_1 \Sigma^{-1} \tag{8}$$

Thus once V_1 is chosen U_1 is obtained by Eq. (8). The matrices U_2 and V_2 may be any matrices with orthonormal columns spanning the null spaces of L^T and L, respectively. It is worthwhile to mention that the null space of L is the space of all vectors x such that

$$Lx = 0 \tag{9}$$

With the orthogonal decomposition given by Eq. (6), an nxm matrix L^+, called the pseudoinverse of L, is defined by

$$L^+ = V \begin{bmatrix} \Sigma^{-1} & \vdots & 0 \\ 0 & \vdots & 0 \end{bmatrix} U^T \tag{10}$$

L^+ is uniquely defined by L; it does not depend on the particular orthogonal decomposition of L.

<u>Application of SVD to Dynamic System with Constraints</u>

If the rank of matrix A in Eq. (2) is $r \leq m$ then r of the kinematical variables in q are related by Eq. (2), and there are only NS-r independent generalized coordinates in Eqs. (2) and (3). In other words, the dynamic system possesses NS-r degrees of freedom.

The SVD of the mxn matrix A provides

$$A = USV^T \tag{11}$$

The orthogonal matrices U and V (of dimension mxm and NSxNS, respectively) are partitioned as

$$U = \begin{bmatrix} U_1 & \vdots & U_2 \end{bmatrix} \tag{12}$$

$$V = \begin{bmatrix} V_1 & \vdots & V_2 \end{bmatrix} \tag{13}$$

where U_1 and V_1 are, respectively, mxr and NSxr matrices; U_2 and V_2 are, respectively, mx(m-r) and NSx(NS-r) matrices. Note that r is the rank of A.

Because $AV_2 = 0$, Eq. (2) is satisfied by

$$\dot{q} = A^{+}B + V_2\dot{z} \tag{14}$$

for any vector $\dot{z}$, where A^{+} is the pseudoinverse of A. We shall refer to z as the reduced set of (NS-r) coordinates.

Differentiation of Eq. (2) with respect to time yields

$$A\ddot{q} = -\dot{A}\dot{q}+\dot{B}$$

or,

$$A\ddot{q} = B' \tag{15}$$

Following Eq. (14), express $\ddot{q}$ in terms of $\ddot{z}$ as

$$\ddot{q} = A^{+}B' + V_2\ddot{z} \tag{16}$$

Note from Eq. (14) or Eq. (16) that V_2 maps the NS kinematic variables $\dot{q}$ (or $\ddot{q}$) to NS-r variables $\dot{z}$ (or $\ddot{z}$). Thus a consistent set of equations of motion in $\ddot{z}$ is given as

$$V_2^T M V_2 \ddot{z} = V_2^T F' + V_2^T F^c - V_2^T MA^{+}B' \tag{17}$$

The coefficient of $\ddot{z}$ is a symmetric positive definite matrix with the characteristic of an "inertia matrix" for the reduced set of coordinates z.

With the Lagrange multiplier method, F^c is established (as in [4]) via

$$F^c = A^T\alpha \tag{18}$$

where α is the column vector of Lagrange multipliers.

Premultiply Eq. (18) by V_2^T to obtain the following

$$\begin{aligned} V_2^T F^c &= V_2^T A^T \alpha \\ &= (AV_2)^T \alpha \\ &= 0 \end{aligned} \tag{19}$$

Thus it is seen that the nonworking constraint forces make no contribution to the equations of motion (Eq. (17)) and need not be recorded.

Employing the transformations given by Eqs. (14) and (16), the minimum dimension governing differential equations of motion are given by

$$V_2^T M V_2 \ddot{z} = V_2^T F' - V_2^T MA^+ B' \tag{20}$$

and

$$\dot{q} = A^+ B + V_2 \dot{z} \tag{21}$$

This method eliminates the forces of constraints, which when included not only enlarge the dimension of the dynamical system but also quite often introduce computational problems.

Computer Program

The simulation consists of three major parts: (1) an arbitrary topology of flexible structures, (2) a controller, (3) a set of sensors and actuators.

(1) For a mechanical system in an arbitrary topology (for example, the system of Figure 1A), the user, with the aid of an interactive set up program, defines an equivalent tree topology (Figure 1B). Constraint equations arising due to loop closure(s) and prescribed motions of selected points on the bodies are built in the program. The user needs only select the type (from a menu), node points and body indices. Based on this information, Eqs. (2) and (3) are formed internally in the simulation, ignoring F^c (which will be eliminated anyway). Based on Eqs. (20) and (21), the simulation numerically forms a set of 2NS-r first order differential equations in the variables q and $\dot{z}$, ready for integration.

(2) The intent of the simulation effort is to build a control system design tool. To this end sensors and actuators are built into the simulation along with a controller specified in a linear, block-diagram format or a user supplied subroutine.

Linear, block-diagram controller format: The controller is a multi-input, multi-output system composed of transfer functions and summing junctions with gains included in the interconnections between elements. An example of this type of controller is shown in Figure 3. The user defines this controller with

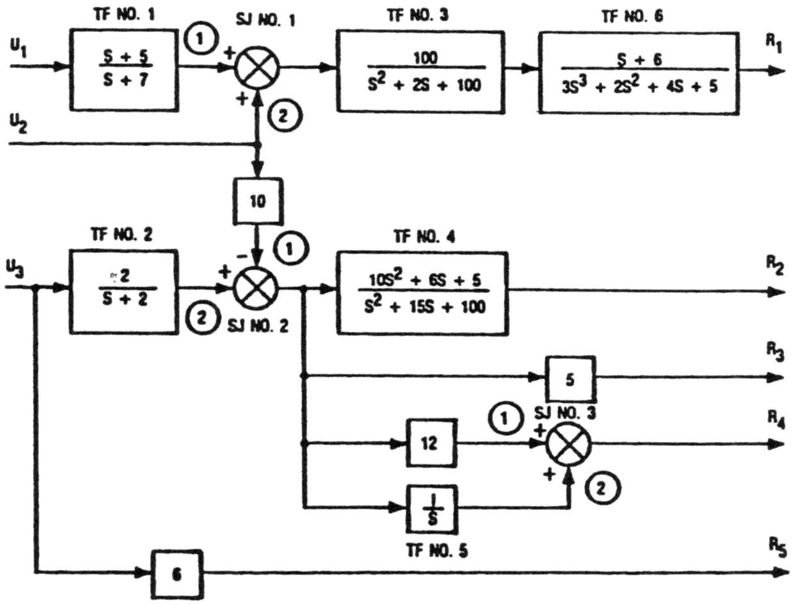

Figure 3. A Typical Linear Controller

the aid of the interactive program. The block diagram is reduced within the simulation to a set of matrix quadruples A,B,C,D corresponding to

$$\dot{x} = Ax + Bu \tag{22}$$

$$y = Cx + Du \tag{23}$$

where x, u, and y are the controller state vector, input and output vectors, respectively, and A,B,C,D are constant coefficient matrices. When the linear controller is continuous Eq. (22) is integrated simultaneously with Eqs. (20)-(21). A discrete linear controller may also be defined in which case the discrete equations corresponding to Eqs. (22)-(23) (formed internally in the program) are updated only at the sample time.

User Supplied Control Subroutine: In the program provision is made for the user to supply his own subroutine but the burden of software development is on the user. Various function generators (e.g. step, ramp, pulse, sawtooth, etc.) are built in the program and the user may use these function generators to define his controller. It should be noted that the continuous linear controller, the discrete controller and the user supplied subroutine may all be used simultaneously and interconnected as desired.

(3) A set of eight sensors has been built into the simulation.

Rate Gyro: The rate gyro is defined by a mounting point and an input axis. The input axis is fixed in the body to which the sensor is attached.

Resolver: A resolver measures the Euler angle associated with a specific hinge axis.

Angular Acceleration Sensor: This sensor is like a rate gyro except it measures the components of the angular accelerations.

Velocity Sensor: A velocity sensor measures the rate of change of the distance between any two points.

Position Sensor: This sensor measures the distance between any two points.

Single Axis Accelerometer: The output of this sensor is the same as a conventional accelerometer.

Tachometer: Measures the Euler rate associated with a specific hinge.

Line of Sight Sensor: Measures the orientation of a line (passing through two points) with respect to a prescribed frame.

The user needs only define the type of sensor, the mounting point node(s) and axis orientation. Likewise a set of four actuators--reaction jet, hydraulic cylinder, momentum wheel, and torque motor--has been defined. The user can generate disturbance or control forces on the structure by simply specifying the actuator type, mounting point node, output axis orientation and actuator command input.

The simulation has a linearization option. A second order

Lagrange interpolating polynomial is used to linearize about a user specified point. The coefficient matrices are available for use in an off-line control design software.

Software Overview

The CONTOPS software consists of three separate computer programs and a number of files as shown in Figure 4.

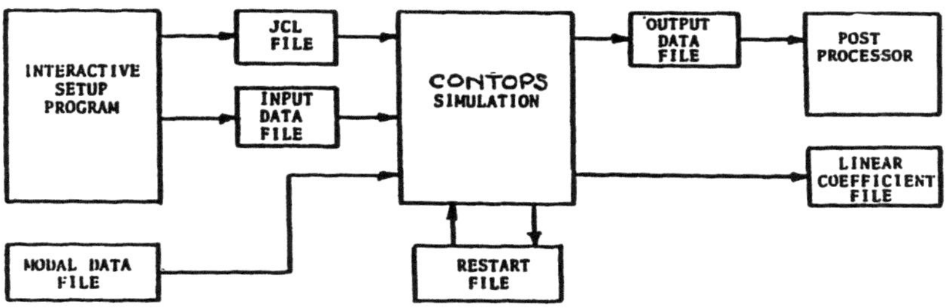

Figure 4. CONTOPS Software Overview

The motivation for having three separate programs is that the set-up program and post-processor run in an interactive mode while the simulation program which requires more execution time is normally run in a batch mode.

The interaction set-up program provides a user-friendly interface with the simulation via the following features:

- o The user is prompted for input so that essential elements are not omitted.
- o Free format inputs are accepted so a knowledge of data format and column number is not required.
- o Editing capability is provided for high level changes (delete or add bodies) and low level changes (change a data value).
- o The file structure and JCL (Job Control Language) files are set up automatically so the user doesn't need a high level knowledge of the computer operating system.
- o Error checking is performed in the interactive session reducing the possibility that the simulation batch job will bomb-out.

The simulation program is the heart of CONTOPS. It formulates the structure equations and performs numerical integration to obtain the time history response of the system. It also computes and outputs the linearized model coefficients. A restart file is created at the end of each simulation run so that subsequent runs can be started at that point at a later date.

The post-processor is interactive so that data can be examined and output modified based on simulation results. This makes it possible to obtain all the simulation output but doesn't make it necessary. Since the output data files have a standard format, it is possible to cross plot data from different runs for comparison purposes. By making the post-processor a stand alone program it enables the addition of special print or plot capabilities for special simulation cases.

References

1. Singh, R. P., VanderVoort, R. J., and Likins, P. W., "Dynamics of Flexible Bodies in Tree Topology--A Computer Oriented Approach," AIAA Journal of Guidance Control and Dynamics, V8, No. 5, Sept. 1985, pp. 584-590. Also Proceedings of the AIAA Dynamics Specialist Conference, Palm Springs, CA, May 1984, pp. 327-337.

2. Singh, R. P., and Likins, P. W. "Manipulator Interactive Design with Interconnected Flexible Elements," Proceedings of 1983 Automatic Control Conference, San Francisco, CA, June 1983, pp. 505-512.

3. Singh, R. P., and Likins, P. W., "Singular Value Decomposition for Constrained Dynamical Systems," to appear in the Journal of Applied Mechanics, Transactions of ASME.

4. Bodley, C., Devers, A., Park, A., and Frisch, H., "A Digital Computer Program for Interaction Simulation of Controls and Structures (DISCOS)," NASA Technical Paper 1219, May 1978.

Dynamics of a High-Speed Rotor Touching a Boundary

W. SZCZYGIELSKI and G. SCHWEITZER

Institute of Mechanics, ETH Zurich

Summary

When a rotor touches upon a boundary its motion depends on impact, friction, and gyroscopic effects. They result in a typical self-excitation leading to a variety of phenomena. For a gyro-pendulum touching a rigid plane wall the equations for small motions are derived. They are piecewise linear, globally however, strongly nonlinaer. An energy balance for the different phases of the motion helps to explain stationary, quasistationary and chaotic solutions, containing cascades of period doubling and Feigenbaum structures. The solutions can be classified for a whole set of parameters.

1. Introduction

A rotating body coming into contact with a second one, its housing or some other boundary, can exhibit a quite fascinating dynamical behaviour. Technically such a "two-body system" is given when a Kaplan-turbine rubs along its labyrinth seals, when a magnetically suspended rotor touches upon its emergency bearing or when the resonance vibrations of an elastic rotor are limited by some caging device. The technical relevance of such problems has been known for some time. There is the mentioning of selfexcited rubbing vibrations by Den Hartog /1/, the investigation of the nonlinear synchronous rubbing vibrations caused by a rotating unbalance by Black /2/ and many others, the recent paper of Muszynska /3/ showing subharmonic vibrations, and the suggestions of Zippe and Meinke /4/ on caging mechanisms for rotors passing through critical speeds. Our objective is somewhat different from the technical papers and goes back to the basic questions of how the modeling of the dynamics can be structured, how does the energy transfer within the various phases of motion work, what are adequate methods for analyzing the nonlinear equations of motion, and what kind of vibration phenomena are to be expected. The model for our investigation consists of a well-balanced gyropendulum which due to initial

Dynamics of Multibody Systems
IUTAM/IFToMM Symposium Udine 1985
Editors: G. Bianchi and W. Schiehlen

disturbances can touch a rigid wall. It will be shown that there are four obviously distinct phases of motion, the free motion, the impact, and the gliding and the rolling contact phase, and there is an additional complex chattering phase describing the transition from impact to gliding. The rotational energy of the rotor is transferred to lateral motions by impact and gliding supported by gyroscopic forces thus enabling self- excitation. The physical insight into the energy balance facilitates the interpretation of the equations of motion. The equations are piecewise linear for each phase but globally they are strongly nonlinear. For their solution analytical methods are combined with numerical simulations. Thus it is possible to classify quite a spectrum of vibration phenomena, like self-excited vibrations with multiple limit cycles and chaotic motions. First experimental results on a test rig corroborate main theoretical results and encourage extensions of it to technical applications.

2. Model and Equations of Motion

The gyropendulum as given in fig. 1 has two degrees of freedom. Its deviations from the vertical equilibrium position are the small angles α, β. The mass and geometric parameters are obvious from the figure.

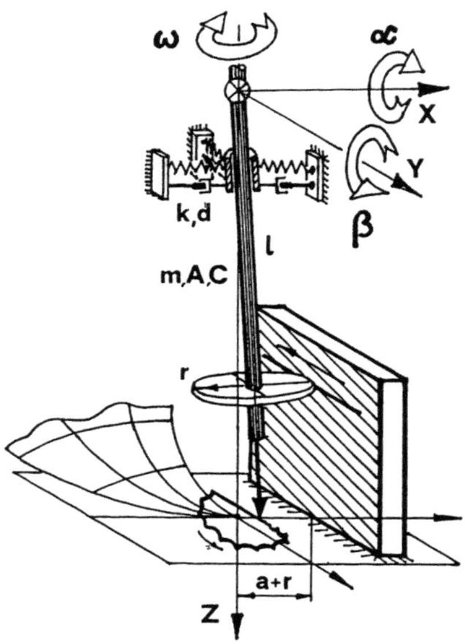

Fig.1. Mechanical model of the gyropendulum and its boundary

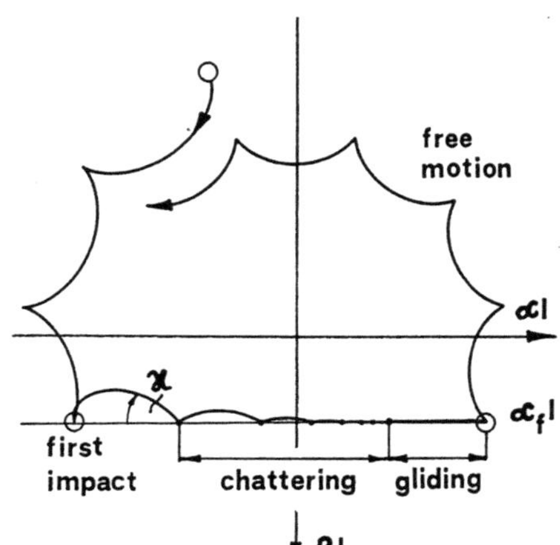

Fig. 2. Phases of motion and its characteristic positions

An additional isotropic spring and damping is attached to the shaft of the pendulum. When the rim of its wheel touches upon the rigid plane wall the normal velocities of the skew impact will be described by classical theory with a restitution coefficient ε being one for elastic and zero for unelastic impact. The gliding motion is governed by the Coulomb friction μ. The angular velocity ω of the rotor is constant, at least in the average, and any variation of speed caused by circumferential impact or gliding forces will be balanced in a sufficiently short time by the rotor drive. Due to the small angles α, β the spherical motion of the pendulum is represented by a trajectory traced out of a plane by the tip of the rotor axis. Fig.2 shows such a trajectory, indicating the various phases of motion, which will be dealt with now.

The free motion of the gyropendulum is described by

$$(1)\qquad \begin{aligned} A\ddot{\alpha} + C\omega\dot{\beta} + d\dot{\alpha} + k\alpha &= 0 \\ B\ddot{\beta} - C\omega\dot{\alpha} + d\dot{\beta} + k\beta &= 0 \end{aligned}$$

The gliding motion along the plane implies the constraint $\beta = a/l$, and assuming $\mu r/l \ll 1$ equation (1) has to be altered into

$$(2)\qquad \begin{aligned} A\ddot{\alpha} \qquad\qquad + d\dot{\alpha} + k\alpha &= \operatorname{sign}(\omega r - \dot{\alpha} l) l F_R \\ -C\omega\dot{\alpha} \qquad\qquad + k a/l &= -l F_N \end{aligned}$$

where the friction force F_R and the normal force F_N are related by Coulomb`s law

$$(3)\qquad F_R = \mu F_N \qquad \text{with } F_N > 0$$

Assuming a high rotor velocity $\omega \gg \dot{\alpha}, \dot{\beta}$ implies a distinct direction of the friction force, and then equ. (2) reduces to

$$(4)\qquad A\ddot{\alpha} + (d - \mu C\omega)\dot{\alpha} + k\alpha = -\mu k a/l$$

Obviously for $d < \mu C\omega$ the gliding motion will be unstable. The gliding turns into free motion when the normal force $F_N = 0$, and then following from equ. (2), the velocity $\dot{\alpha}_f$ always has the same value

$$(5)\qquad \dot{\alpha}_f = ka/C\omega l$$

That the spin of the gyro $C\omega$ plays an important role during the free and the contact phase has been demonstrated by Magnus /5/ when he shows how a gyroaxis cann roll along an arbitrary contour. In our case a rolling motion cannot develop from the gliding as long as we assume a fast rotor where the relative velocity between rim and wall does not vanish. It would not be too difficult, however, to include the rolling phase as it is the least complicated one. Rolling may be of interest for technical reasons and especially for a closed circular boundary. It will be investigated separately /7/.

The impact is described by integrating the equations of motion over the time of impact, assuming that the position does not change during that time, leading to relations between the velocities before and after the impact

$$(6)\qquad \begin{aligned} A(\dot{\alpha}^{+} - \dot{\alpha}^{-}) &= \int 1F_R \, dt \\ A(\dot{\beta}^{+} - \dot{\beta}^{-}) &= -\int 1F_N \, dt \end{aligned}$$

Using equ. (3) and the kinematic relation for the normal velocities

$$(7)\qquad \dot{\beta}^{+} = -\varepsilon \dot{\beta}^{-}$$

we obtain from (6) the tangential velocity

$$(8)\qquad \dot{\alpha}^{+} = \dot{\alpha}^{-} + \mu(1 + \varepsilon)\dot{\beta}^{-}$$

The restitution coefficent ε applies to the normal direction. The consideration of restitution effects for the tangential direction does not seem to be necessary /6/, as the tangential forces are more adequately characterized by the friction coefficient.

The chattering phase describes the transition from impact to gliding as shown in the measured trajectory of fig. 3. Chattering is modeled by a converging sequence of impacts and free motions (fig. 4).

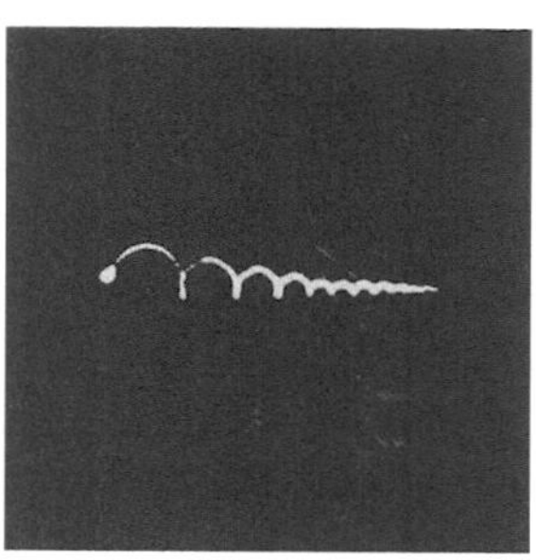

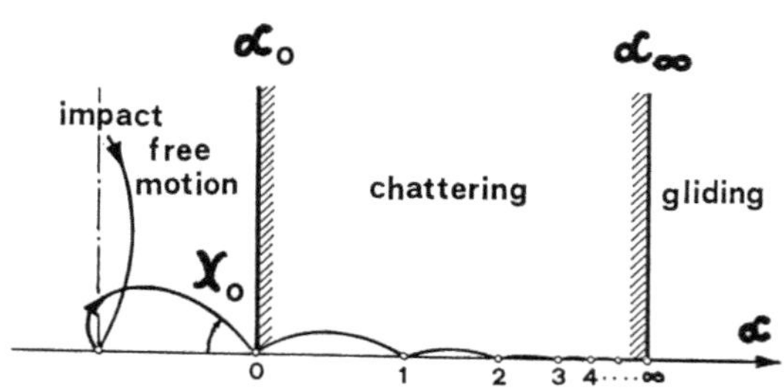

Fig. 3. Measured trajectory of the gyro axis during chattering

Fig. 4. Model for the chattering transition from free motion to gliding

The initial state of the phase is given by an angle α_0 at the position of a suitable impact which has an angle of approach $\varkappa_0$ with

(9) $$0 < \dot{\beta}_0^- / \dot{\alpha}_0^- = \tan \varkappa_0 < 1$$

After each impact the motion is at least approximately a free nutation circle, with the nutation frequency $\omega_N = \omega C/A$, and the total phase can be expressed piecewise by closed form solutions. The infinite sequence converges to the final gliding state /7/ for $\varepsilon \ll 1$

(10) $$\alpha_\infty = \alpha_0 + \frac{2\varepsilon}{\omega_N(1-\varepsilon)} \dot{\beta}_0^- \quad , \quad \dot{\alpha}_\infty = \dot{\alpha}_0^- + \mu \frac{1+\varepsilon}{1-\varepsilon} \dot{\beta}_0^-$$

The chattering motion of the gyro-axis can be illustrated (fig. 5), too, by looking at the finite displacements $M_1, \ldots M_i$ of the axis of the moment of momentum vector under a sequence of impacts, each of the M_i being the center for a new nutation circle /8/.

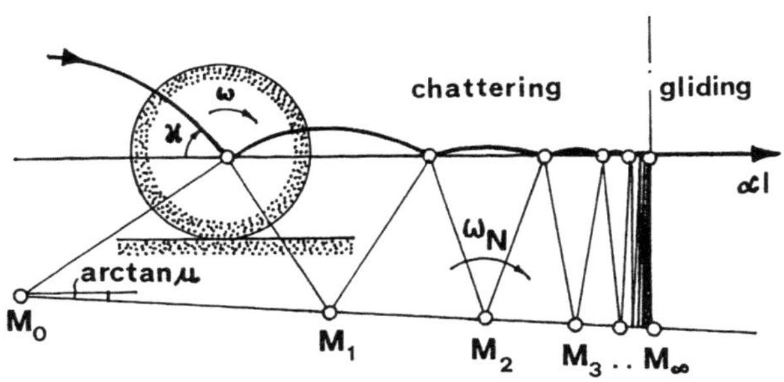

Fig. 5. Shifting of the moment of momentum axis and resulting trajectory of the gyro-axis during chattering

Further analysis even leads to a nice equivalence between the chattering and an immediate gliding after an anelastic impact ($\varepsilon = 0$). This equivalence substantially facilitates numerical solutions.

3. Energy Balance

The free motion is conservative or damped ($d \geq 0$). During the gliding phase, however, energy is transmitted to the lateral α-motion, which can even become unstable as equ. (4) shows. The impact phase usually will be dissipative but it can be receptive as well. This is the case when the kinetic energy after the impact $A(\dot{\alpha}_+^2 + \dot{\beta}_+^2)/2$ is larger than before the impact $A(\dot{\alpha}_-^2 + \dot{\beta}_-^2)/2$. Such a "superball-effect" occurs when the angle of approach $\varkappa$, which can be derived from the energy condition and equs. (7,8) is determined by

$$\text{(11)} \qquad \tan\varkappa = \dot{\beta}^-/\dot{\alpha}^- < 2\mu/\left[1 - \varepsilon - \mu^2(1+\varepsilon)\right]$$

Physically this means that the energy transmitted by the friction impact and coming from the rotation of the gyro is larger than the energy dissipated by the classical impact mechanism. And this energy transfer from the fast to the slow modes of the motion still works, when the gyro is not driven at constant velocity ω as long as the variation of ω through an impact remains sufficiently small.

During the chattering phase condition (7) usually is fulfilled, so this phase, too, will be receiving energy and can contribute to self-excitation. Now the question is of how do these energies contribute to the global motion, and how does any periodic or asymptotic behaviour look like?

4. Solution Techniques

Physical, mathematical and numerical methods have been used to find solutions. Physical reasoning for energy and force balance essentially helps to structure the problem, to predict and to explain special solutions, and to reduce the domain of possible inital conditions. The latter is quite an important issue and it works as follows. Assuming the existence of a gliding phase, we know from equ. (5) that the end of the gliding phase is described by a kinematically very simple, reduced state. All possible motions, the phase flow, only differ in the angle α_f where the rotor leaves the wall. The gliding phase channels the phase flow and reduces the dimension of the state space. That means that an arbitrary initial state $[\alpha_0, \beta_0, \dot{\alpha}_0, \dot{\beta}_0]$ is reduced to $[\alpha_f]$ and the constants $\dot{\alpha}_f$, $\beta_f = a/l$, $\dot{\beta}_f = 0$, and in the next step α_f is written in the dimensionless form $\Psi_f = \alpha_f l/a$. For a = 0 a simple special solution can be derived. Besides the initial conditions the other parameters can be reduced to only four essential ones: μ, ε, d, and $\Omega_0 = C^2\omega^2/kA$ which means the ratio of nutation and precession frequency and characterizes the gyro velocity.

Analytical solutions are easily obtained for the various phases. Due to the unsteady transitions between the phases, however, analytical methods for the global motion, like averaging or asymptotic methods, fail. So analytical solutions are only used to support numerical procedures, mainly by avoiding very cumbersome integrations.

Numerical solutions now can be obtained for an already highly restricted set of parameters. They are represented by trajectories of the gyro tip in the figures 6a to 6d showing typical examples that will be discussed now.

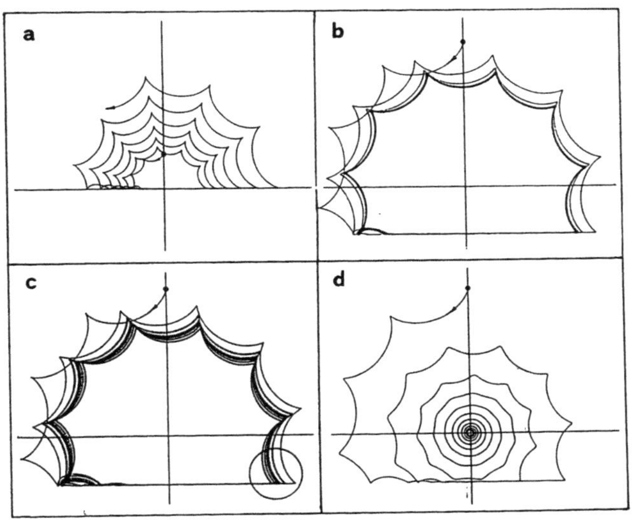

Fig. 6. Trajectories a) unstable motion, b) limit cycle with one loop, c) chaotic limit cycle, d) asymptoticlly stable

5. Discussion of Solutions

The trajectories in fig. 6 all show the first part of the motion from the initial condition (α_0, β_0) to the reduced initial condition α_f which marks the end of the gliding phase. Depending on this initial condition the figures show four different typical types of behaviour, described in the legend. For the special case, where the gap a=0 (fig. 6a) some characteristics of the trajectory can be proven generally. The velocity at the break-off point α_f is always zero (equ. 5), the trajectory of the free motion is an epicycle with sharp edges. Using complex variables the motion

can be analysed, and once one loop is known the other ones follow from it by a mere scaling factor and are self-similar. The scaling factor already decides on the stability, being > 1 for the unstable motion of fig. 6a. A stability chart is shown in fig. 7, the stability depending on the gyro speed Ω_o and the friction μ. There are certain speeds where the instability areas in a repetitive way are very large. At these gyro speeds the free trajectory contacts the wall with an edge of the epicycle, i.e. with a zero velocity and therefore the length of the gliding phase and the energy input is largest.
The chaotic motion of fig. 6c will be dealt with next.

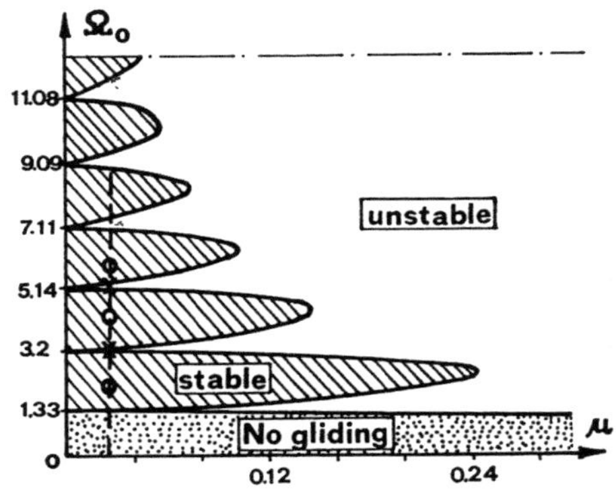

Fig. 7. Stability chart (a=0, d=0, ε=0,6), measured values o x , μ=0,025
Fig. 8. Fractal structure of the trajectories near the break-off point α_f

6. Chaotic motions

The trajectory in fig. 6d converges to an area that looks like a broad limit cycle, forming a socalled strange attractor /9, 10/. The area is not covered uniformly but shows a fractal structure (fig. 8). The motion is not periodic even when it shows orbital stability and some analogy to a limit cycle. What does distinguish this attractor from others known in literature, is its cyclic variation of state space dimension. It piecewise has dimension 1, 2 and 4 corresponding to the different phases of motion. Furthermore there are two independent mechanisms of dissipation, impact and viscous damping. The erratic looking motion, it is deterministic however, depends on the sensitivity with respect to the initial condition α_f. The reason for it is, that small variations of the position of the peaks of the free epicycles near the wall have a large influence on the conditions of the impact and the further motion. In order to more clearly investigate the chaotic motion we sample the motion in a strategic or in a stroboscopic way. This Poincaré mapping is done by determining the sequence of end points α_f for all the consecutive gliding phases of a unique trajectory. For a periodic motion with one loop per period this obviously always gives one point $\alpha_{fi} = \alpha_{f1}$. Varying an essential parameter, for example ε, this one-periodic motion can bifurcate into a two-periodic motion. By further variation of ε a cascade of period doubling leads to a chaotic behaviour. Periods with up to $2^7=128$ loops have been simulated. Even bifurcations of the class $3*2^N$ and $5*2^N$ occur within the windows of the chaotic area. Figure 9 shows such a cascade, followed by chaotic areas, a socalled "Feigenbaum". For each parameter value 1000 loops were calculated and the corresponding 1000 breakoff points α_{fi} have been plotted. The Feigenbaum shows the typical structure with windows and an abrupt end. There it collides with an unstable limit cycle and jumps to another domain of attraction. The Feigenbaum number

$$\delta_K = \frac{\varepsilon_K - \varepsilon_{K-1}}{\varepsilon_{K+1} - \varepsilon_K} = 4{,}669201\ldots$$

where ε_K is the parameter value at the K-th bifurcation, could be approximated by $\delta_5 = 4{,}38$ and $\delta_6 = 4{,}86$.

Of course the variation of the friction coefficient μ leads to a similar behavior (fig. 10), with, however, an inverse cascade.

Experiments

First experimental results have been obtained with the test rig shown in fig. 11. It allows well defined arbitrary initial conditions ($\alpha_0, \beta_0 < 0.1$) for the gyropendulum, a constant rotor speed ω at arbitrary level up to 50 Hz, a simple mounting of different wall materials as steel, brass or teflon, and the measuring of trajectories of the shaft by proximity sensors. One quantitative result is shown within fig.7, the points indicating measured stable and unstable motions, depending on the gyro speed Ω_0.

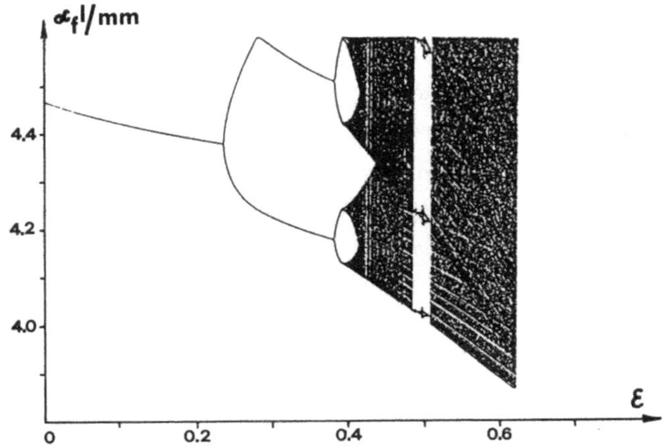

Fig. 9. Feigenbaum for $\alpha_f(\varepsilon)$

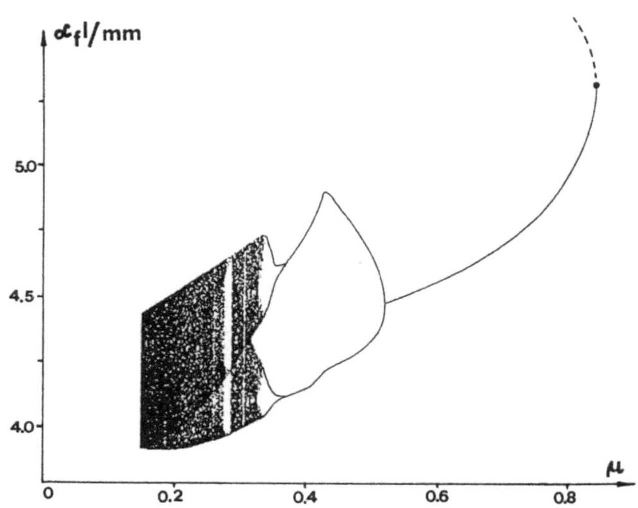

Fig. 10. Feigenbaum for $\alpha_f(\mu)$

8. Conclusions and Outlook.

The simple model of a gyropendulum contacting a plane wall gives a deep insight into the self-excitation mechanism of rubbing contacts. A special value is attributed to the physical explanations of the various phenomena. There is reason to believe that the methods can be used to solve more technical problems as well. Using conformal mapping some of the current results have already been transferred to circular concave und convex boundaries (fig. 12). Further investigations will include elastic and damped boundaries, different material combinations, and the consideration of unbalance forces. The final objective is to get some control on the asymptotic state of the complex contact dynamics.

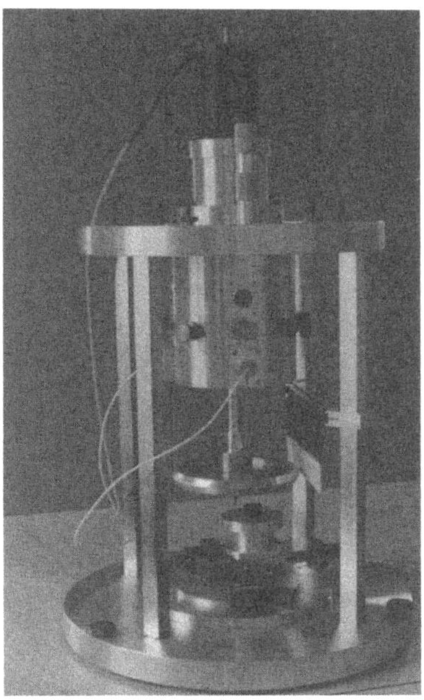

Fig. 11. Experimental set-up

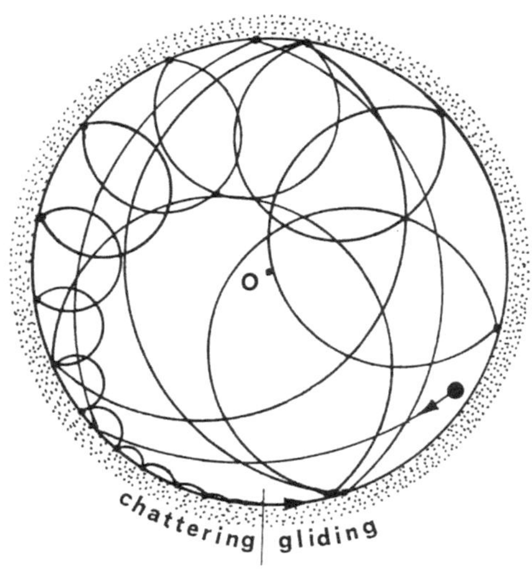

Fig. 12. Trajectory for circular boundary

References

1. Den Hartog, I.P.: Mechanical Vibrations. Mc Graw Hill (1956) 291-293

2. Black, H.F.: Interaction of whirling rotor with a vibrating stator across a clearance annulus. J. Mech. Eng. Sc. 10 (1968) 1-12.

3. Muszynska, A.: Partial lateral rotor to stator rub. Third Int. Conf. on Vibrations in Rotating Machinery. I Mech E Publ. C 281/84, York, Sept. 1984.

4. Zippe, G.; Meinke, P.: Verfahren und Vorrichtung zum Durchlaufen kritischer Drehzahlen langgestreckter Rotoren. Patentschrift DE 2632586 C2.

5. Magnus, K.: Beiträge zur Kinetik des Kurvenkreisels. Ingenieur-Archiv 43 (1974) 145-157.

6. Lehmann, T.: Elemente der Mechanik III. Braunschweig: Vieweg 1977.

7. Szczygielski, W.: Dynamisches Verhalten eines schnell drehenden Rotors bei Anstreifvorgängen. Diss. ETH Zürich (to appear).

8. Magnus, K.: Kreisel,Theorie und Anwendungen.Berlin-Heidelberg-New York: Springer-Verlag 1971.

9. Eckmann, J. P.: Roads to turbulence in dissipative dynamical systems. Rev. Modern Phys. 53 (1981) 643-653.

10. Ott, E.: Strange attractors and chaotic motions of dynamical systems. Rev. Modern Phys. 53 (1981) 655-671.

Computer Analysis of Multibody Systems with »Natural Coordinates«

J. UNDA, J.M. JIMENEZ, A. AVELLO & J. GARCIA DE JALON

University of Navarra
Department of Applied Mechanics
Urdaneta 7, 20006 San Sebastián, SPAIN

Summary

In this paper a new method for the numerical analysis of multi-rigid-body systems is described. This method uses a new system of non independent coordinates formed by the cartesian coordinates of some points of the mechanism, and -in the three dimensional case- by the cartesian components of some unitary vectors fixed to the elements. All of them determine the position and the motion of the multi-rigid-body system. The constraint equations arise from the rigid body condition of each element and from the constraints introduced by the joints. In the 3-D case, the inclusion of unitary vectors as mechanism coordinates allows an easy formulation of pair constraints when the pair is related to a particular direction, as in revolute (R), cylindrical (C), or prismatic (P) pairs. The constraint equations are always linear or quadratic in the problem coordinates. The differential equations of motion are obtained easily through the application of the Theorem of Virtual Power. Some examples of dynamic analysis of planar and three-dimensional multibody systems are presented.

Introduction

Many mechanical systems can be effectively modeled as formed by a set of interconnected rigid bodies. The dynamic analysis of such systems can be performed with the highest generality using numerical methods. These methods have been mostly developed in the 70s and 80s, and they have given birth to general purpose computer programs as DRAM[1-2], IMP[3], ADAMS[4] and DADS-3D[5].

The distinctive characteristics of the numerical methods of analysis on which these programs are based are the nature of the coordinates chosen for the definition of the position and motion of every element of the mechanism, the constraint equations that relate these non-independent coordinates, and finally, the formulation of the differential equations of motion whose nume

Dynamics of Multibody Systems
IUTAM/IFToMM Symposium Udine 1985
Editors: G. Bianchi and W. Schiehlen

rical integration gives the time evolution of the mechanism.

This article describes a method for the kinematic and dynamic analysis of multi-rigid-body systems, based on the use of a new system of dependent coordinates called "natural coordinates". The natural coordinates are easy to select and handle, they have a clear and direct physical meaning, and their constraint equations can be formulated very easily. With them the need of pre- and post processing work is reduced to a minimum.

Mechanism coordinates

Numerical methods for the analysis of multibody systems diversify mainly by the kind of coordinates that they choose for the definition of the mechanism motion. These coordinates, which we will call "mechanism coordinates", must define unequivocally the position of every element of the mechanism. Usually, their number is greater than the number of degrees of freedom, so they are not independent but related by the kinematic constraint equations. From the mechanism coordinates, a set of independent ones can be chosen in order to define the motion of the input elements or to integrate the differential equations of motion. The computation of the mechanism coordinates from the independent ones implies the solution of the position problem through the constraint equations, which are usually non linear.

There are mainly three different kinds of mechanism coordinates: relative coordinates[1-3], that define the position of every element with respect to the previous one in the kinematic chain; reference point coordinates[4-7], that define independently the absolute position of every element by means of the cartesian coordinates of a point and Euler's Angles or Euler's Parameters; and natural coordinates[8-9], which describes the position of each element through the cartesian coordinates of two or more of its points and -in the 3-D case- with the cartesian components of some unitary vectors, as it will be seen later.

Relative coordinates are the lowest in number, they imply loop constraint equations, and they demand complicate topological preprocessing. Reference point coordinates are greater in number, but they don't need any preprocessing because their constraint equations arise from the kinematic pairs. Both relative and reference point coordinates lead to nonlinear constraint equations involving trascendental functions. Natural coordinates are moderate in number, and they use rigid body constraints (between points and unitary vectors that belong to the same element) and pair constraints (only in some cases,because the constraints introduced by some kinematic pairs are automatically taken into account without any equation). In Figure 1 a planar and a three-dimensional mechanism, using natural coordinates as mechanism coordinates, are shown. The arrows show the points

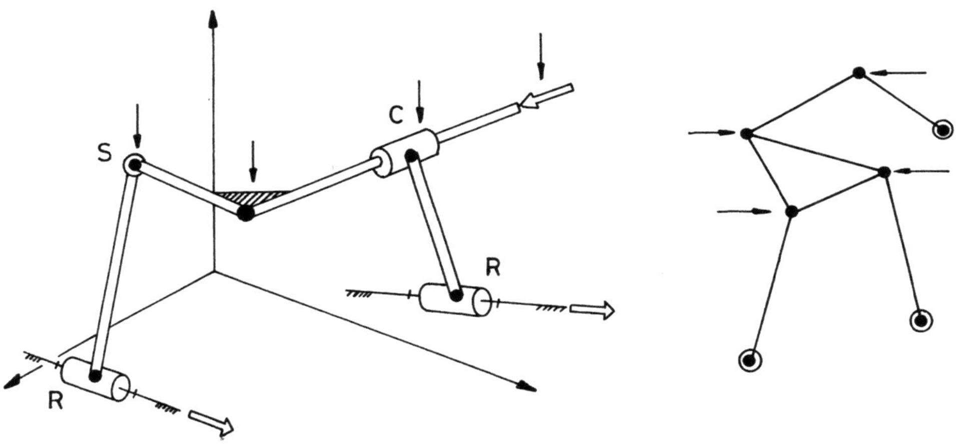

Fig.1. Examples of mechanisms with natural coordinates.

and vectors whose cartesian coordinates and components are the mechanism coordinates. With natural coordinates the constraint equations are always quadratic or linear; then, neither angular coordinates nor trascendental functions appear in the formulation.

Constraint equations with natural coordinates

In order to define appropiately the motion of the multibody elements each one must have at least two points in the planar case, and three points or unitary vectors in the 3-D one. With natural coordinates, points and vectors are optimally chosen when located in kinematic pairs, because the mechanism coordinates are then shared by contiguous elements, so the number of coordinates are minimized, and also because the pair constraint equations are derived easier or completely eliminated. Later it will become clearer how natural coordinates must be chosen for a particular mechanism.

Let us consider several disconnected planar and three-dimensional elements as can be seen in Fig.2. These elements have several points and vectors that must move according to the rigid body condition.

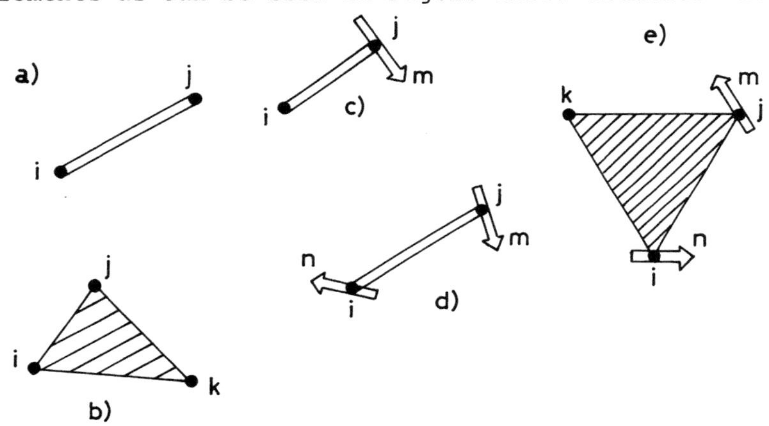

Fig. 2. Elements modeled with natural coordinates.

This can be guaranteed by imposing the rigid body constraint equations. For instance, for the planar binary link in Fig. 2a the condition of constant distance between points i and j must be imposed. This condition can be formulated in the form,

$$(x_i-x_j)^2 + (y_i-y_j)^2 = d_{ij}^2 \tag{1}$$

where d_{ij} is the distance between points i and j. In the three-dimensional case the problem is not much more complicated. For instance, for the element in Fig. 2d the following equations must be imposed,

$$(x_i-x_j)^2 + (y_i-y_j)^2 + (z_i-z_j)^2 = d_{ij}^2 \tag{2}$$

$$(x_i-x_j)m_x + (y_i-y_j)m_y + (z_i-z_j)m_z = C_2 \quad (3)$$

$$(x_i-x_j)n_x + (y_i-y_j)n_y + (z_i-z_j)n_z = C_3 \quad (4)$$

$$m_x n_x + m_y n_y + m_z n_z = C_3 \quad (5)$$

$$m_x^2 + m_y^2 + m_z^2 = 1 \quad (6)$$

$$n_x^2 + n_y^2 + n_z^2 = 1 \quad (7)$$

where C_2, C_3 and C_4 are constants. Eq.(2) is a constant distance condition. Eqs. (3)-(5) are constant angle conditions, and Eqs.(6) and (7) impose the unit module condition to the unitary vectors. In an analogous way the rigid body constraint equations can be generated for every element.

Once the rigid body conditions have been formulated, it is necessary to impose the pair constraint equations that guarantee that the elements move according with the possibilities of relative motions allowed by the kinematic pairs that join them. With natural coordinates this can also be done very easily. In Fig. 3 we can see a Spherical (S), a Revolute (R) and a Cylindrical (C) pair. The Spherical pair is automatically taken into account when the contiguous elements share a point, because then the

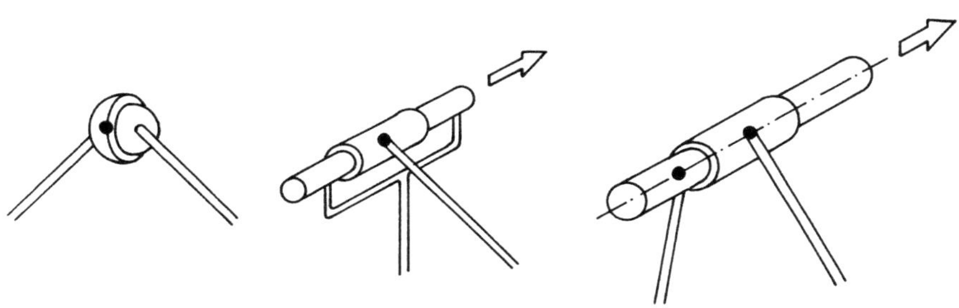

Fig.3. Spherical (S), Revolute (R) and Cylindrical (C) pairs.

only possible relative motion between the two elements is a rotation with unspecified axis. Analogously the Revolute pair is introduced when two elements share a point and an unitary vector, because only a rotation about the axis determined by this vector is allowed. The Cylindrical pair is taking into account sharing an unitary vector and imposing the condition of alignement between this vector and two points on the axis, each one belonging to a different element. Other kinematic pairs like the Prismatic (P) one or the Universal Joint (U) can be introduced in an analogous way.

Kinematic analysis

The kinematic constraint equations arising from the rigid body conditions and the kinematic pairs can be expressed vectorially in the form

$$\{\Phi(\{x\})\} = \{0\} \tag{8}$$

where $\{x\}$ is the vector of mechanism coordinates. These equations are quadratic of linear. Differentiating them with respect to time we obtain the velocity equation

$$[A(\{x\})]\ \{\dot{x}\} = \{0\} \tag{9}$$

where the jacobian matrix $[A]$ is linear or constant in the mechanism coordinates. This equation also means that the velocity vector $\{\dot{x}\}$ belongs to the nullspace of matrix $[A]$, which is the space of allowable motions. If $[R]$ is a matrix whose columns constitute a basis of the nullspace of matrix $[A]$, the following expressions can be written

$$[A]\ [R] = [0] \tag{10}$$

$$\{\dot{x}\} = [R]\{\dot{z}\} \tag{11}$$

This last equation means that the mechanism velocities can be expressed in terms of a lower dimension vector $\{\dot{z}\}$ of independent velocities. These independent velocities can be chosen in a large variety of ways, the most simple of them is selecting a subset or an extraction of $\{\ddot{x}\}$. To compute accelerations, Eq. (9) can be differentiated again giving

$$[A(\{x\})]\{\ddot{x}\} = -[\dot{A}(\{\dot{x}\})]\ \{\dot{x}\} \tag{12}$$

the relationship between the mechanism accelerations and the independent accelerations can be obtained differentiating Eq.(11)

$$\{\ddot{x}\} = [R]\{\ddot{z}\} + [\dot{R}]\{\dot{z}\} \tag{13}$$

Dynamic Analysis

To establish the differential equations of motion the Principle of Virtual Power can be used. It can be shown that the balance of virtual power between inertial and external forces with a virtual velocity $\{\dot{x}\}$ can be expressed in the form

$$\{\dot{x}\}^T [M] \{\ddot{x}\} = \{\dot{x}\}^T \{f\} \tag{14}$$

where $[M]$ is a matrix that can be formed by assembling the mass matrices of the elements, which depend on the mass, the position of center of gravity, the inertia tensor and the position of the element. In Eq.(14) virtual velocities can not be eliminated because they aren't independent. Using Eq. (11) to introduce independent virtual velocities we arrive at

$$[R]^T [M] \{\ddot{x}\} = [R]^T \{f\} \tag{15}$$

Eq. (15) completed with eq. (12) provide the mechanism accelerations, but this is not enough to integrate the differential equations of motion, because eq.(12) is the second derivative of the constraint equations, and it can be shown that their solution tends to violate the constraint equations (8). This problem can be solved via the integration of the mixed system of differential-algebraic equations (15) and (8) or with the stabilization technique due to Baumgarte[10], but it is our experience that it is better to formulate the differential equations of motion in independent coordinates. So, substituting Eq.(13) in Eq.(15) we obtain

$$[R]^T[M][R]\{\ddot{z}\} = [R]^T\{f\} - [R]^T[M][\dot{R}]\{\dot{z}\} \tag{16}$$

in this equation the term $[\dot{R}]\{\dot{z}\}$ can be identified with the help of Eq. (13) as the mechanism accelerations computed with the true velocities and null independent accelerations. If the mechanism velocities $\{\dot{x}\}$ are integrated numerically jointly with the independent accelarations $\{\ddot{z}\}$, then the new position of the mechanism is obtained without having to solve the position problem. According to the experience of the authors of this paper this is the best method for the dynamic analysis of multibody systems. A symbolic formalism leading to an equation similar to eq. (16) has been derived by Schiehlen[11].

The formulation described in this paper for the kinematic and dynamic analysis of multibody systems can be easily accomodated to include reonomic and nonholonomic constraints, nonlinear springs and dampers, impacts, etc. Actually, all of these characteristics are being implemented in a computer program called COMPAMM (COMPuter Analysis of Machines and Mechanisms), in the Department of Applied Mechanics of the University of Navarra.

Computer results

First of all in Table 1 we present the CPU time obtained in a VAX 11/780 computer in a dynamic simulation of three planar mechanisms using reference point coordinates (two cartesian coordinates and an angle for each element) and natural coordinates. In both cases the dynamic simulation has been done using inde-

		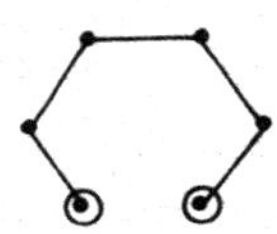	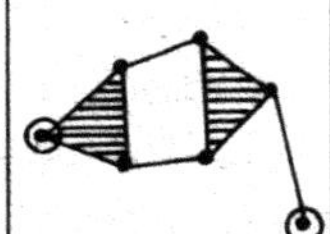
Reference Point Coord.	193.76	397.56	578.17
Natural Coordinates	113.65	208.70	438.70

Table 1. Numerical comparison between reference point coordinates and natural coordinates.

pendent coordinates and the "projected" equation (16). These results indicate that natural coordinates can be a competitive alternative to conventional methods.

An interesting characteristic of natural coordinates is the little postprocessing work needed to represent graphically the simulation results. In the planar case, as the mechanism coordinates are cartesian coordinates of points, it is enough to represent directly these points and to draw the lines corresponding to constant distance conditions. In Fig.4 two examples are presented: a bar that falls on a staircase and a triple pendulum that falls in a vertical plane colliding with an obstacle. In the three-dimensional case the method also produces clearly displayable results. In Fig. 5 we can see two views of the mo-

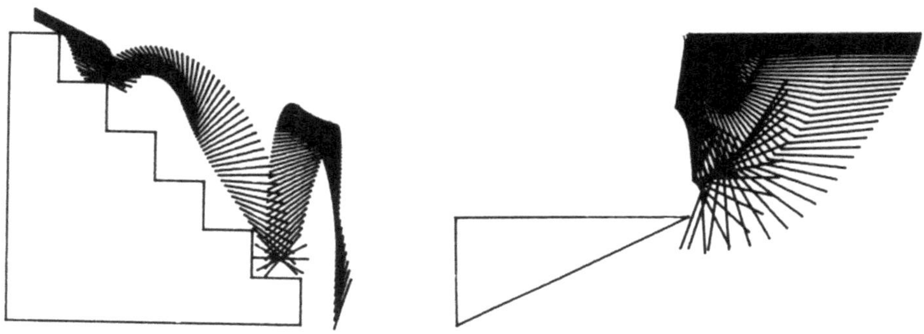

Fig. 4. Two examples of dynamic simulation of planar mechanisms in natural coordinates.

tion of a solid with a fixed point, showing constant nutation angle and precession velocity. It can be observed that the inclusion of unitary vectors as mechanism coordinates facilitates considerably the representation of motion.

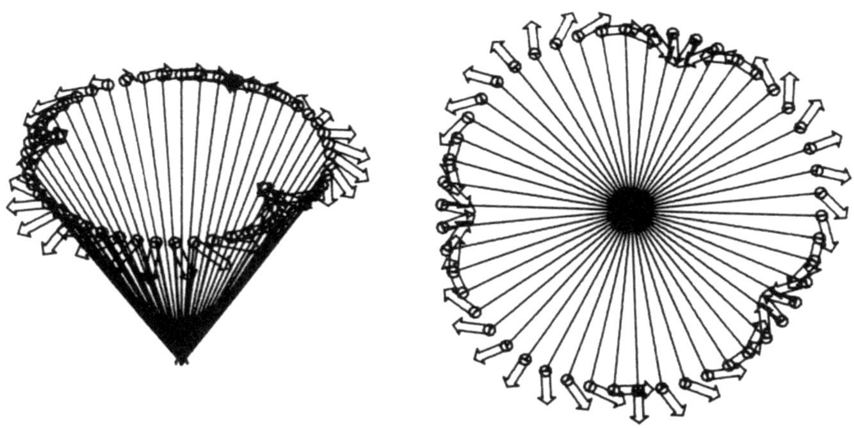

Fig. 5. Computer simulation of the motion of a solid with a fixed point.

Conclusions

A new method for the computer simulation of planar and three-dimensional multibody systems has been presented. This method is based in the use of "natural coordinates", which are constituted by cartesian coordinates of points and cartesian components of unitary vectors. This method can take into consideration a large variety of kinematic pairs. The constraint equations are at most quadratic, and very easy to formulate and evaluate. Open and closed loop multibody systems are considered in the same formulation. The differential equations of motions are formulated very easily and in a very efficient way in independent coordinates. The postprocessing work has been reduced to a minimum due to the particular mechanism coordinates used. The whole method is conceptually very simple, of very easy implementation and numerically very efficient.

References

1. Chace, M.A. and Smith, D.A., DAMN - Digital Computer Program for the Dynamic Analysis of Generalized Mechanical Systems, Trans. SAE 80, 969-991, 1971.

2. Smith, D.A., Chace, M.A. and Rubens, A.C., The Automatic Generation of a Mathematical Model for Machinery Systems, ASME J. of Eng. for Ind., 629-635, 1973.

3. Sheth, P.N. and Uicker, J.J., IMP (Integrated Mechanism Program): A Computer-Aided Design Analysis System for Mechanisms and Linkages, ASME J. of Eng. for Ind., 454-464, 1972.

4. Orlandea, N., Chace, M.A. and Calahan, D.A., A Sparsity Oriented Approach to the Dynamic Analysis and Design of Mechanical Systems (Parts 1 and 2), ASME J. of Eng. for Ind., 773-784, 1977.

5. Wehage, R.A. and Haug, E.J., Generalized Coordinate Partitioning for Dimension Reduction in Analysis of Constrained Dynamic Systems, ASME, J. of Mech. Design, 104, 247-255, 1982.

6. Nikravesh, P.E. and Chung, I.S., Application of Euler Parameters to the Dynamic Analysis of Three-Dimensional Constrained Mechanical Systems, ASME J. of Mech. Design, 104, 785-791, 1982.

7. Nikravesh, P.E.; Wehage, R.A. and Kwon, O.K., Euler Parameters in Computational Kinematics and Dynamics (Parts 1 and 2), ASME Papers 84-DET-112 and 94-DET-113, 1984.

8. Unda, J., Gimenez, J.G. and García de Jalón, J., Computer Simulation of the Dynamic Behaviour of Road Vehicles, 8th IAVSD Sysmposium, Cambridge, USA, August 1983.

9. Vilallonga, G., Unda, J. and García de Jalón, J., Numerical Kinematic Analysis of Three-Dimensional Mechanisms Using a "Natural" System of Lagrangian Coordinates, ASME Paper 84-DET-199, 1984.

10. Nikravesh, P.E., Some Methods for Dynamic Analysis of Constrained Mechanical Systems: A Survey, in Computer Aided Analysis and Optimization of Mechanical System Dynamics (Ed. by E.J. Haug), Springer-Verlag, 1984.

11. Schiehlen, W.O., Dynamics of Complex Multibody Systems, SM Archives, 9, 159-195, 1984.

Earthquake Vibrational Response of Turbo-Machines

F. ZIEGLER

Department of Civil Engineering
Technical University of Vienna
Austria

H.L. HASSLINGER

HEBAG-Vienna, Austria

Summary

Substructure synthesis method is employed to the gyroscopic system of a turbo-machine mounted on a table-like foundation. A single-point excitation through a stochastic earthquake process is assumed which is modeled by filtered white noise. The time dependent spectral method is adapted to frequency response functions given in a discrete form. R.m.s.-values of deformations and forces are calculated for spinning and non-spinning rotors, respectively.

Introduction

Safety of structures and large turbo-machines of power stations during earthquakes, requires the study of the vibrational characteristics with respect to random excitations.

Slightly damped linear elastic structures with symmetric stiffness and mass matrices are analyzed by standard procedures,e.g. by modal analysis combined with the response spectrum method and SRSS-superposition, see for example Clough and Penzien /1/, Newmark and Rosenblueth /2/.

Hybrid structures, like dam-reservoir systems /3/,/4/, liquid filled storage tanks /5/, and the flexible rotor (with horizontal axis) mounted on a table foundation /6/, the problem under consideration, require more sophisticated models for dynamic analysis. Due to the gyroscopic effects of the spinning rotor, see e.g. Magnus /7/, non-standard methods are required. Thus, we use the probabilistic formulation of the earthquake excitation and employ a substructure technique by isolating the rotor from its foundation /8/. Bearings are modeled by anisotropic linear

Dynamics of Multibody Systems
IUTAM/IFToMM Symposium Udine 1985
Editors: G. Bianchi and W. Schiehlen

elastic springs. The linear elastic table foundation is a substructure of standard (modal) type. Since the horizontal dimensions are small enough with respect to wave lengths, a single--point excitation of the table footings by a stochastic process is assumed whose samples are the free-field seismograms of the site. Thus, interaction ground effects are not considered here. Extension of the substructure method to include a linear elastic homogeneous or layered half-space model of the ground is possible without any theoretical difficulties.

Methods developed for rotational systems with axial symmetry are not directly applicable, since the whole structure has usually only one plane of symmetry through the rotor axis. We solve the linearized equations of motion of the gyroscopical system by a real matrix procedure and we represent the dynamical system by proper frequency response functions. Hence, we employ spectral methods in the probabilistic evaluations. Attention is paid mainly to the reaction forces and deformations in the rotor bearings, when considering the rotor spinning and non-spinning.

Substructuring in time and frequency domains

The table foundation shown in Fig.1 is a modally vibrating conventional substructure. We assume the deck to be rigid with four degrees of freedom. Absolute coordinates are the inertial displacements of the center of mass and the Newton-Euler angular rotation about the vertical principal axis. They are sampled in the column matrix $\underset{\sim}{x}_F(t)$. Single-point translational motion of the ground is the three-component vector $\underset{\sim}{z}(t)$. The vector $\underset{\sim}{x}_{FO}$ contains some constant lengths of the system.

The gyroscopical subsystem consists of the rotor, properly discretized, as shown in Fig.1, and the coordinates are sampled in $\underset{\sim}{x}_R(t)$. The stress free configuration of that part of the coordinates which are independent of the joint displacements at the bearings (tensionless springs) is denoted by $\underset{\sim}{x}_{R1}$, and the remaining coordinates are given by $\underset{\sim}{x}_{R2}$. The vector of the coupling coordinates at tensionless springs is $\underset{\sim}{x}_{CO}$ and the specifically defined coupling vectors are $\underset{\sim}{x}_{C1}$ and $\underset{\sim}{x}_{C2}$. In the following two subsections each of the two systems is considered individually.

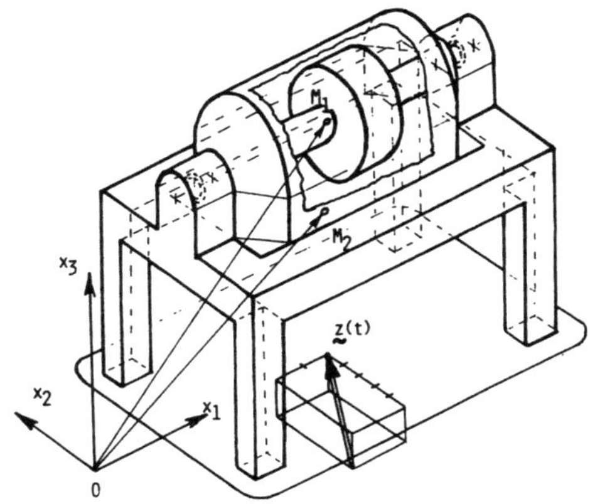

Fig.1 Rotor connected by anisotropic(linear) springs to the table foundation. Single-point excitation $\underset{\sim}{z}(t)$.

The table foundation

Slight damping is added to the modally separated equations of motions. All external forces and moments acting on that part of the structure, building-up the column matrix $\underset{\sim}{f}_F(t)$ and the coupled equations of motion in absolute coordinates become

$$\underset{\sim}{m}_F \ddot{\underset{\sim}{x}}_F + \underset{\sim}{k}_F(\underset{\sim}{x}_F - \underset{\sim}{x}_{FO} - \underset{\sim}{z}_F) = \underset{\sim}{f}_F \quad . \tag{1}$$

In relative coordinates

$$\underset{\sim}{y}_F = \underset{\sim}{x}_F - \underset{\sim}{x}_{FO} - \underset{\sim}{z}_F \;, \tag{2}$$

ground acceleration becomes a driving term

$$\underset{\sim}{m}_F \ddot{\underset{\sim}{y}}_F + \underset{\sim}{k}'_F \underset{\sim}{y}_F = -\underset{\sim}{m}_F \ddot{\underset{\sim}{z}}_F + \underset{\sim}{f}_F \quad . \tag{3}$$

Using

$$\underset{\sim}{x}_C = \underset{\sim}{t}_F \underset{\sim}{x}_F + \underset{\sim}{r}_F \;, \tag{4}$$

where $\underset{\sim}{t}_F$ is a proper transformation matrix. Sampling the generalized coupling forces in $\underset{\sim}{f}_C(y_R)$, the reaction forces in the bearing joints can be expressed as

$$\underset{\sim}{f}_F = \underset{\sim}{t}_F^T \underset{\sim}{f}_C \;, \tag{5}$$

$\underset{\sim}{t}_F^T$ is the transpose of $\underset{\sim}{t}_F$.

Solution of the associated eigen-problem

$$\underset{\sim}{m}_F\underset{\sim}{\Psi}_F\underset{\sim}{\omega}_F^2 = \underset{\sim}{k}_F\underset{\sim}{\Psi}_F \tag{6}$$

renders the diagonal matrix of eigen-frequencies $\underset{\sim}{\omega}_F$ and the modal matrix $\underset{\sim}{\Psi}_F$. Therefore, the free-vibrations due to initial conditions $\underset{\sim}{y}_{FO}$ and $\dot{\underset{\sim}{y}}_{FO}$ become

$$\underset{\sim}{y}_F = \underset{\sim}{\Psi}_{F1}(t)\underset{\sim}{y}_{FO} + \underset{\sim}{\Psi}_{F2}(t)\dot{\underset{\sim}{y}}_{FO} \quad . \tag{7}$$

Modal analysis

$$\underset{\sim}{y}_F = \underset{\sim}{\Psi}_F\underset{\sim}{q}(t) \tag{8}$$

renders the set of non-coupled oscillator equations

$$\ddot{q}_j + \omega_{Fj}^2 q_j = -\ddot{z}_{pj} + f_{pj} \quad , \quad \ddot{z}_{pj} = \underset{\sim}{\Psi}_{Fj}^T\underset{\sim}{m}_F\ddot{\underset{\sim}{z}}_F/m_j^* \quad ,$$
$$f_{pj} = \underset{\sim}{\Psi}_{Fj}^T\underset{\sim}{f}_F/m_j^* \quad , \quad m_j^* = \underset{\sim}{\Psi}_{Fj}^T\underset{\sim}{m}_F\underset{\sim}{\Psi}_{Fj} \quad . \tag{9}$$

Frequency response due to time harmonic excitation

$$\underset{\sim}{z}_F = \underset{\sim}{z}_{Fe}\exp(i\nu t) \quad , \quad \underset{\sim}{f}_F = \underset{\sim}{f}_{Fe}\exp(i\nu t) \quad , \tag{10}$$

where amplitudes are complex, becomes

$$\underset{\sim}{y}_{Fe}(i\nu) = \underset{\sim}{F}_{Fz}\underset{\sim}{z}_{Fe} + \underset{\sim}{F}_{Ff}\underset{\sim}{f}_{Fe} \quad . \tag{11}$$

The matrices

$$\underset{\sim}{F}_{Fz} = (\nu^{-2}\underset{\sim}{m}_F^{-1}\underset{\sim}{k}_F - \underset{\sim}{I})^{-1} \quad , \quad \underset{\sim}{F}_{Ff} = (\underset{\sim}{k}_F - \nu^2\underset{\sim}{m}_F)^{-1} \tag{12}$$

are related by

$$\underset{\sim}{F}_{Fz} = -\nu^2 \underset{\sim}{F}_{Ff}\underset{\sim}{m}_F \quad . \tag{13}$$

Their modal decomposition is

$$\underset{\sim}{F}_{Fz} = \underset{\sim}{\Psi}_F(\nu^{-2}\underset{\sim}{\omega}_F^2 - \underset{\sim}{I})^{-1}\ \underset{\sim}{m}^{*-1}\underset{\sim}{\Psi}_F^T\underset{\sim}{m}_F \quad ,$$

$$\underset{\sim}{F}_{Ff} = \underset{\sim}{\Psi}_F(\underset{\sim}{k}^* - \nu^2\underset{\sim}{m}^*)^{-1}\ \underset{\sim}{\Psi}_F^T \quad , \tag{14}$$

where $\underset{\sim}{m}^*$ and $\underset{\sim}{k}^*$ denote the diagonalized mass and stiffness matrices, respectively.

The gyroscopic subsystem

Linearized equations of motion of the discretized and undamped rotor have the matrix form in absolute coordinates,

$$\underset{\sim}{m}_R\ddot{\underset{\sim}{x}}_R + \underset{\sim}{g}_R\dot{\underset{\sim}{x}}_R + \underset{\sim}{k}_R(\underset{\sim}{x}_R - \underset{\sim}{x}_{R1}) - \underset{\sim}{k}_C(\underset{\sim}{x}_C + \underset{\sim}{x}_{C1}) = \underset{\sim}{0} \ . \tag{15}$$

Substitution of

$$\underset{\sim}{x}_{C1} = -\underset{\sim}{x}_{CO} + \underset{\sim}{x}_{C2} \tag{16}$$

and

$$\underset{\sim}{x}_{RO} = \underset{\sim}{x}_{R1} + \underset{\sim}{x}_{R2} \tag{17}$$

and relating

$$\underset{\sim}{k}_C\ \underset{\sim}{x}_{C2} = \underset{\sim}{k}_R\ \underset{\sim}{x}_{R2} \tag{18}$$

renders the final form

$$\underset{\sim}{m}_R\ddot{\underset{\sim}{x}}_R + \underset{\sim}{g}_R\dot{\underset{\sim}{x}}_R + \underset{\sim}{k}_R(\underset{\sim}{x}_R - \underset{\sim}{x}_{RO}) = \underset{\sim}{k}_C(\underset{\sim}{x}_C - \underset{\sim}{x}_{CO}) \ , \tag{19}$$

with gyroscopic forces $\underset{\sim}{g}_R\dot{\underset{\sim}{x}}_R$. Introducing the relative coordinates

$$\underset{\sim}{y}_R = \underset{\sim}{x}_R - \underset{\sim}{x}_{RO} \quad , \quad \underset{\sim}{y}_C = \underset{\sim}{x}_C - \underset{\sim}{x}_{CO} \tag{20}$$

renders

$$\underset{\sim}{m}_R \ddot{\underset{\sim}{y}}_R + \underset{\sim}{g}_R \dot{\underset{\sim}{y}}_R + \underset{\sim}{k}_R \underset{\sim}{y}_R = \underset{\sim}{k}_C \underset{\sim}{y}_C \quad . \tag{21}$$

Following the representation /6/, /12/, with the numerical advantage of small real matrices, $\underset{\sim}{g}_R \underset{\sim}{h}$ is a symmetric matrix, the associated eigen-problem is determined through

$$\underset{\sim}{k}_R \underset{\sim}{\Psi}_R - \underset{\sim}{m}_R \underset{\sim}{\Psi}_R \underset{\sim}{\omega}_R^2 - \underset{\sim}{g}_R \underset{\sim}{h} \underset{\sim}{\Psi}_R \underset{\sim}{\omega}_R = \underset{\sim}{O} \quad . \tag{22}$$

An efficient solution technique is developed in /6/ and /12/, and may be compared to that of Meirovich /9/. Hence, analogously to Eq.(7), free vibrations are represented as

$$\underset{\sim}{y}_R = \underset{\sim}{\Psi}_{R1}(t) \underset{\sim}{y}_{RO} + \underset{\sim}{\Psi}_{R2}(t) \dot{\underset{\sim}{y}}_{RO} \quad . \tag{23}$$

Forced vibrations are given by the convolution

$$\underset{\sim}{y}_R = \int_0^t \underset{\sim}{\Psi}_{R2}(t-\tau) \underset{\sim}{m}_R^{-1} \underset{\sim}{k}_C \underset{\sim}{y}_C(\tau)\, d\tau \quad , \tag{24}$$

where

$$\begin{aligned} \underset{\sim}{\Psi}_{R2}(t) = (-\underset{\sim}{\Psi}_R \cos \underset{\sim}{\omega}_R t + \underset{\sim}{h} \underset{\sim}{\Psi}_R \sin \underset{\sim}{\omega}_R t) \underset{\sim}{H}_R^{-1} \underset{\sim}{\omega}_R^{-1} \underset{\sim}{\Psi}_R^{-1} + \\ (\underset{\sim}{h} \underset{\sim}{\Psi}_R \cos \underset{\sim}{\omega}_R t + \underset{\sim}{\Psi}_R \sin \underset{\sim}{\omega}_R t) \underset{\sim}{H}_R^{-1} \underset{\sim}{\omega}_R^{-1} \underset{\sim}{\Psi}_R^{-1} \underset{\sim}{h} \quad , \end{aligned} \tag{25}$$

and

$$\underset{\sim}{H}_R = \underset{\sim}{\Psi}_R^{-1} \underset{\sim}{h} \underset{\sim}{\Psi}_R + \underset{\sim}{\omega}_R^{-1} \underset{\sim}{\Psi}_R^{-1} \underset{\sim}{h} \underset{\sim}{\Psi}_R \underset{\sim}{\omega}_R \tag{26}$$

is the sum of two similarity transformations.

Frequency response functions follow from the time-harmonic excitation of the bearing joints

$$\underset{\sim}{y}_C = \underset{\sim}{y}_{Ce} \exp(i\nu t) \quad , \quad \underset{\sim}{f}_C = \underset{\sim}{f}_{Ce} \exp(i\nu t) \quad , \tag{27}$$

$$\underset{\sim}{y}_{Re}(i\nu) = \underset{\sim}{F}_R(i\nu) \underset{\sim}{k}_C \underset{\sim}{y}_{Ce}(i\nu) \quad . \tag{28}$$

The inversion of the following large complex matrix is cumbersome

$$\underset{\sim}{F}_R(i\nu) = (\underset{\sim}{k}_R-\nu^2\underset{\sim}{m}_R+i\nu\underset{\sim}{g}_R)^{-1} \tag{29}$$

thus, the expansion

$$\begin{aligned}\underset{\sim}{F}_R(i\nu)=\{&-(\underset{\sim}{\Psi}_R\underset{\sim}{\omega}_R+i\nu\underset{\sim}{h}\underset{\sim}{\Psi}_R)(\nu^2\underset{\sim}{I}-\underset{\sim}{\omega}_R^2)^{-1}\underset{\sim}{H}_R^{-1}\underset{\sim}{\omega}_R^{-1}\,\underset{\sim}{\Psi}_R^{-1}\underset{\sim}{h} + \\ &(-\underset{\sim}{h}\underset{\sim}{\Psi}_R\underset{\sim}{\omega}_R-i\nu\underset{\sim}{\Psi}_R)(\nu^2\underset{\sim}{I}-\underset{\sim}{\omega}_R^2)^{-1}\underset{\sim}{H}_R^{-1}\underset{\sim}{\omega}_R^{-1}\underset{\sim}{\Psi}_R^{-1}\}\underset{\sim}{m}_R^{-1}\end{aligned} \tag{30}$$

is used as a much more efficient representation, where $\underset{\sim}{H}_R$ of Eq.(26) is understood.

The dynamical system

Coupling forces $\underset{\sim}{f}_C$ in Eq.(5) are the combination of two vectors

$$\underset{\sim}{f}_C = -\underset{\sim}{k}_{\Theta}(\underset{\sim}{x}_C+\underset{\sim}{x}_{C1}) + \underset{\sim}{k}_C^T\underset{\sim}{x}_R \quad . \tag{31}$$

Substitution of relative displacements according to Eq.(20) and of Eq.(16), renders

$$\underset{\sim}{f}_C = -\underset{\sim}{k}_O\underset{\sim}{y}_C+\underset{\sim}{k}_C^T\underset{\sim}{y}_R-(\underset{\sim}{k}_O\underset{\sim}{x}_{C2}-\underset{\sim}{k}_C^T\underset{\sim}{x}_{RO}) \quad . \tag{32}$$

The bracketed term vanishes since coupling forces are proportional to the relative displacements. Considering Eqs.(20),(4) and (2), those coordinates are expressible in the form

$$\underset{\sim}{y}_C=\underset{\sim}{x}_C-\underset{\sim}{x}_{CO}=\underset{\sim}{t}_F(\underset{\sim}{y}_F+\underset{\sim}{z}_F)+(\underset{\sim}{t}_F\underset{\sim}{x}_{FO}+\underset{\sim}{r}_F-\underset{\sim}{x}_{CO}) \quad , \tag{33}$$

and since the relative displacements vanish with the coordinates, the bracketed second term is put to zero. Finally, the coupling forces become

$$\underset{\sim}{f}_C = -\underset{\sim}{k}_O\underset{\sim}{t}_F(\underset{\sim}{y}_F+\underset{\sim}{z}_F) + \underset{\sim}{k}_C^T\underset{\sim}{y}_R \quad . \tag{34}$$

Substitution of Eq.(5) in Eq.(3), together with inserting Eq.(33) in (21), renders the coupled dynamical set of system equations

$$\underset{\sim}{m}_F\ddot{\underset{\sim}{y}}_F + (\underset{\sim}{k}_F+\underset{\sim}{t}_F^T\underset{\sim}{k}_O\underset{\sim}{t}_F)\underset{\sim}{y}_F = \underset{\sim}{t}_F^T\underset{\sim}{k}_C^T\underset{\sim}{y}_R - \underset{\sim}{m}_F\ddot{\underset{\sim}{z}}_F \quad , \tag{35}$$

$$\underset{\sim}{m}_R \ddot{\underset{\sim}{y}}_R + \underset{\sim}{g}_R \dot{\underset{\sim}{y}}_R + \underset{\sim}{k}_R \underset{\sim}{y}_R = \underset{\sim}{k}_C \underset{\sim}{t}_F \underset{\sim}{y}_F \quad , \tag{36}$$

if $\underset{\sim}{t}_F \ddot{\underset{\sim}{z}}_F = \underset{\sim}{O}$ is considered.

Combination of the subsystems can be quite efficiently performed in the frequency domain. Considering Eq.(27) and Eq.(10) together with Eq.(5), and inverting Eq.(8),

$$\underset{\sim}{q}_e(i\nu) = \underset{\sim}{\Psi}_F^T \underset{\sim}{y}_{Fe} \quad , \tag{37}$$

renders by the separation of Eq.(35) the input-output relation of relative amplitudes

$$\begin{aligned} &\{\underset{\sim}{I} + \underset{\sim}{\Psi}_F^T \underset{\sim}{F}_{Ff}(i\nu) \underset{\sim}{t}_F^T (\underset{\sim}{k}_O - \underset{\sim}{k}_C^T \underset{\sim}{F}_R(i\nu) \underset{\sim}{k}_C) \underset{\sim}{t}_F \underset{\sim}{\Psi}_F\} \underset{\sim}{q}_e(i\nu) \\ &= \underset{\sim}{\Psi}_F^T \{\underset{\sim}{F}_{Fz}(i\nu) - \underset{\sim}{F}_{Ff}(i\nu) \underset{\sim}{t}_F^T (\underset{\sim}{k}_O - \underset{\sim}{k}_C^T \underset{\sim}{F}_R(i\nu) \underset{\sim}{k}_C) \underset{\sim}{t}_F\} \underset{\sim}{z}_{Fe} \quad . \end{aligned} \tag{38}$$

Solution is simplified by using Eqs.(12) and (13) for the hermitic matrix before inversion is performed:

$$\begin{aligned} \underset{\sim}{q}_e(i\nu) = &\{\underset{\sim}{k}^* - \nu^2 \underset{\sim}{m}^* + \underset{\sim}{\Psi}_F^T \underset{\sim}{t}_F^T (\underset{\sim}{k}_O - \underset{\sim}{k}_C^T \underset{\sim}{F}_R(i\nu) \underset{\sim}{k}_C) \underset{\sim}{t}_F \underset{\sim}{\Psi}_F\}^{-1} \underset{\sim}{\Psi}_F^T \{\nu^2 \underset{\sim}{m}_F - \\ &\underset{\sim}{t}_F^T (\underset{\sim}{k}_O - \underset{\sim}{k}_C^T \underset{\sim}{F}_R(i\nu) \underset{\sim}{k}_C) \underset{\sim}{t}_F\} \underset{\sim}{z}_{Fe} \quad . \end{aligned} \tag{39}$$

Splitting the solution into real and imaginary part renders a hypermatrix. In their main diagonal are symmetric matrices, those off-diagonal are the transpose of each other due to the hermitic property of $\underset{\sim}{F}_R(i\nu)$. Thus, the real system matrix is symmetric.

There are two efficient methods available for the consideration of slight damping in the frequency response functions: The first is a numerical, quadrature type of the elastic-viscoelastic correspondence of time-harmonic vibrations for homogeneously smeared damping in the whole structure, recently extended to include resonances by Ziegler /10/. The second is a perturbational procedure of Meirovich and Ryland /11/ which can handle also local damping. The latter is employed in the numerical example in a form due to Hasslinger /6/. Figs.2,3 and 4 show typical frequency response functions of the undamped system.

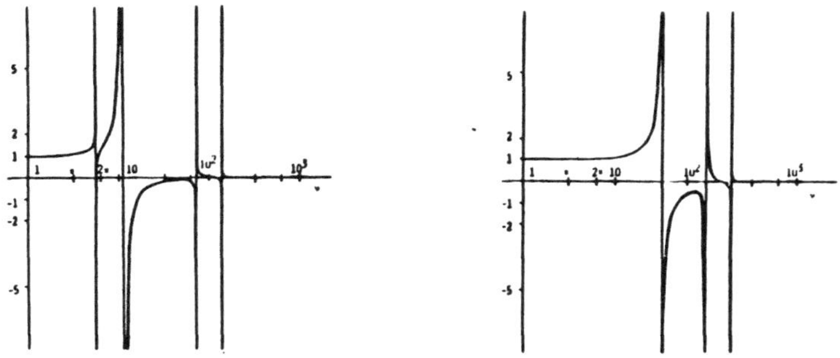

Fig.2 Frequency response function of bearing force, undamped system. Non-spinning rotor.
(a) horizontal component (b) vertical component

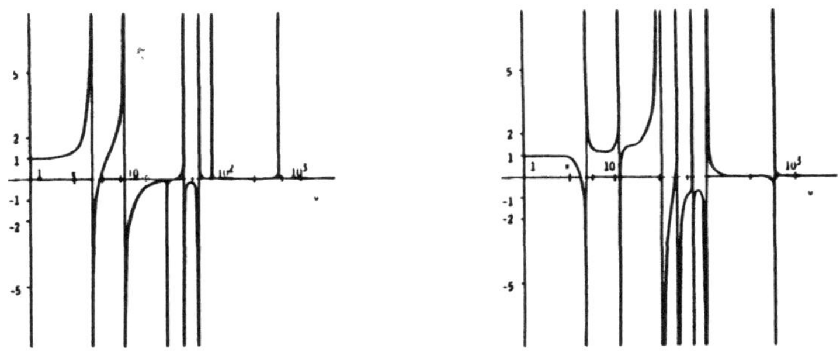

Fig.3 Real part of frequency response function of bearing force, undamped system. Spinning rotor 100π/s.
(a) horizontal component (b) vertical component

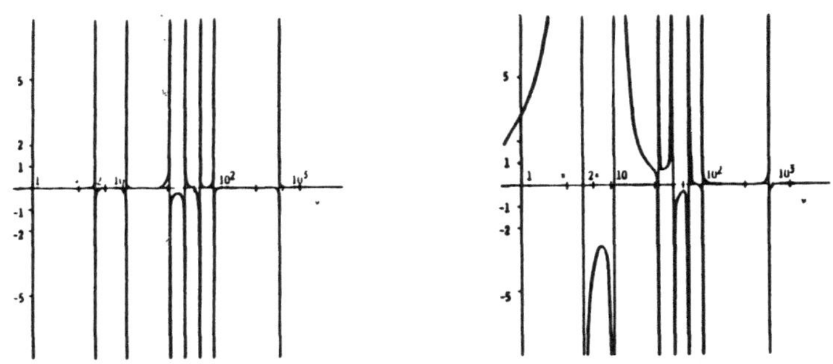

Fig.4 Imaginary part of frequency response function of bearing force, undamped system. Spinning rotor 100π/s.
(a) horizontal component (b) vertical component

Stochastic response: The time-dependent spectral method

Since the frequency response functions of the slightly damped system are given in discrete form at equidistant points of the forcing frequency axis within a properly wide frequency band, we employ the non-stationary spectral method developed by Priestley /13/ in the improved version of /3/ for envelope type stochastic forcing functions.

To generate the vectorial single-point earthquake process we use non-stationary white noise of zero mean with envelope function $\Phi(t)$,

$$\eta = \Phi(t)w(t) \quad , \quad t \geqslant 0 \quad , \tag{40}$$

and we add a Kanai-Tajimi filter (parameters $\omega_0=4\pi/s$, $\zeta=0.6$) with stationary frequency response,

$$F_G(i\nu)=\{1-(\nu/\omega_0)^2+(2\zeta\nu/\omega_0)^2-2i\zeta\nu^2/\omega_0^2\}/\{(1-\nu^2/\omega_0^2)^2+(2\zeta\nu/\omega_0)^2\} \tag{41}$$

to the gross(gyroscopic) system to account for ground effects of the site. The directional distribution of the vector $\underset{\sim}{z}(t)$ in principal system axis is assumed constant by 2:5:1, i.e.parallel, perpendicular to rotor axis and vertical, respectively. Hence, the non-stationary spectral input-output relation of the total system (including the ground filter) becomes

$$\underset{\sim}{S}_{yy}(i\nu,t) = \underset{\sim}{F}_t(i\nu,t)\underset{\sim}{S}_{ww}\underset{\sim}{F}_t^T(-i\nu,t) \quad . \tag{42}$$

The time dependent frequency response matrix of filter plus gross system is determined by the double integral, where integration has been interchanged,

$$\underset{\sim}{F}_t(i\nu,t) = \frac{1}{2\pi}\int_{-\infty}^{\infty}\underset{\sim}{F}_S(i\nu)f(i\nu,\xi,t)d\xi \quad , \tag{43}$$

$$f(i\nu,\xi,t) = \int_0^t \Phi(t-\tau)\exp\{i(\xi-\nu)\tau\}d\tau \quad , \tag{44}$$

where $\underset{\sim}{F}_S(i\nu)$ is the stationary frequency response function of the total system including the filter. In case of simple envelopes $\Phi(t)$, Eq.(44) is analytically integrable:
$\Phi(t)=H(t)$, the Heaviside step-function, switching at t=0,renders

$$f(i\nu,\xi,t)=i\{1-\exp i(\nu-\xi)t\}/(\nu-\xi) \quad . \tag{45}$$

Modulation by exponential functions,

$$\Phi(t) = e^{\alpha_1 t} - e^{\alpha_2 t} \quad , \quad \alpha_2 < \alpha_1 < 0 \quad , \quad t \geqslant 0 \quad , \tag{46}$$

renders

$$f(i\nu,\xi,t)=i\sum_{k=1}^{2} (-1)^{k-1}\{e^{\alpha_k t} - e^{i(\nu-\xi)t}\}\{(\nu-\xi)-i\alpha_k\}/\{(\nu-\xi)^2+\alpha_k^2\}. \tag{47}$$

Thus, the single integral of Eq.(43) is left for numerical integration over a sufficiently wide band of frequencies. A second numerical integration renders the evolution of the variance matrix

$$\underset{\sim}{\sigma}^2_{YY}(t) = \int_{-\infty}^{\infty} \underset{\sim}{S}_{YY}(\nu,t)\,d\nu \tag{48}$$

which is the main result of this paper. Only the stationary values are reported in Table 3.

Numerical example and stationary response

According to a gross-system of nine degrees of freedom, Fig.1, the eigenfrequencies have been calculated in the sequence: 5.47, 11.05, 16.39, 32.67, 51.58, 70.86, 99.29, 136.18, 190.28 s^{-1}, in case of non-spinning rotor. The associated 9x9 modal matrix is presented in Table 1. When the rotor is spinning at the angular speed 100π/s, the eigenfrequencies are changed according to the sequence: 4.80, 10.95, 16.39, 32.65, 49.13, 51.58, 72.30, 103.46, 570.49 s^{-1}, with the modal matrix as shown in Table 2.

Intensity matrix of the three-dimensional white noise excitation process with the assumption of $|\underset{\sim}{z}|=1$ and randomly chosen cross-spectral densities, is

$$\begin{bmatrix} 0.365 & 0.522 & 0.243 \\ 0.522 & 0.913 & 0.304 \\ 0.243 & 0.304 & 0.183 \end{bmatrix} \tag{49}$$

Table 1 Modal matrix, non-spinning rotor.

0.	0.	0.630	0.	-0.653	0.	0.	0.	0.
-0.376	0.615	0.	0.	0.	-0.566	0.	0.081	0.
0.	0.	0.	0.625	0.	0.	-0.648	0.	-0.096
0.899	0.230	0.	0.	0.	-0.271	0.	-0.134	0.
0.	0.	0.776	0.	0.757	0.	0.	0.	0.
0.028	0.752	0.	0.	0.	0.771	0.	-0.115	0.
0.	0.	0.	0.780	0.	0.	0.738	0.	0.130
0.	0.	0.	-0.016	0.	0.	-0.187	0.	0.987
0.223	0.061	0.	0.	0.	0.105	0.	0.981	0.

Table 2 Modal matrix, rotor spinning at $100\pi/s$.

0.	0.	0.630	0.	0.	-0.653	0.	0.	0.
-0.362	0.643	0.	0.005	0.032	0.	-0.579	-0.018	0.003
0.000	0.000	0.	0.623	0.197	0.	0.033	-0.649	-0.006
0.903	0.158	0.	-0.053	0.788	0.	-0.177	0.022	-0.005
0.	0.	0.776	0.	0.	0.757	0.	0.	0.
0.045	0.748	0.	-0.011	-0.008	0.	0.792	0.025	-0.004
0.010	0.005	0.	0.780	-0.048	0.	-0.029	0.754	0.008
-0.017	-0.008	0.	-0.019	0.372	0.	0.059	0.015	0.717
0.226	0.045	0.	0.006	-0.445	0.	-0.042	-0.096	0.696

Stationary results are presented in Table 3.

Table 3 Stationary r.m.s.-values of nine deformations and deformation rates for non-spinning (a) and spinning rotor (b), angular speed $100\pi/s$.

	y_{11}	y_{12}	y_{13}	y_{14}	y_{21}	y_{22}	y_{23}	y_{24}	y_{25}
(a)	0.037	0.112	0.004	0.120	0.046	0.126	0.006	0.0001	0.030
(b)	0.037	0.110	0.004	0.120	0.046	0.118	0.006	0.003	0.030
	$\dot{y}_{11}$	$\dot{y}_{12}$	$\dot{y}_{13}$	$\dot{y}_{14}$	$\dot{y}_{21}$	$\dot{y}_{22}$	$\dot{y}_{23}$	$\dot{y}_{24}$	$\dot{y}_{25}$
(a)	0.600	1.160	0.141	0.759	0.739	1.390	0.176	0.005	0.192
(b)	0.600	1.120	0.140	0.630	0.739	1.280	0.176	0.019	0.163

Conclusions

Spectral method proved to be efficient for response calculations of the gyroscopic system. From Table 3 and from typical r.m.s-values of the system with spinning rotor, it can be concluded that the earthquake with dominating horizontal excitation renders strong vibrations and large bearing forces in orthogonal directions. Dominating is the vertical displacement and the yaw rotation in comparison to those of the system with non-spinning rotor. The table foundation is stabilized with respect to translational motions in its plane.

Acknowledgement

This project is supported by a grant within the national research project S30/03 "Structural Mechanics" of the Austrian FWF, 1985-1989, through the final stage.

References

1. Clough, R.W.; Penzien,J.: Dynamics of structures. New York: McGraw Hill 1975.

2. Newmark, N.M.; Rosenblueth, E.: Fundamentals of earthquake engineering. Englewood Cliffs, N.J.: Prentice-Hall 1971.

3. Höllinger, F.;Ziegler, F.: Nonstationary random vibrations of linear elastic gravity dam with naturally sloped reservoir. (In German) ZAMM 63 (1983) 49-54.

4. Höllinger, F.: Time-harmonic and nonstationary stochastic vibrations of arch dam-reservoir-systems. Acta Mechanica 49 (1983) 153-167.

5. Ziegler, F.: Random vibrations of liquid-filled containers (excited by earthquakes). In: Hennig, K.(ed.): Random vibrations and reliability. Proc. IUTAM Symposium Frankfurt/Oder 1982. Berlin: Akademie Verlag 1983, 359-367.

6. Hasslinger, H.L.: Oscillatory systems with rotors subjected to earthquake excitation (In German), Dissertation, T.U.Vienna 1982.

7. Magnus, K.: Gyrodynamics. CISM Courses and Lectures No.53, Udine. Wien, New York: Springer 1974.

8. Hasslinger, H.L.: Substructure synthesis method for gyroscopic systems (In German). ZAMM 65 (1985) T56-T58.

9. Meirovich, L.: A new method of solution of the eigenvalue problem for gyroscopic systems. AIAA 12 (1974) 1337-1342.

10. Ziegler, F.: The elastic-viscoelastic correspondence in case of numerically determined discrete elastic response spectra. ZAMM 63 (1983) T135-T137.

11. Meirovich, L.; Ryland, G.: Response of slightly damped gyroscopic systems. J. Sound and Vibration 67 (1979) 1-19.

12. Hasslinger, H.L.: Vibrations of a conservative, gyroscopic system (In German). ZAMM 63 (1983) T53-T55.

13. Priestley, M.B.: Power spectral analysis of non-stationary random processes. J. Sound and Vibration 6 (1967) 86-97.

MIX
Papier aus verantwortungsvollen Quellen
Paper from responsible sources
FSC® C105338

If you have any concerns about our products,
you can contact us on
ProductSafety@springernature.com

In case Publisher is established outside the EU,
the EU authorized representative is:
Springer Nature Customer Service Center GmbH
Europaplatz 3, 69115 Heidelberg, Germany

Printed by Libri Plureos GmbH
in Hamburg, Germany